T0275896

INTRODUCTION TO THE CHEMISTRY OF FOOD

INTRODUCTION TO THE
CHEMISTRY OF FOOD

MICHAEL ZEECE
Professor Emeritus
Department of Food Science
University of Nebraska
Lincoln, Nebraska, United States

ACADEMIC PRESS

An imprint of Elsevier

Academic Press is an imprint of Elsevier
125 London Wall, London EC2Y 5AS, United Kingdom
525 B Street, Suite 1650, San Diego, CA 92101, United States
50 Hampshire Street, 5th Floor, Cambridge, MA 02139, United States
The Boulevard, Langford Lane, Kidlington, Oxford OX5 1GB, United Kingdom

Notices
Knowledge and best practice in this field are constantly changing. As new research and
experience broaden our understanding, changes in research methods, professional practices, or
medical treatment may become necessary.

Practitioners and researchers must always rely on their own experience and knowledge in
evaluating and using any information, methods, compounds, or experiments described herein.
In using such information or methods they should be mindful of their own safety and the safety
of others, including parties for whom they have a professional responsibility.

To the fullest extent of the law, neither the Publisher nor the authors, contributors, or editors,
assume any liability for any injury and/or damage to persons or property as a matter of products
liability, negligence or otherwise, or from any use or operation of any methods, products,
instructions, or ideas contained in the material herein.

Library of Congress Cataloging-in-Publication Data
A catalog record for this book is available from the Library of Congress

British Library Cataloguing-in-Publication Data
A catalogue record for this book is available from the British Library

ISBN: 978-0-12-809434-1

For information on all Academic Press publications visit our
website at https://www.elsevier.com/books-and-journals

Publisher: Charlotte Cockle
Acquisitions Editor: Nina Rosa de Araujo Bandeira
Editorial Project Manager: Laura Okidi
Production Project Manager: Selvaraj Raviraj
Cover Designer: Christian Bilbow and original art Megan Mclaughlin

Typeset by TNQ Technologies

Contents

9. Food systems and future directions

TEC URL: http://textbooks.elsevier.com/web/Manuals.aspx?isbn=9780128094341

Acknowledgments

I wish to thank my wife, Pauline Davey Zeece, for her comments and suggestions regarding the contents of this book. Her expertise in developmental psychology contributed to summarizing research regarding food additives and hyperactivity in children. I wish to thank our daughter, Megan Mclaughlin (9speedcreative.com), for the artwork on the cover of this book. I also wish to thank our sons, Michael Zeece and Eric Zeece, for their ongoing encouragement and support.

CHAPTER ONE

Chemical properties of water and pH

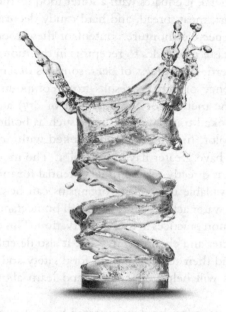

Learning objectives

This chapter will help you describe or explain:

- Water's structure
- The hydrogen bond and its importance to water
- What a food acid is, including examples
- pH and titratable acidity
- The importance of water to food color, taste, and texture
- Why oil is not soluble in water
- Water activity and its importance to food quality and safety

Introduction to the Chemistry of Food
ISBN: 978-0-12-809434-1
https://doi.org/10.1016/B978-0-12-809434-1.00001-3

1

Introduction

Water is the major component of all living things and therefore an important part of food. Water affects the texture, taste, color, and microbial safety of everything we eat. The moisture content of food is a good indicator of its texture. In general, it equates with a softer food texture. For example, the texture of yogurt, meat, bread, and hard candy decreases in that order and parallels the respective moisture content of these foods. Water is the vehicle that carries taste molecules to receptors in the mouth. For example, the sweetness of cherries, bitterness of beer, sourness of lemons, saltiness of pretzels, and pungency of peppers results from compounds (tastants) dissolved in water. The method of cooking (wet or dry) affects food flavor and color. Food cooked using wet methods, such as boiling, are generally low in flavor and color. In contrast, foods cooked with dry methods, such as frying or grilling have greater flavor and color. The moisture content of foods, such as milk, is directly related to its potential for microbial spoilage. Control of water available to spoilage organisms can be accomplished by lowering the food's water activity level (a_w) with humectants or by dehydration. Both are common practices in food preservation. This chapter describes the properties of water and chemistry in food. It also describes the chemical concepts of acids and their relationships to food safety and spoilage.

These questions will help you explore and learn about water and its effects on food.
- How can surface tension be demonstrated using a cup of water and a paperclip?
- Why did my can of pop explode in the freezer?
- Why does it take longer to boil potatoes in Denver than in Chicago?
- What is a pKa?
- Gee fizz, what makes soda pop so tasty?
- Why did the biscuit dough package explode in the refrigerator? Hint: The answer involves acid-base chemistry.
- So, what happens when oil is added to water? Why doesn't it dissolve?
- What is the acid-ash hypothesis and does alkaline water make my bones stronger?

Structure of water

Before considering the effects of water in food, it is necessary to understand its unique molecular properties. The physical and chemical properties of water directly result from its molecular composition and structure. Water is a

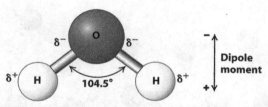

Fig. 1.1 Water molecule bond angle. *Permission source https://alevelbiology.co.uk/notes/water-structure-properties/.*

simple compound containing only three atoms: one oxygen and two hydrogens. Hydrogen atoms in water are bonded to the oxygen atom with precise spacing and geometry. The length of the oxygen bond to hydrogen is exactly 0.9584 A° and the angle formed between all three atoms is 104.45°. A more visual interpretation of a water molecule's structure is shown as a ball and stick model (Fig. 1.1).

The bond between oxygen and hydrogen is a true covalent bond, but electrons in this bond are not shared equally due to the difference in electronegativity between oxygen and hydrogen atoms. Oxygen is a highly electronegative atom and hydrogen is weakly electronegative. As a result of the difference in negativity, electrons spend more time on the oxygen end of the bond, giving it a slightly negative charge. Conversely, electrons spend less time at the hydrogen atom giving it a slightly positive charge. The asymmetrical distribution of electrons between hydrogen and oxygen is termed a dipole. Dipoles are noted by Greek letter delta (δ) and indicates a partial positive or negative charge exists in the bond. The letter δ together with the appropriate sign (positive, δ^+ or negative, δ^-) indicates the direction of bond polarity. The dipoles between hydrogen and oxygen atoms are responsible for the force that holds water molecules together, called hydrogen bonding. Water molecules have a V shape, providing optimal geometry for hydrogen bonding between water molecules. Each water molecule is hydrogen bonded to four others and this extensive interaction is responsible for its unique physical properties (Fig. 1.2, Yan, 2000).

While water molecules are linked by hydrogen bonding, their position is not fixed. Water molecules in the liquid state rapidly exchange their bonding partners.

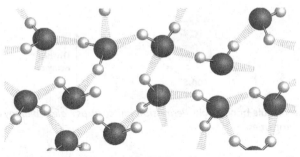

Fig. 1.2 Hydrogen bonding of water molecules. Permission source Shutterstock ID: 350946731.

Physical properties of water

Surface tension is a surface property of liquids that allows resistance to external forces. Water's surface tension results from the attractive forces (hydrogen bonding) between molecules. Surface tension also enables insects (e.g., water spiders) to walk on water and unusual objects to float on the surface of water (Fig. 1.3).

How can surface tension be demonstrated using a cup of water and a paperclip?

Floating a metal paperclip on the surface of water is often used to demonstrate its surface tension properties. Adding a drop of dish washing detergent to the water immediately causes the paperclip to sink. The explanation for its

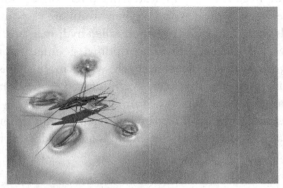

Fig. 1.3 Water Strider Insect walking on water Permission source Shutterstock ID: 276367415.

sinking is that detergents are surfactants that disrupt hydrogen bonds between water molecules.

Surfactants: Surfactants are substances containing both polar and nonpolar properties. They disrupt hydrogen bonding between water molecules and destroy its surface tension. Droplet formation is another example of water's surface tension property. Water exiting an eye dropper or sprayer forms discrete spherical droplets because molecules near the surface have fewer hydrogen bonding partners. Those in the interior have greater hydrogen bonding. Water molecules are thus pulled to the center of the droplet, resulting in a spherical shape (Labuza, 1970; Yan, 2000).

Specific heat capacity: The amount of energy required to raise the temperature of one gram of water (one degree centigrade) is known as the specific heat capacity. The specific heat of water is higher than other similarly sized molecules (e.g., methane), due to extensive hydrogen bonding. The high specific heat capacity of water enables it to absorb or lose large amounts of heat without undergoing a substantial change in temperature. For example, the temperature of water is slow to increase as it is heated, until it reaches 100°C. Water's specific heat capacity regulates the temperature of the planet because large bodies of water act as a buffer to changes in air temperature. Water's specific heat explains why the temperature in Hawaii stays within a relatively small range.

Phase changes of water

Water undergoes reversible state transitions from solid to liquid to gas depending on conditions of temperature and pressure. The structure and mobility of water molecules differ in these states. In the gas state, water molecules have the highest mobility because the hydrogen bonding force weakens as temperature increases. Conversely, the mobility of water molecules is lower in liquid and solid states because the strength of hydrogen bonding is higher at lower temperature. Water's physical properties are unique compared to molecules of similar size. Water exists in the solid state (ice) at 0 °C and below. It melts and transitions to the liquid state as the temperature increases from 0 °C to 100 °C, above 100 °C water exists in the gas state. In contrast, methane is a molecule of similar size and weight. However, the melting and boiling points of methane are very different from water. Methane exists in the solid state at −182.6 °C and transitions to the gas state at −161.4 °C (Table 1.1).

Table 1.1 Physical properties of methane and water.

Physical property	Methane (CH$_4$)	Water (H$_2$O)
Molecular Weight	16.04	18.01
Melting Point	−182.6 °C	0 °C
Boiling Point	−161.4 °C	100 °C

Water as a solid

At 0 °C, water becomes a solid (ice) with structural and physical properties that are substantially different from the liquid state. Freezing water is an exothermic (heat liberating) process. While that statement may seem incorrect, heat is removed during the transition from liquid to solid state. At 0 °C water exists as crystalline lattice, variably composed of nine distinct forms. The bond angle between oxygen and hydrogen atoms is different for water molecules in the liquid and solid states. Specifically, the angle increases from 104.5° (liquid state) to 106.6° in ice. The thermal conductivity of ice is greater because water molecules in the liquid state absorb some energy through their motion.

Why did my can of pop explode in the freezer?

When water forms a crystal lattice, the space between molecules becomes larger and its density is lowered. The increased bond angles and greater distance between water molecules in ice means that a given amount of water occupies a larger volume as ice and thus has lower density. Water expands about 9% in volume in the frozen state. This change in volume is the reason why a can of pop left in the freezer looks like it is about to explode.

Melting point of water: When ice melts, heat is absorbed from the environment. This transition is an example of an endothermic process. Approximately 80 calories of heat are absorbed per gram of ice as it melts. The transfer of energy in melting ice is known as the latent heat of fusion. It is a measure of the amount of heat required to convert a solid to a liquid. Making ice cream at home takes advantage of water's high latent heat of fusion. The ice cream mix is placed in a bucket of ice to which salt is added. Salt causes ice to melt and the resulting endothermic process absorbs heat from the liquid ice cream mix causing it to solidify. Latent heat of fusion can be observed when ice is added to a glass of pop. The temperature of the beverage is lowered to about 0 °C and remains steady until the ice is melted.

Water as a gas

Water has a high boiling point compared to molecules of similar size and composition (e.g., methane). The reason for water's higher boiling point is that greater amounts of heat must be added to overcome its attractive forces (hydrogen bonds). Water's heat of vaporization (about 540 calories per gram) is very large for a molecule of its size. The transition of liquid water to the gas phase is called vaporization. There are two types of vaporization: evaporation and boiling. Evaporation is a transition from liquid to gas occurring at its surface and at a temperature below the boiling point. The amount of evaporation is directly related to the exposed surface area and pressure of the air above it. Boiling results from formation of gas bubbles below the surface of the liquid water that rise to its surface. The boiling point is also dependent on the pressure of air above the liquid. Water boils at less than 100 °C when the pressure is reduced. Conversely, the boiling point of water is greater than 100 °C when the pressure is increased.

Why does it take longer to boil potatoes in Denver than in Chicago?

Denver's elevation (approximately 5,000 ft) results in lower air pressure above the water. Consequently, water boils at a lower temperature (above 95 C) compared to sea level and a longer time is required to cook potatoes. Similarly, a closed system, such as a pressure cooker, operates at higher than ambient pressure and potatoes are cooked in less time.

Sublimation occurs when ice is converted directly to the gas state without going through the liquid state. Water molecules in ice require energy (heat) and sublimation, like melting, is an endothermic process. Dry ice (solid carbon dioxide) is excellent for keeping foods frozen because it absorbs large quantities of heat during sublimation. Sublimation is also responsible for that shrunken ice cube found in the back of the freezer. Sublimation is the physical basis of lyophilization that is the process used to make a variety of shelf stable foods. Lyophilized foods are made by placing frozen product in a chamber and removing water by a strong vacuum. Under these conditions, water in the product undergoes direct solid to gas transition with little or no rise in temperature.

Chemical properties of water

For a simple molecule composed of only three atoms, water has many properties that can only be explained by considering its chemistry. The water with which we interact in everyday life is highly concentrated. Its concentration (about 55 Molar) results from the extensive hydrogen bonding between molecules. A mole is the notation used in chemistry to describe how many atoms or molecules of a substance are present. Molarity (abbreviated as M) is the term used to indicate the concentration. A 1 M solution contains 1 mol of a substance dissolved in 1 L of liquid. Moles and molarity are terms used in chemistry and cooking to identify how many atoms or molecules we are working with in the laboratory or kitchen. The definition of a mole is based on the carbon atom. Carbon has a mass of 12.000 g, corresponding to 6.02×10^{23} atoms of carbon per mole of carbon. The number 6.02×10^{23} is a constant known as Avogadro's number in honor of the 18th century Italian physicist, Amadeo Avogadro, who first proposed the concept. A mole of any element contains 6.02×10^{23} atoms of that element and its mass corresponds to its atomic weight. Using the Periodic Table of Elements, we find that one mole of iron (Fe) contains 6.02×10^{23} atoms and weighs about 56 g. The same rule applies to molecules. One mole of water (H_2O) contains 6.02×10^{23} molecules. Since its atomic weight is 18, water weighs about 18 g. Similarly, the atomic weight of salt (NaCl) is 58. One mole of NaCl contains 6.02×10^{23} molecules and weighs 58 g.

Acid-base chemistry

The introduction to the principles of acid–base chemistry and measurement is included in this chapter because of its importance to food quality and safety. The level of acid in food is critical to its preservation. Foods high in acid store well and typically do not require refrigeration. The centuries old process of fermentation creates acids that inhibit the growth of spoilage microorganisms and convert perishable commodities (milk and meat) into stable foods (cheese and sausage). In contrast, foods low in acid and high in moisture spoil quickly and can promote the growth of pathogenic microorganisms. pH is one way to measure of the level of acidity in food. For example, pH is critical in determining the extent of thermal processing needed to can foods safely.

What is an acid or base?

The oldest and least accurate description of acid and base was founded on experiential observation. Acids were associated with a sour or sharp taste and the ability to turn litmus paper red. Acids react with some metals (e.g., iron and zinc) to liberate hydrogen gas. Bases were associated with a bitter taste and the ability to turn litmus paper blue. Bases in water solution give it a slippery feel. More precise definitions of acids and bases were provided in the late 19th and early 20th centuries. The first was proposed by Swedish chemist Svante Arrhenius in 1877. His definition stated that an acid is anything that produces hydrogen ions (H^+) in water. Similarly, anything that produces hydroxide ions (OH^-) in water is a base. For example, hydrochloric acid (HCl) is an acid according to the Arrhenius definition because HCl dissolves in water to form hydrogen ions (H^+) and chloride ions (Cl^-).

$$HCl + H_2O \rightarrow H^+ + Cl^-$$

Hydrogen ion is very reactive and once formed quickly adds to a water molecule to form the hydronium ion (H_3O^+), as shown in the following equation. The hydronium ion is preferred way to view the acidic form of water.

$$H^+ + H_2O \rightleftharpoons H_3O^+$$

What is a proton? A proton is a hydrogen atom separated from its electron. Hydrogen is the simplest element made up of only two elemental particles, one proton (positively charged) and one electron (negatively charged). When a hydrogen atom loses its electron, it becomes the positively charged species, called a proton.

While the Arrhenius definition is valid for describing acids and bases in water, it does not apply to acid–base reactions that take place in nonaqueous environments. The second and broader definition of acid or base was provided independently by Johannes Brønsted and Martin Lowry in 1923. According to their definition, anything that can donate a proton is an acid and anything that can accept a proton is a base. The advantage of this broader definition is illustrated in the simple reaction between ammonia (NH_3) and hydrochloric acid (HCl). In this reaction, the acid (proton donor) is HCl and the base (proton acceptor) is NH_3. The product of the reaction between an acid and a base is a salt named ammonium chloride (NH_4Cl).

$$NH_3 + HCl \rightarrow NH_4Cl$$

The neutralization of an acid by a base, or vice versa, results in formation of a salt. An example of neutralization is the reaction between hydrochloric acid (HCl) and sodium hydroxide (NaOH) resulting in formation of salt, sodium chloride (NaCl).

$$HCl + NaOH \rightarrow NaCl + H_2O$$

Ionization of water

Water's ability to ionize has a substantial impact on food. Ionization is the process by which an atom or molecule becomes charged by gaining or losing an electron to form an ion. Water molecules continually dissociate to form ions and re-associate to form water in a rapid and dynamic process. Water spontaneously ionizes to form two species: hydronium and hydroxide ions (H_3O^+ and OH^-). The equations for this process are:

$$H_2O \rightarrow H^+ + OH^-$$

$$H^+ + H_2O \rightarrow H_3O^+$$

A summary equation is:

$$2H_2O \rightleftarrows H_3O^+ + OH^-$$

In pure water at 25 °C, ionization produces equal amounts of hydronium and hydroxide ions. Because the rates of dissociation and re-association of water molecules are equal under these conditions, the process is at equilibrium. It should be kept in mind that the total amount of ionization in water is small. In pure water at 25 °C, the concentration of hydronium ion is only 0.0000001 M. Similarly, the concentration of hydroxide ion is the same, 0.0000001 M. Writing numbers this way is tedious and leads to mistakes (e.g., adding or leaving out a zero) that are avoided by using exponential (scientific) notation as shown below.

$$0.0000001M\ H_3O^+ = 1.0 \times 10^{-7}M\ H_3O^+$$

pH and measuring acidity

pH is the term used to express the measurement of hydronium ion concentration in solution. The chemical definition of pH is given as "the negative log of the hydronium concentration" and corresponds to the following equation.

$$pH = -Log[H_3O^+]$$

This equation of acidity is very useful and often employed to find the pH of materials (in solution) including water, soil, and food. The hydronium ion concentration of pure water at 25 °C is 1.0×10^{-7} M. Using the equation for pH, the value for water under these conditions is 7.

$$pH = -Log[1.0 \times 10^{-7}M]$$

$$pH = -(-7)$$

$$pH = 7$$

This equation can be used to find the pH of soda pop and egg white, as shown below. Soda pop contains phosphoric acid and is an acidic food. The hydronium ion concentration of a typical pop is 1.0×10^{-4} M, therefore its pH is:

$$pH = -Log(1.0 \times 10^{-4}M)$$

$$pH = -(-4)$$

$$pH = 4$$

Egg white, in contrast to soda pop, represents one of the few alkaline foods. This means it contains a enough base to raise its pH above 7. The hydronium ion concentration of egg white is 1.0×10^{-8} M. Therefore, its pH is:

$$pH = -Log(1.0 \times 10^{-8}M)$$

$$pH = -(-8)$$

$$pH = 8$$

The pH scale ranges from 1 to 14, with pH 1 being very acidic and pH 14 being strongly basic (alkaline). One of the only places to find a substance with a pH of 1 is inside the stomach. Its digestive fluid contains HCl and the pepsin enzyme essential to digestion. The highly acidic environment denatures proteins and makes their breakdown by pepsin more effective. In contrast, the pH of the small intestine is alkaline (about pH 8), and promotes further digestion by other enzymes.

What about the pH scale?

Because the pH scale is a logarithmic representation of the hydronium ion concentration, a change of 1 pH unit represents a 10-fold change in hydronium ion concentration. Examples of pH and corresponding hydronium ion concentration for a variety of foods is shown in Table 1.2.

Strong and weak acids

Strong acids dissociate completely in water. Strength of an acid is indicated by a molecule's ability to donate protons. Hydrochloric acid (HCl), for example, is a strong acid because it dissociates completely in water to form

Table 1.2 pH of Common Foods.

pH	$[H_3O^+]$		Source
1	0.1	1.0×10^{-1}M	Digestive fluid
2	0.01	1.0×10^{-2}M	Lemon Juice
3	0.001	1.0×10^{-3}M	Vinegar
4	0.0001	1.0×10^{-4}M	Tomatoes
5	0.00001	1.0×10^{-5}M	Beer
6	0.000001	1.0×10^{-6}M	Meat
7	0.0000001	1.0×10^{-7}M	Drinking water
8	0.00000001	1.0×10^{-8}M	Egg White

hydronium and chloride ions. This is shown in the equation below. Addition of 0.01 Mole of HCl to one liter of water results in a concentration of 0.01 M for both H_3O^+ ions and 0.01 M Cl^- ions. The pH of this solution is obtained from its hydronium ion concentration (1.0×10^{-2} M), or 2.0.

$$HCl + H_2O \rightarrow Cl^- + H_3O^+$$

In contrast to strong acids, weak acids are poor proton donors because they only partially dissociate. The generalized equation below for a weak acid HA, shows that it dissociates to form H^+ and A^- ions. Dissociation of weak acid HA to its products H^+ and A^- is shown in the equation proceeding from left to right (right pointing arrow). Conversely, formation of HA from H^+ and A^- is shown in the equation proceeding from right to left (left pointing arrow). An equilibrium is quickly established between the two competing processes:

$$HA \rightleftharpoons H^+ + A^-$$

Note: In this generalized equation, the hydrogen ion is designated as the simple proton H^+ and not the hydronium ion. Additionally, the A^- ion is the conjugate base of the acid HA. A conjugate base is what remains after an acid has donated a proton. The equilibrium expression used to show the dissociation of weak acids is written as:

$$K_a = \frac{[H^+][A^-]}{HA}$$

Terms in this equation represent the concentration of the species and the term K_a represents the dissociation constant. This equation illustrates that acids with greater dissociation to products (H^+ and A^-) have a higher K_a value. In other words, that acids with a high K_a are more acidic than those with a low K_a.

Using K_a to calculate pH

The pH of a solution of a weak acid can be calculated knowing its concentration and dissociation constant, K_a. For example, the K_a of acetic acid (e.g. the acid in vinegar) with its assumed concentration of 0.1 M can be used to find the pH of vinegar. In this example, the K_a of acetic acid is 1.8×10^{-5} and its concentration is 0.1 M. The equation for dissociation of acetic acids is:

$$CH_3COOH \rightleftharpoons CH_3COO^- + H_3O^+$$

The equation shows that acetic acid dissociates to form equal amount of acetate and hydronium ions. The data for this calculation are given below. The initial concentration for acetic acid as 0.1 M. Since there is no dissociation at first, the concentrations of CH_3COO^- and H_3O^+ are zero. The second line gives concentrations of CH_3COO^- and H_3O^+ at equilibrium. Here the acetic acid concentration is 0.1 M minus the amount that dissociated (x) or 0.1-x. Similarly, the concentrations of CH_3COO^- and H_3O^+ are equal, but unknown (x).

Calculating pH of 0.01 M Acetic acid.

	CH_3COOH	CH_3COO^-	H_3O^+
Initially	0.1 M	0.0	0.0
At Equilibrium	0.1-x	x	x

Using the data above, the pH for this acetic acid solution is calculated as follows:

$$K_a = \frac{[CH_3COO^-][H_3O^+]}{CH_3COOH}$$

$$1.8 \times 10^{-5} = \frac{[x][x]}{0.1 - x}$$

$$1.8 \times 10^{-5} = \frac{[x^2]}{0.1}$$

$$X^2 = 1.8 \times 10^{-6}$$

$$x = [H^+] = \sqrt{1.8 \times 10^{-6}} = 1.33 \times 10^{-3}$$

$$pH = -Log[1.32 \times 10^{-3}] = 3-Log\ 1.32$$

$$\mathbf{pH} = 2.88$$

What is pKa?

pKa is an expression denoting the degree of acid ionization. pKa like pH, is a log-based representation of a number. pKa is defined as the negative log of the K_a value. For example, it is more convenient to use the pKa of acetic acid as 4.7 as opposed to writing 1.8×10^{-5}. It is important not to confuse pH and pKa. pH is a direct measure of hydronium ion concentration and pKa represents the ability of an acid to dissociate. Additionally, it should be noted that there is an inverse relationship between Ka and pKa values. The smaller the Ka, the larger the pKa.

Weak acids contribute to the flavor of food and are responsible for its resistance to spoilage. Carbonated beverages represent an extreme example of an acidic food. These beverages are made by dissolving carbon dioxide (CO_2) in liquid. Dissolved CO_2 becomes Carbonic Acid. Many of these carbonated beverages contain additional acids such as citric acid and/or phosphoric. Cheese and yogurt contain propionic, butyric, and lactic acid. Wine contains malic and lactic acids. Fruits such as blueberries and cherries contain malic, lactic, and tartaric acids. Table 1.3 contains additional examples of weak acids found in food.

There is considerable difference in the strength of weak acids. The examples in Table 1.3 are listed with strongest acids at the top and weakest at the bottom. The K_a and pKa values show an inverse relationship for the examples given. The asterisks (*) indicate acids that are polyprotic. Polyprotic acids have more than one carboxylic acid (COOH) group and only the lowest K_a is given.

Table 1.3 Common Food Acids K_a and pKa.

Acid	Formula	K_a	pKa
Phosphoric	H_3PO_4	6.0×10^{-3}	2.2
Tartaric	$H_2C_4H_4O_6$	$6,8 \times 10^{-4}$	3.0
Citric	$H_3C_6H_5O_7$	7.4×10^{-4}	3.1
Malic	$C_4H_6O_5$	1.4×10^{-3}	3.4
Lactic	$CH_3(CHOH)COOH$	1.4×10^{-4}	3.9
Acetic	CH_3COOH	1.8×10^{-5}	4.7
Butyric	$CH_3(CH_2)_2COOH$	1.5×10^{-5}	4.8
Propionic	CH_3CH_2COOH	1.3×10^{-5}	4.9
Carbonic	H_2CO_3	4.5×10^{-7}	6.3

Weak acids and buffering

A buffer is a solution that can resist change in pH when an acid or base is added to it. Buffers can be weak acids or weak bases. The concept of buffering can be demonstrated using acetic acid as a model. It has been shown that a 0.1 M acetic acid solution has a pH of about 2.9. At pH 2.9 there is an equilibrium between the acid and its dissociation to hydronium and acetate ions.

$$CH_3COOH + H_2O \rightleftarrows CH_3COO^- + H_3O^+$$

Acetic Acid   water   Acetate ion   Hydronium ion

When a small amount of acid (H_3O^+) is added to this solution, a proton is added to the acetate ion (CH_3COO^-) causing a shift in the equilibrium and a slight decrease in pH. Continuing to add acid will result in a decrease of pH until all the acetate has been neutralized. At that point, the buffering effect ends and pH of the solution declines sharply.

$$CH_3COO^- + H_3O^+ \rightleftarrows CH_3COOH + H_2O$$

Conversely, when a small amount of base (OH^-) is added to the solution, a proton is taken from acetic acid (CH_3COOH) causing a shift in the equilibrium and increase the pH. Continuing to add base results in an increase of pH until all the acetic acid has been neutralized. At that point, the buffering effect ends and pH of the solution increases sharply.

$$CH_3COOH + OH^- \rightleftarrows CH_3COO^- + H_2O$$

Without a buffer, small additions of acid or base cause a large decrease, or increase, respectively, in pH.

Henderson-Hasselbalch equation

Weak acids and bases play important roles in the chemistry of food and life. Since these molecules only partially dissociate in water, their strength as an acid or base varies considerably. As stated above, pKa is a measure of an acid's ability to ionize and pH is a direct measure of acidity (i.e., hydronium ion concentration). The relationship between pH and pKa can be combined in an useful expression known as the Henderson–Hasselbalch equation. For example, this equation can be used to illustrate the buffering effect of weak acids and bases. Briefly, the equation states that pH of a weak acid is equal to its pKa plus the log of ratio between the basic and acidic forms.

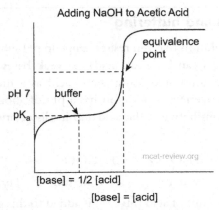

Fig. 1.4 Titration of a weak acid with a strong base. *Permission source https://chemistry. stackexchange.com/questions/75525/what-is-causing-the-buffer-region-in-a-weak-acid- strong-base-titration.*

$$pH = pKa + Log \frac{[Base]}{[Acid]}$$

Using acetic acid as the model system the equation becomes:

$$CH_3 COOH + OH^- \quad \rightleftarrows \quad CH_3COO^- + H_2O$$

$$\text{Acetic Acid} \qquad\qquad \text{Acetate ion (conjugate base)}$$

$$pH = 4.7 + Log \frac{[CH_3COO^-]}{[CH_3 COOH]}$$

The graph in Fig. 1.4 illustrates the titration (neutralization) of a weak acid (acetic acid) with a strong base: sodium hydroxide (NaOH). The pH is indicated on the vertical axis and the amount of NaOH added is indicated on the horizontal axis. Starting at the left-hand portion of.the curve (no added NaOH), the pH is at its lowest and directly corresponds to the concentration of acetic acid. As NaOH is added, some acetic acid is converted to acetate and the pH increases substantially. Continued addition of NaOH does not have the same affect on pH in the middle portion of the curve. In fact, there is relatively little change in pH in this middle portion even though an increasing amount of base is added. The flat region of the curve in this graph corresponds to the buffering capacity of acetic acid. At the exact midpoint of this curve the concentrations of acetic acid and its con-jugate base, acetate ion, are equal. Maximum buffering capacity is reached when the pH of the solution is equal to the pKa of the acid. This point is verified by substituting equal values for the concentrations of acetic acid and acetate ion into the Henderson–Hasselbalch equation. Addition of

NaOH beyond the point when 90% of the acetic acid has been neutralized, causes the pH to increase sharply (as shown in the right hand portion of the curve).

Acid-base chemistry in food

Now that the essential principles of acid and base chemistry have been described, we can explore the importance of these concepts in food.

Gee fizz, what makes soda pop tasty?

Fizziness and a taste described as being sharp or crisp are desirable properties of carbonated soft drinks. The preference for fizziness and sharp flavor of carbonated beverages is rooted in chemistry and biology.

Chemistry produces acid. Fizziness in soda is created by dissolving carbon dioxide (CO_2) at high pressure, in water. Carbon dioxide dissolves in water to form the weak acid, carbonic acid (H_2CO_3). Carbonic acid subsequently forms hydronium ions in water and lowers the pH as illustrated in the following equations.

$$CO_2 + H_2O \rightleftharpoons H_2CO_3 (\text{carbonic acid})$$

$$H_2CO_3 + H_2O \rightleftharpoons H_3O^+ + HCO_3^- (\text{Bicarbonate})$$

Releasing the pressure reverses the reaction and liberates carbon dioxide

$$H_2CO_3 + H_3O^+ \Rightarrow CO_2 \uparrow + H_2O$$

Biology, acid stimulates taste receptors: The sharp and desirable tastes of carbonated beverages results from activation of a taste bud receptor. Specifically, the sensation is caused by the enzyme, carbonic anhydrase (Chandrashekar et al., 2009). This enzyme is fixed to sour taste bud cells on the tongue. Carbonic anhydrase quickly catalyzes the conversion of CO_2 to Carbonic Acid (H_2CO_3), serving as a stimulus that triggers the pleasant sensory experience in the brain. Phosphoric acid (H_3PO_4), also present in most soft drinks, increases the level of CO_2 and intensifies the taste response.

Chemical leavening

Leavening agents are widely used in baking applications and consist of mixtures of acids and bases (Lindsay, 2007). Leavening agents produce gas (CO_2) by a chemical reaction instead of a yeast fermentation. They provide

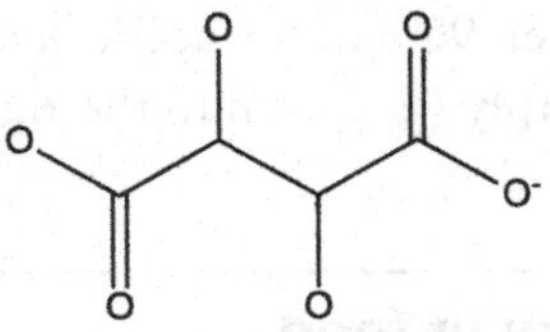

Potassium Bitartrate

Fig. 1.5 Potassium bitartrate.

the advantages of convenience and added texture to baked goods. Chemical leavening is used in quick breads, cakes, cookies, and refrigerated biscuit dough. Chemical leaveners produce carbon dioxide gas when water is added and/or in response to heat. Sodium bicarbonate is the base and source of CO_2 gas produced following a chemical reaction with acid. For example, addition of vinegar to baking soda produces carbon dioxide bubbles and is a common demonstration of this reaction. A variety of acids can be used in combination with sodium bicarbonate to produce gas, but potassium bitartrate (more commonly known as cream of tartar) is widely used in food systems. Potassium bitartrate (shown in Fig. 1.5) is the acidic salt of tartaric acid.

Acidic salts

Potassium bitartrate, the acid salt widely used for leavening applications, is made from tartaric acid. Tartaric acid contains two carboxylic acid groups. Each is capable of donating a proton to a base (proton acceptor). Potassium bitartrate is made by neutralizing one of the carboxylic acid groups with potassium hydroxide, resulting in the potassium (K) salt.

Baking soda, baking powder, and double acting baking powder

Baking soda is the common name for sodium bicarbonate ($NaHCO_3$). Baking soda liberates carbon dioxide (CO_2) gas by the addition of an acid, as shown in the equation.

$$NaHCO_3 + H_3O^+ \rightarrow CO_2 \uparrow + Na_2CO_3 + H_2O$$

Alternatively, baking soda liberates CO_2 gas when heated.

$$NaHCO_3 + Heat \rightarrow CO_2 \uparrow + Na_2CO_3 + H_2O$$

Baking powder is a dry mixture of baking soda (sodium bicarbonate) and an acid salt. This combination liberates carbon dioxide by the same chemical process as shown in the above reactions between sodium bicarbonate and potassium bitartrate. In the example below, potassium bitartrate reacts with sodium bicarbonate (baking soda) to form Carbonic Acid.

$$NaHCO_3 \text{ (bicarbonate)} + KC_4H_5O_6 (\text{K bitartrate})$$
$$\rightarrow NaKC_4H_4O_6 \text{ (NaK tartarate)} + H_2CO_3$$

Carbonic Acid then decomposes to produce the carbon dioxide gas responsible for expanding the dough.

$$H_2CO_3 (\text{carbonic acid}) \rightarrow H_2O + CO_2 \uparrow$$

When the reaction liberating carbon dioxide is initiated by the addition of water alone, the mixture is referred to as a single-acting baking powder. There are advantages to having an acid in the mixture beyond generating carbon dioxide gas. The alkaline nature of sodium bicarbonate (baking soda) alone can give quick breads and other baked goods bitter flavors and a yellowish color. The acid contained in a baking powder mixture neutralizes some of the carbonate, reducing the negative effects caused by alkalinity. Double-acting baking powder is also a mixture of baking soda and acids. While the chemistry of gas production is the same for both single- and double-acting baking powders, there is a difference in how it occurs. Double-acting powder contains two types of acid: one that functions as soon as water is added and a second that functions when heat is applied. The first acid, potassium bitartrate, quickly produces a relatively small amount of CO_2 gas when water is added, allowing time for mixing and pouring operations. The second acid produces additional CO_2 gas when the mixture is heated to about 140 °F/60 °C. Compounds such as sodium aluminum sulfate, sodium aluminum phosphate, or sodium acid pyrophosphate are examples that produce acid when heated.

Titratable acidity

Determination of pH is a convenient way to measure the level of the hydronium ion concentration. But pH determination does not represent the total amount of acid (hydronium ion) potentially available from all weak acids in the food. Grapes contain several weak acids (i.e., citric, tartaric, and malic) whose individual levels are subject to change during ripening and with variety. Titratable acidity is the method of choice to assess grape acidity. It is defined as the percentage of acid in a food as determined by titration with a standard base (Sader, 1994). In this procedure, a known amount of food sample is titrated with a strong base (NaOH) to an endpoint of pH 8.2. Knowing the precise volume and concentration of NaOH used in the titration enables calculation of the total acid present. The following example illustrates calculation of percent acidity in wine using the above equation and provides percent acid values for both tartaric and malic acids, the principle acids in wine grapes.

$$\% \text{ Acid } = \frac{\text{Vol of NaOH} \times [\text{NaOH}] \times \text{Eq wt of acid} \times 100}{\text{wine sample wt(grams)}}$$

Since the equation calls for the weight of wine sample in grams, it is assumed that each mL of wine has a weight of 1 g, therefore, 20 mL of wine equals 20 g. The volume of 0.1 M NaOH required to titrate this wine sample to a pH of 8.2 is 25 mL ($= 0.025$ L). The equivalent weight (Eq) is an acid, such as acetic acid 60.5 g. The equivalent weight of malic and tartaric acids is 67 g and 75 g, respectively.

Substituting the value of 75 g tartaric into the equation for % acid in the wine sample as tartaric acid, the calculation is:

$$\% \text{ Acid} = \frac{0.025\text{L} \times 0.1\text{M} \times 0.075\text{g} \times 100}{0.020\text{g}} = 0.94\%$$

Similarly, substituting the value of 67 g as malic acid into the equation for % acid in the wine sample, the calculation is:

$$\% \text{ Acid } = \frac{0.025\text{L} \times 0.1\text{M} \times 0.067\text{g} \times 100}{0.020\text{g}} = 0.84\%$$

Titratable acidity is used in wine making to measure acidity at various times during ripening of the grapes. It is used to determine the optimum time for harvesting. It is also used in evaluating wine quality.

What is the acid-ash hypothesis and does alkaline water make my bones stronger?

The acid–ash hypothesis holds that "acid" diets contribute to increased bone loss, potentially leading to osteoporosis. This idea has been adopted by some in the lay community who promote the alkaline diet. However, the evidence to date does not support the therapeutic value of alkaline diets. In particular, drinking alkaline water (pH 8—9) has not been shown to reduce or prevent bone demineralization.

The acid–ash hypothesis holds that diets high in acid producing foods (e.g., meat, poultry, fish, dairy products, eggs and alcohol) result in high acid level when metabolized. Conversely, diets low in acid producing foods (e.g., fruits, vegetables, nuts, and legumes) result in low acid level when metabolized. Foods high in protein, for example, are termed acidic because of their potential to increase acid load. However, a high acid load does not alter blood pH because mineral-containing molecules in bone are mobilized to neutralize the acid. The concern is that over time, diets high in acid-producing foods increase the potential renal acid load (PRAL) and decrease bone mineral density, leading to osteoporosis. Research data linking alkaline diets with reduced risk for osteoporosis and other conditions (kidney stone formation, high blood pressure, cognitive function) are highly mixed. The evidence to date suggests that high acid-producing diets combined with low calcium intake increases the risk of bone erosion. However, diets supplemented with calcium and potassium may provide a protective effect (Fenton et al., 2009; Nicolli and McLearn, 2014; Granchi et al., 2018). Additionally, diets high in fruits in vegetables are well documented to provide a range of health-promoting benefits.

Water in food

Water is the single most important component of food and the largest constituent of milk, meat, and most fruits and vegetables. It contributes the largest percent of composition in many processes. Water is also important in a wide range of dry foods such as flour, powdered milk, cocoa powder, and dried fruits. It is important to understand water's role in the complex matrix of food.

Solutes, solubility, and solutions

In simple terms, a solution is defined as a homogenous mixture composed of two or more substances. The substance dissolved is called the solute and the solution it is dissolved in, is called the solvent. Solutes can be either hydrophilic (water-loving) or hydrophobic (water hating). The extent to which a solute interacts with water dictates its solubility. Solutes can be classified as is ionic, polar, or non-polar. Ionic solutes are atoms or compounds that have lost or gained an electron resulting in a net charge (positive or negative). Ionic compounds, such as sodium chloride (NaCl), are composed of positively and negatively charged atoms joined by an ionic bond. Polar solutes are molecules in which the electric charge is unevenly distributed between two atoms in bond and results in formation of a dipole. Ethanol and glucose are examples of polar solutes that interact well (dissolve) with water without being truly ionized. Non-polar solutes are not ionized and do not contain dipoles. Methane, ethane, fat, and oil are examples of non-polar compounds. They do not dissolve in water.

Ion-dipole interaction

Mixing of NaCl (table salt) and water represents an example of an ion-dipole interaction. Na and Cl atoms are joined together by ionic bonds in salt molecules and tightly packed to form the dry crystalline solid. Dissolving NaCl in water is a process initiated by the attraction of water molecules to salt atoms in the solid. Specifically, negative and positive ends of water's dipole are attracted to Na and Cl atoms in a process called hydration (Fig. 1.6). Hydrated atoms then dissociate from salt molecules and become Na^+ and Cl^- ions in a true solution. Once dissolved in water, Na^+ and Cl^- ions remain strongly associated with water molecules and approximately

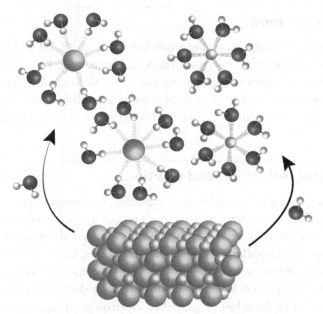

Fig. 1.6 Hydration of salt by water.

6—7 water molecules are attracted to each ion. The solubility of larger solutes, such as proteins, can be greatly increased when hydrated salt ions are attracted to their ionic surfaces (Fig. 1.6).

While NaCl is highly soluble in water, there is considerable variability in solubility of salts. Barium sulfate ($BaSO_4$) is extremely insoluble in water. It is essentially inert under human or animal physiological conditions and has medical applications in CAT scans and other diagnostic tools.

Dipole-dipole interaction

The hydration of glucose with water is an example of dipole-dipole interaction. In this example, glucose hydroxyl (OH) groups attract water molecules forming a complex (Fig. 1.7). Interactions between glucose OH groups and water, represent the strongest form of dipole-dipole interaction (i.e., hydrogen bonds). Proteins rely upon the polar R groups of serine and threonine, for example, to provide the interaction with water needed for solubility and structural stability. Starch and other polysaccharides rely upon their numerous hydroxyl groups to interact with water and provide hydration, viscosity and other functional properties in food.

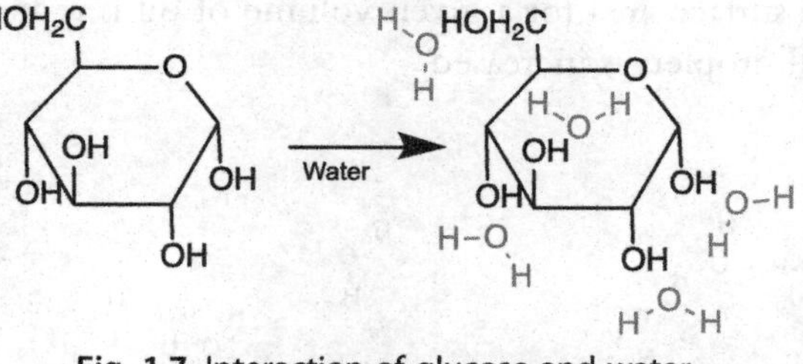

Fig. 1.7 Interaction of glucose and water.

Interaction of non-polar substances with water

While water is often called the universal solvent, not everything dissolves in it. The old adage, "oil and water do not mix" is very true. The interaction of oil (a non-polar solute) with water is different from interactions of solutes containing dipoles or ionic groups. Non-polar molecules like methane, ethane, and food oils are composed almost entirely of carbon and hydrogen atoms. Bonds between these atoms are covalent bonds in which the electrons are shared almost equally. The difference in electronegativity between hydrogen and carbon atoms is small, resulting in a weak dipole.

So what happens when olive oil is added to water?

Oils are lighter (lower density) than water and float on top of the liquid. Oil will not dissolve in pure water because its molecules do not ionize and have extremely weak dipoles. A blender can be used to vigorously mix the oil and water together to form a suspension. Immediately after blending is stopped, the solution is cloudy because the oil has been dispersed into thousands of tiny droplets. In a short time, oil droplets will coalesce and reform into the original pool that floats on the surface.

Why does this happen? Tiny oil droplets formed by blending are pushed in pockets (cavities) within water's extensively hydrogen bonded structure. Formation of cavities requires water molecules immediately surrounding the oil droplet to be stretched out of their optimal shape. Water molecules under these conditions are constrained in mobility and resemble a fixed cage (Fig. 1.8). The combination of creating a cavity in water and constraining the mobility of water molecules results in an energetically unfavorable situation. However, the amount of unfavorable energy in can be reduced by clustering numerous oil droplets into a smaller number of larger ones. This can be demonstrated by considering the geometry of spheres. The amount of exposed surface area in spheres decreases as the diameter increases. The surface area for a given volume of oil is reduced when the diameter of oil droplets is increased.

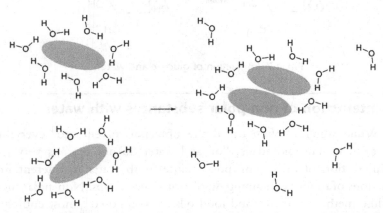

Fig. 1.8 Interaction of water with non-polar substances with water. *Permission source:*
https://www.researchgate.net/publication/320062418/figure/fig3/AS:543321181638657@
1506549272392/The-hydrophobic-effect-Water-molecules-form-an-ordered-structure-
around-a-hydrophobic.png.

Water dispersions

Surfactants are compounds that lower the surface tension of a liquid (water in this case), enabling desirable functionalities in food materials like foaming, emulsifying, and suspension of non-polar molecules in water. Food surfactants are molecules that contain both polar and non-polar groups. This dual character with respect to polarity enables them to form stable mixtures of non-polar materials and water.

Emulsification

Naturally occurring surfactants, such as lecithin, are excellent emulsifiers used in a variety of foods such as chocolates, dressings, desserts and pudding. Lecithin is the generic name for several similar compounds found in plant and animal tissues. It is typically derived from egg yolk or soybean oil. Lecithin is an amphiphilic molecule, meaning that it contains both polar and non-polar groups. Lecithin is a phospholipid composed of a glycerol backbone, two fatty acids and phosphocholine (Fig. 1.9). The fatty acids contribute to the hydrophobic property of lecithin while its strongly polar property is contributed by phosphate (negatively charged) groups.

Examples of lecithin ability to act as an emulsifier include salad dressings and chocolates. Mayonnaise is a dressing made by blending oil with egg yolks containing lecithin. The end product a is stable emulsion, despite the fact that this mixture contains about 80% oil and 8% water. Vinaigrette is a common salad dressing prepared by adding vinegar (acetic acid) and oil to water followed by vigorous mixing. After a short time, the dispersed mixture separates into oil and water fractions. However, if a small bit of liquid egg

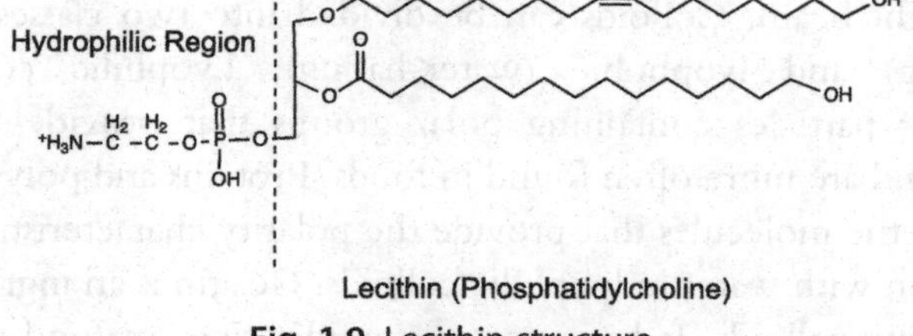

Fig. 1.9 Lecithin structure.

yolk (lecithin) is added, a continuous dispersion is formed. Lecithin is also widely used in chocolate products to improve functional properties. Cocoa butter, the fat in chocolate, contains very little water. Making chocolate requires large amounts of sugar, a very hydrophilic molecule. Lecithin's combination of polar and non-polar groups makes it possible to emulsify the fat and sugar components of chocolate making it easier to flow in coating and molding applications. Lecithin is added to cocoa powder to enhance its dispersion in hot chocolate and chocolate milk beverages. Proteins can also be emulsifiers. Milk contains substantial amounts of fat that can separate and float on the surface, a defect known as creaming. Milk fat is made into a stable emulsified system by homogenization (physical process). The process of homogenization uses high pressure to force milk through very small holes, creating a new population of smaller and more stable, fat globules with milk proteins serving as the emulsifying agent.

Colloids

The term colloid in food systems typically refers to a dispersion of particles in a liquid medium such as water. Colloidal particles are very small and cannot be seen with the unaided eye. They range in diameter from one nm (nanometer) to about one thousand nm. One nm equals 10^{-9} or 0.000000001 m. Particles with dimensions in this range are larger than individual water molecules. When dispersed in a colloidal system, these particles are termed the dispersed phase and the medium in which it is dispersed in is the continuous phase (Petrucci et al., 2007). Particles in stable colloids do not sediment by gravity nor can they be removed by filtration. It is possible to determine whether a mixture is true colloid by shining a light through it. Colloidal solutions will look cloudy or turbid with a light beam passing through them because suspended particles diffract and scatter light. In contrast, true solutions contain only dissolved solutes and will remain clear in a light beam. Colloids can be divided into two classes: lyophilic (water-loving) and lyophobic (water-hating). Lyophilic colloids are composed of particles containing polar groups that provide interaction with water and are more often found in foods. Proteins and polysaccharides are generally the molecules that provide the polarity characteristics essential for interaction with water in lyophilic colloids. Gelatin is an ingredient that forms lyophilic colloids. It has numerous applications in food such as ice cream, yogurt, and gummy bears. Gelatin is composed of collagen that is a protein derived from animal connective tissue and bone. Collagen hydrates well in water and forms reversible gel solutions in response to heating or cooling. When cooled, collagen solution becomes a soft gel by re-establishing intermolecular hydrogen bonds that cross link the molecules. Lyophobic colloids have greater affinity for non-polar materials like oil, are less stable, and eventually coagulate unless a stabilizer is added. Lyophobic colloids have little application in foods, unless first coated with a hydrophilic material such as a polysaccharide. Colloidal dispersions in food systems can fail. Homogenized milk is a colloidal dispersion stabilized by milk proteins. If milk is frozen, crushing caused by expansion of water destabilizes the colloid and causes coalescence of its proteins. The result after thawing milk is a watery mess with coagulated protein on the bottom and separated fat on top.

Water activity (a_w)

Water activity is a measurement used for assessing the water available for chemical and biochemical reactions in food (Yan, 2000). Water activity differs from moisture content. The latter refers to the amount of water present in a material. Typically, moisture content is determined by weighing a food sample, drying it thoroughly in a desiccator, and then reweighing it. The difference in the two weight determinations, expressed as a percent of the sample weight, is its moisture content. In contrast, water activity determines available moisture in a food material by measuring vapor pressure. Some water molecules in food are very tightly associated with its components (e.g., proteins and carbohydrates) and cannot be removed by drying. Water molecules that are strongly absorbed in a monolayer on the solute's surface are called bound water. Water activity determination is the method used by food producers and regulatory agencies to determine shelf life and ensure the safety of foods. It is defined by the ratio of the vapor pressure in the food to the vapor pressure of pure water, under the same conditions (Reid and Fennema, 2007). Specifically, calculation of water activity (a_w) involves measuring vapor pressure of the food material relative to the vapor pressure of pure water and incorporating those values into the following equation.

$$a_w \cong \rho / \rho_0$$

In this equation:
- ρ is the partial pressure of water above the sample
- ρ_0 is the partial pressure of pure water, at the same temperature
- The range of a_w values is from 0 to 1.0

Importance of a_w to food spoilage and safety

Determinations of water activity are relatively easy to perform. Numerous studies have established a link between water activity and the growth of various microorganisms. This relationship enables prediction of a food's susceptibility to microbial spoilage and the potential growth of pathogens. In general, the growth of most bacteria, yeasts, and molds is controlled by water activity levels of less than 0.9, 0.7 and 0.5, respectively. Examples of foods and minimum water activity levels that support microbial growth are listed in Table 1.4.

The rationale for linkage between water activity and ability of a microorganism to grow is based on the obvious need for water to support essential biochemical reactions in the organism (i.e., bacteria, mold, or yeast) (Beuchat, 1981). Microorganisms need to transfer out nutrients in and waste materials of their cells. If available water in the food is insufficient for these purposes, growth is inhibited. It is important to note that microorganisms are not destroyed by water activity levels below their cutoff for growth. Most will resume multiplying when a favorable level returns.

Table 1.4 Foods, water activity, and microbial growth.[1]

Food examples	a_w	Microorganisms that can grow
Highly perishable foods e.g., fresh and canned vegetables, meat, fish, milk, cooked sausages	0.95	Escheria coli Pseudomona
Cured meat (ham), cheddar, swiss and provolone cheese	0.91	Salmonella, Clostridium botulinum,
Fermented sausage (salami), dry cheese, margarine	0.87	Many yeasts, Candida, Torulopsis
Fruit Juice concentrates, maple and chocolate syrup, condensed milk, flour	0.80	Most molds, Saccharomyces spp., Staphylococcus aurous
Jam, marshmallows, glace fruits	0.75	Most halophilic bacteria, toxic aspergilli molds
Dried fruits, caramel, toffee, honey, whole egg powder	0.50 to 0.60	Osmophilic yeasts, some molds

[1]Water Activity of Some Foods and Susceptibility to Spoilage by Microorganisms.
Adapted from Beuchat, LR., 1981. Microbial stability as affected by water activity. Cereal Foods World 26, 345−349.

Relationship between water activity and moisture content

A closer look at water activity and its relationship to moisture content can be seen in the graph of these two variables in a model food system. This type of graph is known as an isotherm (Fig. 1.10). The horizontal axis in this figure corresponds to water activity level in the food. Similarly, the vertical axis corresponds to moisture content in the food. A curve, representing the sorption isotherm (hydration), is obtained by plotting water activity determinations for increasing levels of moisture content (green line). Conversely, a curve representing the desorption isotherm (dehydration) is obtained by plotting water activity determinations for decreasing levels of moisture content (blue line). Dashed vertical lines in the graph define three distinct types of water existing in the food system. Starting at the left, the strongly bound monolayer represents those water molecules that have reduced mobility and cannot be removed by drying. The less strongly bound water represents water molecules that are attracted by chemical forces, such hydrogen bonding, and ion-dipole bonding to solutes in the food, but perhaps at a greater distance. The less strongly bound category includes molecules entrapped in capillary spaces. The third group refers to water molecules that are not associated with the solute molecules and termed free water.

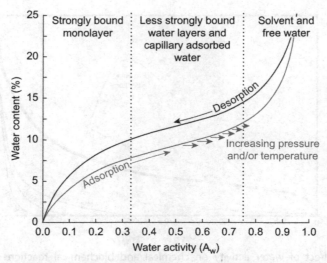

Fig. 1.10 Moisture Content versus Water Activity (a_w) (For interpretation of the references to color in this figure, see the color plate.). *Adapted from Bell, L.N., Labuzza, T.P., 2000. Practicle aspects of moisture sorption isotherm measurement and use. Second Ed.AACC. Egan Press Egan MN. http://www1.lsbu.ac.uk/water/water_activity.html [Accessed March 2019].*

Relationship between water activity and temperature

Water activity isotherms vary considerably with the composition of the food and environmental conditions. A reliable prediction for the growth of microorganisms in a real food material requires that determination be made for each type of product and temperature. Temperature can have a large effect on water activity. Generally, an increase in food temperature will increase its water activity. This means that for a low moisture content e.g., 5%, the water activity and may reduce shelf-life of the due to microbial growth.

Importance of a_w to chemical and biochemical reactions in food

Water activity can influence chemical and biochemical reactions and affect the quality of foods (Chinachoti, 2000). The graph in Fig. 1.11 represents the relationship between water activity (horizontal axis) and reaction rate (left vertical axis) and moisture content (right vertical axis). The sorption isotherm representing a model food is shown as a solid line in Fig. 1.11. Reactions that are affected by the level of water activity include, Maillard browning, lipid oxidation, and enzyme activity. The rate of Maillard browning is highest at water activity (a_w) levels of 0.5 to 0.6. The rate of lipid

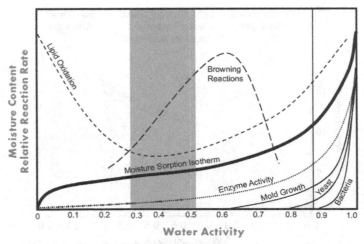

Fig. 1.11 Effect of water activity on chemical and biochemical reactions in food. *Adapted from Labuza, T.P., Tannenbaum, S.R. and Karel, M., 1970. Water content and stability of low moisture and intermediate moisture foods. Food Technol. 24:543–550. Permission Source https://cwsimons.com/determination-of-water-activity-2/.*

oxidation is proportional to the a_w level in a food system (i.e., 0.3 for the low and 0.7 for the high). Water activity can also affect the activity of enzymes in food. For example, lipase (hydrolysis and release of fatty acids) and phenolase (enzymatic browning) are more active at higher a_w levels (e.g., 0.8 to 0.9).

Increasing moisture content without increasing water activity

Humectants provide the advantages of increasing a food's resistance to spoilage, while improving its texture. Salt and sugar are the best known examples of humectants and are generally added to foods as a way of preserving them. The practice of salting meats and making jams with sugar has long been used to extend the self-life of perishable foods. Humectants are compounds that bind water molecules and lower water activity levels. Carbohydrates (i.e., sugars and starches) are good humectants because they have numerous hydroxyl (OH) groups. Glucose contains six hydroxyl groups per molecule and each can hydrogen bond to surrounding water molecules. Molecules containing multiple hydroxyl groups are sometimes called polyols. Foods containing polyols can have relatively high moisture content (20 to 40%), soft texture, and resist spoilage at room temperature. Raisins and other dried fruits, are good examples. The content of sugars (glucose and fructrose) in raisons, decreases water activity to about 0.3. It is possible to change high moisture, perishable foods, such as meat into products that are stable at room temperature by incorporating humectants. This type of product is called an intermediate moisture food. Typically for a meat product, a mixture of salt, polyols, and flavorings are infused into small, cubed meat pieces, followed by cooking. The end product is salami-like sausage with a much softer texture. Intermediate moisture foods are convenience items, but are also important as field rations for feeding the military.

Summary

Water is a key component of most foods. Meat, milk, fruits, and vegetables contain 75% or more water. Water molecules, being composed of hydrogen and oxygen atoms, provide its unique physical and chemical properties (boiling point, surface tension, substance solubility, and ionization of weak acids). Water content affects food texture, those with low moisture level are typically hard and less desirable while high moisture foods are soft but may spoil quickly. Surfactants such as lecithin, proteins, and polysaccharides enable formation of stable dispersions of nonpolar substances (fats) in foods such as chocolate. The acidity of food is important to its quality and safety. Weak acids such as carbonic, malic, and tartaric provide desirable taste and control spoilage. Water supports the growth of microorganisms causing waste and disease. For example, milk with a low pH indicates spoilage. pH is critical to the safety of canned food. Foods with pH of 4.6 and higher must be processed with a higher heat treatment to prevent growth of clostridia botulinum and subsequent production of botulism toxin.

Glossary

Acid 1. A molecule that can donate a proton or accept electrons. 2. A substance that neutralizes alkali and turns litmus red.

Amphiphilic A chemical compound possessing both hydrophilic and hydrophobic properties

Amphiprotic Molecule (or ion) that can donate or accept a proton and act either as an acid or a base. Water, amino acids, hydrogen carbonate ions and hydrogen sulfate ions are common examples of amphiprotic species

Base 1. Molecule that can accept a protons or donor electrons. 2. Substances that are slippery to the touch and taste bitter. Bases react with acids to form salts

Boiling point Temperature at which its vapor pressure is equal to the pressure of the gas above it

Bound water An empirical term used to indicate water molecules that are so tightly bound that conventional drying methods can't remove them

Buffer Substance that resists change in pH when an acid or base is added to it

Colloid Homogenous suspension of one substance in another. Suspended particles do not separate out and are not removed by filtering or centrifugation

Conjugate acid-base pairs In an acid-base reaction, the substance donating a proton is the acid and its conjugate base is the substance receiving that proton

Covalent bond Chemical bond involving the sharing of electrons between atoms in a molecule

Dipole A pair of opposite but equal electrical charges separated by a distance

Endothermic Process or reaction that absorbs heat energy from the environment

Exothermic Process or reaction that liberates heat energy to the environment

Electron A negatively charged component of atoms. Electrons surround the nucleus of atoms and sharing of electrons between atoms is responsible for a major type of chemical bonding (i.e., covalent bond)

Electronegativity The tendency of an atom to attract electrons

Equivalent weight of an acid Weight in grams that will supply one mole of hydrogen ions

Hydrogen Bond Weak bond between two molecules resulting from an electrostatic attraction between a proton (hydrogen) in one molecule and an electronegative atom (oxygen, in the case of water) in the other

Heat of Vaporization The amount of heat required to convert one gram of liquid water to gas

Heat of Fusion/Latent Heat of Fusion Heat per unit mass required for a state change (between a liquid and a solid) to occur without a change in temperature; heat released or absorbed during a constant-temperature process. In the ice cream example, melting ice absorbs heat from the environment and causes the ice cream mix to freeze

Humectant Substance that keeps food moist without increasing water available for microorganism growth; attracts and retains water molecules

Ion Molecule or atom in which the total number of electrons is not equal to the total number of protons, giving it a net positive or negative electrical charge

Ionization Process by which an atom or molecule acquires a negative or positive charge by gaining or losing electrons

Ionic Bond Bond between two groups of opposite charge. Sodium and chloride ions form an ionic bond to make salt, sodium chloride

Melting point Temperature at which the solid and liquid states of a substance are in equilibrium at a specified pressure (usually atmospheric pressure)

Mole Mass of a substance containing the same number of fundimental units as there are atoms in exactly 12.000g of ^{12}C.

Molarity Indicates concentration (i.e., the number of moles of a substance in specific volume 1 L of liquid)

Periodic Table Table of the chemical elements arranged in order of atomic number. Elements with similar atomic structure appear in rows. Elements with similar chemical properties appear in vertical columns

pH A measure of the hydrogen ion concentration in solution and defined as the negative log of that concentration

Polarity A separation of electric charge in a molecule or chemical group (a dipole)

Proton Positively charged component of atoms and is found in the nucleus. Each proton has a mass of approximately one atomic mass unit (amu). The number of protons in an element is equal to its Atomic Number

Neutron An uncharged component of atom and is found in the nucleus. A neutron has a mass that is slightly larger than that of the protein

Solute Substance dissolved in a solution

Solubility Ability of a substance or molecule to form a true solution in water

Solvent Medium used to dissolve a substance

Sublimation Chemical process where a solid turns into a gas without going through a liquid stage

Surface tension Elastic property of liquids resulting from attraction between liquid molecules to each other. In water, the high degree of hydrogen bonding gives water its high surface tension

Surfactant Substances that lower the surface tension of liquids by disrupting attractive forces between liquid molecules. Examples of surfactants include detergents, emulsifiers, proteins and polar lipids

Thermal conductivity Ability to conduct heat. A material with a high thermal conductivity is known as a heat sink. Conversely, a material with a low thermal conductivity is known as an insulator

Titratable acidity Percentage of acid in a sample as determined by titration with a standard base

Vapor pressure Pressure exerted by a vapor in thermodynamic equilibrium with its condensed states (solid or liquid) at a given temperature in a closed system

Water Activity (a_w) Ratio of the vapor pressure of water in solution to the vapor pressure of water, at the same temperature

References

Beuchat, L.R., 1981. Microbial stability as affected by water activity. Cereal Foods World 26, 345–349.

Chandrashekar, J., Yarmolinsky, D., Buchholts, L.V., Oka, Y., Sly, W., Ryba, N.J.P., Zuker, C.S., 2009. The taste of carbonation. Science 326, 443–445.

Chinachoti, P., 2000. Water, water activity. In: Christen, G.L., Smith, S. (Eds.), Food Chemistry: Principles an Application. Science Technology System, West Sacramento CA, pp. p21–34.

Fenton, T.R., Lyon, A.W., Eliasziw, M., Tough, S.C., Hanley, D.A., 2009. Meta-analysis of the effect of the acid-ash hypothesis of osteoporosis on calcium balance. J. Bone Miner. Res. 11, 1835–1840.

Granchi, D., Caudarella, R., Ripamonti, C., Spinnato, P., Bazzocchi, A., Massa, A., Baldini, N., September 12, 2018. Potassium citrate supplementation decreases the biochemical markers of bone loss in a group of osteopenic women: the results of a randomized, double-blind, placebo-controlled pilot study. Nutrients 10 (9). Epub 2018 Sep. 12.

Labuza, T.P., 1970. Properties of water as related to the keeping quality of foods. In: Proceedings of the Third International Congress of Food Science & Technology. Washington, DC, pp. 618–635.

Labuza, T.P., Tannenbaum, S.R., Karel, M., 1970. Water content and stability of low moisture and intermediate moisture foods. Food Technol. 24, 543–550.

Lindsay, R.C., 2007. Flavors. Fennema's Food Chemistry, fourth ed. CRC Press/Taylor & Francis, Boca Raton FL, pp. 639–688.

Nicolli, R., McLearen, H.J., 2014. The acid-ash hypothesis revisited: a reassessment of the impact of dietary acidity on bone. J. Bone Miner. Metab. 32, 469–475.

Petrucci, R.H., Herring, G., Madura, J.D., Bissonnette, C., 2007. General Chemistry: Principles & Modern Applications, ninth ed. Pearson/Prentice Hall, Upper Saddle River, New Jersey.

Reid, D.S., Fennema, O.R., 2007. Fennema's Food Chemistry, fourth ed. CRC Press/Taylor & Francis, Boca Raton FL, pp. 17–82.

Sader, G.O., 1994. Titratable Acidity. In: Nielsen, S.S. (Ed.), Introduction to the Chemical Analysis of Foods, pp. 81–91.

Yan, P., 2000. Food chemistry. In: Christen, G.L., Smith, S. (Eds.), Principles and Application. P9-19. Science Technology System, West Sacramento CA.

Further reading

Properties of water

https://en.wikipedia.org/wiki/Properties_of_water [Accessed Dec 2018].

Acid-base chemistry-in food

https://en.wikipedia.org/wiki/Acid_strength[Accessed Dec 2018].
https://en.wikipedia.org/wiki/Leavening_agent [Accessed Dec 2018].

Titratable acidity in wine

http://www.thevintnervault.com/index.php?p=w_m_tips&id=5801 [Accessed Dec 2018].

Acid-ash hypothesis

Foroutan, R., 2016. Alkaline Diet: does pH affect health and wellness?.
https://foodandnutrition.org/may-june-2016/alkaline-diet-ph-affect-health-wellness/
    [Accessed June 2019].
Frassetto, L., Banerjee, T., Powe, N., Sebastian, 2018. Acid balance, dietary acid load, and
    bone effects-A controversial subject. Nutrients 10 (4). Epub 2018 Apr 21.
https://foodandnutrition.org/may-june-2016/alkaline-diet-ph-affect-health-wellness/
    [Accessed Dec 2018].

Properties of water

Coultate, T., 2009. Food. The Chemistry of its Components, fifth ed. The Royal Society of
    Chemistry. RSC Publishing, Cambridge UK.
Damodaran, S., Parkin, K.L., Fennema, O.R., 2007. Fennema's Food Chemistry, fourth ed.
    CRC Press/Taylor & Francis, Boca Raton FL.

Review questions

1. Define the term hydrogen bond.
2. Why is hydrogen bonding so weak in methane but strong in water?
3. Why does water have such a high boiling point?
4. Why does methane have such a low boiling point?
5. Why isn't oil soluble in water?
6. Why does ice float in water?
7. Why is sucrose so soluble in water?
8. Why did the can of soda pop explode in the freezer?
9. What is responsible for the surface tension of water?
10. Define the term surfactant and give three examples.
11. What is the role of water in microwave cooking?
12. What is responsible for the tingly tastes of soda pop?
13. Define the term weak acid.
14. Define the term pKa and how is it related to the strength of weak acids?
15. Give an example of an acid in food.
16. Describe the acid–ash hypothesis.
17. Describe chemical leavening and what it is used for.
18. How is pH defined?
19. What is a buffer?

20. What is titratable acidity and how is it used?
21. How does addition of an acid change the ionization of water?
22. Define the term water activity.
23. Why does temperature affect the aw value?
24. How does sugar lower water activity?

CHAPTER TWO

Proteins

Learning objectives

This chapter will help you describe or explain:

- Structure of proteins, peptides, and amino acids
- Chemistry of amino acids
- Functional and nutritonal properties of proteins
- Factors that influence protein solubility
- Protein denaturation
- Effects of processing on proteins
- Enzymes and how they work in food

Introduction

Proteins are macromolecules responsible for all biological processes in cells and tissues. Proteins consist of a linear chain of amino acids, the sequence of which is determined by the DNA that encodes for their

Introduction to the Chemistry of Food
ISBN: 978-0-12-809434-1
https://doi.org/10.1016/B978-0-12-809434-1.00002-5

synthesis. Transcription of the DNA blueprint and the subsequent assembly
of amino acids defines a protein's structure and its function. They are varied
by combining the 20 + kinds of amino acid into a polypeptide chain that
assembles and folds into a unique molecule. The forces that stabilize a protein molecule's structure are derived from the chemistry of its constituent
amino acids. In living systems, proteins perform biochemical and structural
functions. As enzymes, proteins are biological catalysts that accelerate the
rate of chemical reactions essential to synthesis, metabolism, regulation,
and other cellular processes. In food, proteins are a source of essential nutrients, textural properties, and flavor. For example, amino acids liberated during ripening of fruits and aging of meat are major sources of flavor in these
foods. Food proteins are derived from a variety of plant and animal sources.
Plant-based protein sources are traditionally found in soy, peas, beans, lentils,
nuts, and wheat. The choice of plant-based proteins is expanding and more
recently includes novel sources such as quinoa, chia, and spirulina (derived
from blue-green algae). Animal protein sources are primarily derived in
meat, milk, egg, and fish. Recently, insect protein has been added to the
list of animal-based protein sources.

There are interesting questions throughout this chapter that will help
you explore and better understand information about proteins.

• How can we use our hands to demonstrate the concept of chirality?
• What is a conditional amino acid?
• Why does egg turn solid when dropped in boiling water?
• What is responsible for the structure of a protein?
• Is a copper bowl better for whipping egg white?
• I just don't crack eggs very well. Will that little bit of yolk in my egg
  whites really matter when I am making meringues?
• Why do my apple slices turn brown?

In the past, proteins were classified on the basis of their solubility as albumins, globulins, glutelins, and prolamines. Albumin proteins are soluble
in water. A principle example is egg white albumin. Globulin proteins are
those soluble in dilute salt solutions. Myofibrillar proteins of muscle are examples of this class. Glutelin proteins are soluble in dilute acid or base solution. High molecular weight glutenin, a type of glutelin, is responsible for
bread dough elasticity and bread texture. Prolamine proteins are soluble
in solutions containing 50%–80% ethanol. Examples of prolamines include
gliadin in wheat, zein in corn, and kafrin in sorghum. Prolamine proteins are
characterized by a high percentage of glutamine and proline amino acids that
are suspected to be linked to celiac disease. Protein science has moved away
from protein solubility-based classification schemes and now relies on a

more precise approach defined by structure and function. Structural classification depends upon comparison of amino acid sequences and structures to classify and determine relationships between proteins. Proteins with the similar amino acid sequences are classified into families that originate from a common ancestor. Legumes, for example, contain seed storage proteins belonging to the vicilin family. Vicilin proteins are found in navy beans, soybeans, peas, lentils, and peanuts. Vicilin proteins have substantial amino acids sequence homology and thus are similar in physical and functional properties.

Proteins and their properties

Amino acids, peptides, and proteins: Amino acids share a common structure as shown in Fig. 2.1A. The central carbon atom of an amino acid is called the alpha carbon. Amino (NH_2) and carboxylic acid (COOH) groups are connected to the alpha carbon via single covalent bond. These groups are often referred to as the alpha amino and alpha carboxylic groups, respectively. A side chain, termed the R group, is also bonded to the alpha carbon and is responsible for differences in the chemical character of amino acids. Imino acids, such as proline and hydroxyproline, are structurally different from other amino acids because their nitrogen atom is contained within a ring that includes the alpha carbon. Three methyl (CH_2) groups join the nitrogen atom to the alpha carbon to complete the ring. The carboxylic acid group is attached to the alpha carbon, but placed outside the ring (Fig. 2.1B). Incorporating proline into a polypeptide backbone substantially restricts its flexibility and makes most secondary structures impossible.

(A)

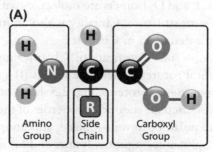

Fig. 2.1A Amino acid structure.

All amino acids, except glycine, occur in chiral forms and are designated by the prefix L or D. The alpha carbon of an amino acids is the chiral center. By definition, each atom bonded to it must be different. Glycine is the only non–chiral amino acid wherein the alpha carbon is bonded to 2 hydrogen atoms. Chirality is biologically important. Many enzymes are stereo specific. This means they will only catalyze reactions with substrate molecules in the correct chiral form. For example, L amino acids are the predominant chiral form found in plant and animal proteins. D amino acids are typically restricted to bacterial proteins.

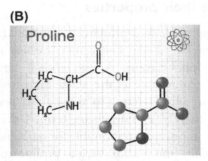

Fig. 2.1B Proline.

How can we use our hands to demonstrate the concept of chirality?

A chiral molecule is indistinguishable from its mirror image but can't be superimposed upon it. You can demonstrate the concept of chirality by placing your hands together palm to palm showing that both are identical (Fig. 2.2). However, the difference is obvious when you place one hand on top of the other. Amino acids are chiral molecules and exist in two stereo–isomeric forms, L and D. Isomers are molecules with the same chemical formula, but different structure. Chirality is also an important property of carbohydrates that is described in Chapter 3.

Amino acid structure and classification: Classification of amino acids based upon the chemical nature of their side chains (R groups) is useful in understanding the properties of proteins. Fig. 2.3 shows amino acids grouped into three categories based on chemical properties of their R groups. Specifically, the classes are polar, polar non–ionized, and nonpolar. Polar amino acids include aspartic, glutamic, arginine, lysine, and histidine. Polar amino

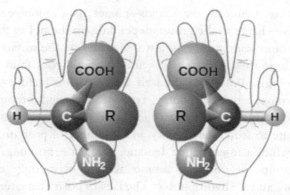

Fig. 2.2 Chiral structure of L- and D-Amino acids. *Permission source https://en.wikipedia. org/wiki/Chirality_(chemistry)#/media/File:Chirality_with_hands.svg.*

Nonpolar R group

CH₃   CH₃ CH₃   CH₃   CH₃   CH₃   CH₃
  | CH   CH   CH |
 Ala | CH₂ | S
            Val   Leu   H₃C–HC   ilu |
                                                           CH₂
            H₂C–CH₂ |
            HN   CH₂   CH₂
              | | |
              H   COOH   Met
             Pro
                          Phe

Polar- Non-Ionized R Group

  H   OH   OH   SH   OH   NH₂   NH₂
  | | | | | | |
 Gly   CH₂   HO–CH   CH₂   CH₂   O=C   O=C
           | | | |
          Ser   Thr   Cys   CH₂   CH₂
                                              CH₂ | |
                                               | Asn   CH₂
                                              Tyr |
                                                                        Gln

Polar-Ionizable R Group

   O   O   NH₂   NH₂
    ‖   ‖ | |
 C–OH   C–OH   CH₂   HN=C   N——CH
  | | | | HC   ‖
 CH₂   CH₂   CH₂   HN   HC   C
  | | | | | N
 Asp   H₂C   CH₂   CH₂   CH₂   H
                       | | | |
                      Glu   CH₂   CH₂   His
                                    | |
                                   Lys   CH₂
                                                   |
                                                  Arg

Fig. 2.3 Classification of amino acids R groups. *Permission source https://www. biologyexams4u.com/2012/09/amino-acids-introduction.html#.XAmaqC3MzAw.*

acid R groups are chemically weak acids or bases that can ionize to give positive or negative charge to the group, depending on the pH of the environment. Polar amino acids are almost always located on the surface of protein molecules because they interact well with water and contribute to the solubility of proteins. Polar non-ionized amino acids include asparagine, glycine, glutamine, serine, threonine, tyrosine, and methionine. R groups of these amino acids are able to hydrogen bond to each other and to water. They contribute to structural stability and solubility of proteins. Nonpolar amino acids include alanine, valine, leucine, isoleucine, proline, and phenylalanine. R groups of nonpolar amino acids contain only carbon and hydrogen atoms and are hydrophobic. The lack of polar character is energetically unfavorable in an aqueous environment. When the nonpolar R groups of a protein are folded into its interior, there is less unfavorable exposure to water. The net effect of clustering nonpolar amino acids into the interior of a protein is a force known as the hydrophobic force (described by the oil-in-water model in Chapter 1). The hydrophobic force contributes substantial energy that stabilizes the three-dimensional structure of proteins.

Acid–base properties of amino acids: Amino acids are weak acid and bases. Specifically, the alpha carboxyl group is a weak acid and the alpha amino group is a weak base. The pK of alpha amino and alpha carboxyl groups of glycine are approximately 9 and 3, respectively. The degree of ionization of the group at a specified pH can be determined using the appropriate pK and the Hederson-Hasselbach equation as described in Chapter 1 and seen below

$$pH = pKa + Log \frac{[Unprotonated]}{[Protonated]}$$

The relationship between pH and charge on an amino acid is demonstrated with glycine in Fig. 2.4. Ionization of amino and carboxyl groups is governed by their pKs. Using the Hederson-Hasselbach equation, a plot illustrating the titration curve of glycine and its charged forms as a function of pH is shown in Fig. 2.4. Beginning at low pH, glycine's alpha carboxyl and alpha amino groups are protonated and the molecule has a positive charge. As the pH is raised above the carboxyl group's pK_1 ($pK_1 = 2.3$), a hydrogen atom is removed and the group becomes negatively charged. However, the alpha amino group retains its positive charge at this pH and the molecule is electrically neutral. The pH at which the positive and negative charges sum to zero is called the isoelectric point or pI (glycine's pI is 5.97). Further increase of the pH above the amino group's pK_2

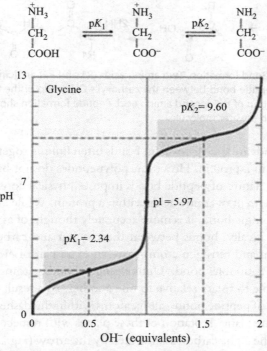

Fig. 2.4 Titration of glycine. *Permission source https://www.biochemden.com/titration-curve-of-glycine/.*

($pK_2 = 9.6$) results in loss of that proton and the amino group is no longer charged. Glycine at pH above 9.6 has a net negative charge.

Primary structure of proteins: Amino acids are the building blocks of proteins. The linear sequence of amino acids in a protein is defined as its primary structure. It is also referred to as a polypeptide chain. The sequence of amino acids in a protein determines its structure and function and is responsible for the huge diversity of proteins found in all forms of life. Random combinations of amino acids in a protein containing 100 amino acids residues is calculated to result in 10^{100} possible polypeptides. Amino acids in polypeptides are joined through a covalent link called the peptide bond. Peptide bonds are formed by a reaction between the carboxylic acid group of the first amino acid (R1) and the amino group of the second amino acid (R2) (Fig. 2.5). During this reaction, a hydroxide (OH) group is split off from the carboxyl end of the first amino acid and a hydrogen is lost from the amino end of the second amino acid. The resulting products are a dipeptide and a water molecule.

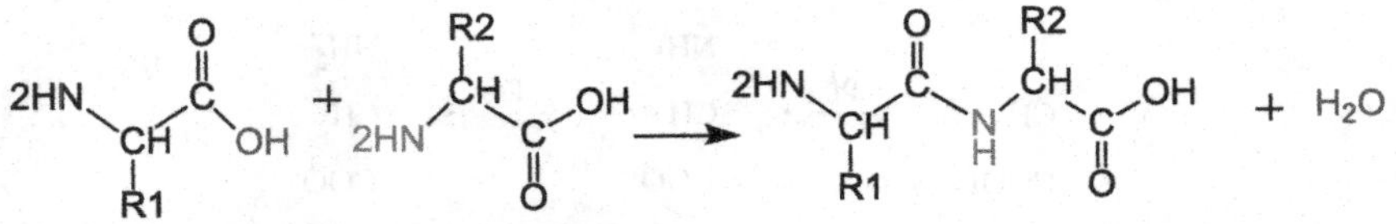

Fig. 2.5 Peptide Bond formation. Two amino acids (R1 and R2) are joined together by formation of a peptide bond between the carboxyl's OH group of the first amino acid and the amine group of the second amino acid. Peptide formation shown here results in a dipeptide and a water molecule.

Peptide bonds are strong covalent bonds often linking together hundreds of amino acids in a protein. However, polypeptides do not have unlimited flexibility. The nature of peptide bonds imposes physical constraints on the rotational movement of amino acids within a protein. While a peptide bond is technically a single bond, it is more accurately thought of as a partial double bond. Specifically, bonds between the oxygen and carbon atoms and between carbon and nitrogen atoms share an extra pair of electrons, effectively making a double bond. Unlike single bonds, atoms in a double bond are not able to rotate relative to one another. As a result of the double bond character of peptide bonds, all six atoms within the dashed line box are fixed in the same plane. Rotation of these planes with respect to each other occurs only at the alpha carbon as indicated by the arrow (Fig. 2.6). Rotation of one plane relative to its neighbor is limited by mutual electron repulsion when atoms are too close together. As an additional consequence of the peptide bond's planar character, amino acids are fixed to each other in either of two configurations, cis or trans. However, the trans configuration is more stable and is found in almost all proteins. Unequal distribution of electrons between these atoms is caused by differences in electronegativity and results in a dipole within the polypeptide backbone. Dipoles (described in

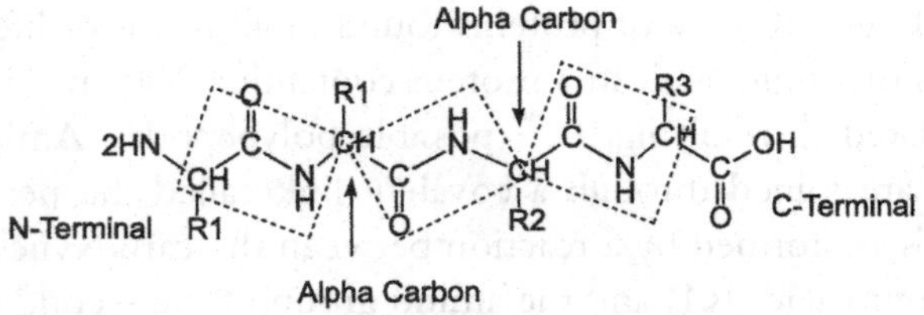

Fig. 2.6 Polypeptide structure. The peptide shown contains four amino acids linked together by three peptide bonds. Individual amino acid side chain positions are indicated as R1 through R4. Alpha carbons in each amino acid are indicated by arrows. The first and last amino acids are designated as its amino (N) and carboxyl (C) terminals. Dashed line boxes illustrate six atoms constrained within the same plane.

Chapter 1) enable hydrogen bonding within and between the structural elements of proteins.

Secondary structure: Within the constraints imposed by the planar nature of peptide bonds, newly synthesized polypeptides self-assemble into secondary structures. Secondary structure is defined as the three-dimensional form of a single polypeptide chain. Several secondary structure conformations are possible, but the alpha helix and the beta sheet forms predominate. Unorganized segments of polypeptide chain also occur and are referred to as a random coil. All forms of secondary structure in proteins are dependent on and defined by patterns of hydrogen bonding in the polypeptide.

The alpha helix is the most common form of secondary structure found in proteins. Several variations of helical structures exist in proteins, but the right-handed alpha helix is most common. In alpha helices, the polypeptide chain appears to be wrapped around a tube (Fig. 2.7). The alpha helix is a coil in which the pitch (the distance along the axis corresponding to one complete turn) is 5.4 Å. Each turn of the coil involves 3.6 amino acids

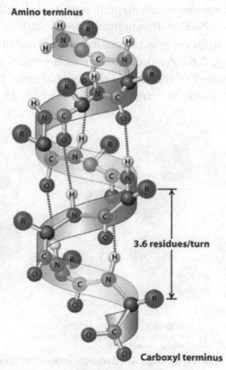

Fig. 2.7 Alpha helix. *WH Freeman.*

and is stabilized by hydrogen bonds in the polypeptide backbone. Every amide (NH) group hydrogen bonds to a C=O (carbonyl) group in the next turn, four amino acids down the chain. All of the hydrogen bonds stabilizing the coil are therefore parallel to the axis of the coil. The R groups of amino acids are oriented perpendicular to the axis of the coil, effectively locating them on the its surface.

Arrangement of amino acids in a polypeptide's primary sequence affects the formation of alpha helices and their overall role in the structure of proteins. Alpha helix formation, for example, is a spontaneous process driven by the nature of its primary structure (amino acid sequence). Alpha helices are likely to a heptad form when the sequence contains a repeat of seven amino acids. Helix forming heptads are characterized by the sequence P–N–P–P–N–N–P, where N and P represent nonpolar and polar amino acids, respectively (Kamtekar et al., 1993). Variations of the heptad repeat and also form alpha helices. Tropomyosin, for example, is a myofibrillar protein that functions in the regulation of muscle contraction. Tropomyosin's structure is unique in that it consists of two completely alpha helical polypeptide subunits. Tropomyosin's heptad repeat is slightly different and consists of the N–P–P–N–P–P–P sequence (Brown, 2010). The importance of heptad repeats to tropomyosin's structure can be seen in the end view of its alpha helices (Fig. 2.8). Amino acids in each heptad are numbered 1 through 7 and 1′ through 7′ for each of the two subunits. Nonpolar amino acids are located at positions 1 and 4 of each heptad. These nonpolar R

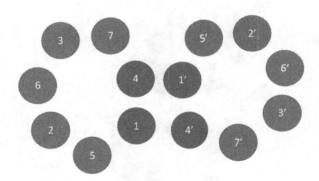

Alpha Helix – end view
Non-polar R groups providing hydrophobic interaction are in red

Fig. 2.8 Alpha helix end view. *https://proteopedia.org/wiki/index.php/Tropomyosin#T ropomyosin.27s_Structure.*

groups form a hydrophobic patch that continues along the length of each tropomyosin subunit and stabilizes its structure. Hydrophobic interaction between subunits is the basis for tropomyosin's molecular structure.

Beta sheet is a common type of secondary structure consisting of beta strands laterally connected by hydrogen bonds. In a beta strand, the carbonyl (C=O) and amide (NH) groups of each amino acid are oriented perpendicular to the direction of the polypeptide chain, enabling extensive hydrogen bonding between chains). Individual beta strands are relatively short in length, but are connected through loop segments called beta turns that contain four amino acids. Multiple beta strand segments are linked together to form a beta sheet in either parallel or anti-parallel direction. Beta strands that are aligned in the same N-to C-terminal direction are described as a parallel beta sheet. Conversely, beta strands that are aligned in the opposite direction (C to N terminal) are termed anti-parallel beta sheets. Hydrogen bonding between beta strands is maximized in the anti-parallel configuration because the bond angle is near zero, creating the shortest distance between carbonyl and amide groups (Fig. 2.9).

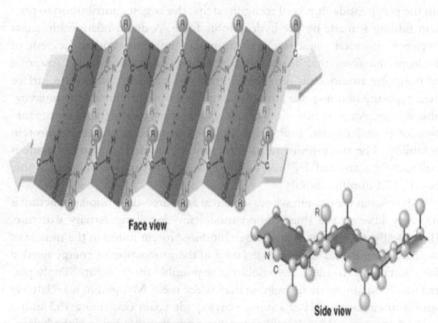

Face view

Side view

Fig. 2.9 Beta sheet-pleated structure. https://www.mun.ca/biology/scarr/Gr09-06.html. *Figure © 2000 by Griffiths et al.; All text material ©2005 by Steven M. Carr.*

Amino acids R groups are shown extending above and below the pleated sheet in Fig. 2.9. This illustrates the potential for them to stick together through complimentary interactions. For example, a beta sheet with predominately nonpolar R groups on one side can hydrophobically interact with the nonpolar side of another sheet. Conversely, a beta sheet with predominately polar R groups can interact via hydrogen bonding with the polar side of an adjacent sheet. Beta sheets are often shown in molecular models of protein structure as flat ribbons with an arrow head at one end indicating direction of polypeptide chain. Beta sheets are also slightly twisted as a result of their peptide bond geometry. The potential to form beta sheet secondary structures is also linked to amino acids sequence of the polypeptide. Beta sheet formation is more likely to occur in polypeptides containing the heptad repeat of P-N-P-N-P-N-P amino acids, where N and P, correspond to nonpolar and polar amino acids, respectively (Luke et al., 2006).

Tertiary structure refers to a polypeptide chain and its secondary elements folded into a three-dimensional molecule. Secondary structures including alpha helix, beta sheet, and random coil are typically bent or stretched in the folded conformation of protein molecules. For most proteins, folding is accomplished through physical and chemical forces acting on the polypeptide. It is well recognized that the largest contribution to protein folding is made by the hydrophobic force. A direct relationship exists between the total surface area of nonpolar molecules and the extent of hydrophobic force that drives folding of the polypeptide chain. Clustering of nonpolar amino acids into the interior of a protein reduces the surface area exposure of non-polar amino acids to water and minimizes the unfavorable energy. Amino acids with polar R groups are typically found on the surface of folded protein molecules. Polar R groups contribute to protein solubility. The three-dimensional structure of a folded protein molecule is collectively stabilized by hydrophobic forces, hydrogen bonding, ionic bonds, and disulfide bonds.

Myoglobin is a protein whose sequence and three-dimensional structure is precisely known and thus a good model for describing tertiary structure. Biologically, myoglobin is an oxygen binding protein found in the muscle of animals. It binds and stores oxygen used in the production of energy needed for contraction. In meat, myoglobin is responsible for its color. Simply put, red meat contains more myoglobin than white meat. Myoglobin is a relatively small protein composed of a single polypeptide chain containing 153 amino acids. Approximately 80% of these amino acids are arranged in alpha helices. Each of its eight alpha helices shown as coiled ribbons are connected to each other by random coil segments (Fig. 2.10). This model illustrates how helical segments are folded and make contact at crossing points. Links formed between helices at the points of contact are stabilized through hydrogen bonding or hydrophobic interaction of amino acid R groups. Myoglobin's polar R groups are located in regions of helices that are exposed to the aqueous environment and provide good water solubility to the protein (Velisek, 2014).

Fig. 2.10 Tertiary structure of myoglobin. https://en.wikipedia.org/wiki/Myoglobin. *Takano T. 1977. Structure of myoglobin refined at 2-0 A resolution. II. Structure of deoxymyoglobin from sperm whale. J. Mol. Bio. 110: 569—84.*

Quaternary structure refers to the three-dimensional organization of protein molecules composed of more than one polypeptide chain or subunit. A subunit is a single polypeptide chain with its folded conformation that can contain secondary structures such as alpha helix or beta sheet. Several types of bonding forces occurring at multiple sites are responsible for stabilizing a protein's quaternary structure. Bonding forces between subunits are provided by hydrogen bonds, electrostatic bonds, and the occasional covalent disulfide bond. A majority of plant and animal proteins are composed of multiple subunits that vary in amino acid composition and/ or in size. Quaternary structure complexity can be an advantage when subunits work cooperatively to perform a biological function. The protein hemoglobin, for example, illustrates the importance of quaternary structure in cooperativity. Hemoglobin is a tetramer composed of 2 types of subunits

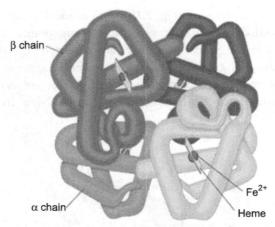

Fig. 2.11 Quaternary structure of hemoglobin. *http://themedicalbiochemistrypage.org/ protein-structure.php#quaternary.*

α and β, arranged as (α$_2$β$_2$) (Fig. 2.11). Each hemoglobin subunit can bind oxygen and is linked by hydrogen bonds and hydrophobic interactions. The oxygen binding function of hemoglobin is an example of cooperatively. Binding oxygen to the first hemoglobin subunit changes the conformation of its neighboring subunits and subsequently increases the molecule's affinity for oxygen.

Forces responsible for protein structure

van der Waals forces are weak forces arising from interactions between induced dipoles of neutral atoms in molecules. A dipole can be induced between two neutral atoms, A and B, by bringing them sufficiently close to cause their electrons to become polarized. At close distances, induced dipoles provide an attractive force, but the attractive force decreases sharply as distance between them increases. Conversely, when the distance between atoms A and B becomes too close, their negatively charged electrons will cause the atoms to be repulsed. Viewed individually, van der Waals forces provide only very weak attractive forces between atoms. However, when viewed collectively for a large molecule like a protein, their sum represents a significant force of attraction.

Hydrogen bonds are bonds between groups of atoms in which a permanent dipole exists. Permanent dipoles occur between atoms with a large difference in electronegativity. Hydrogen and oxygen share a strong dipole.

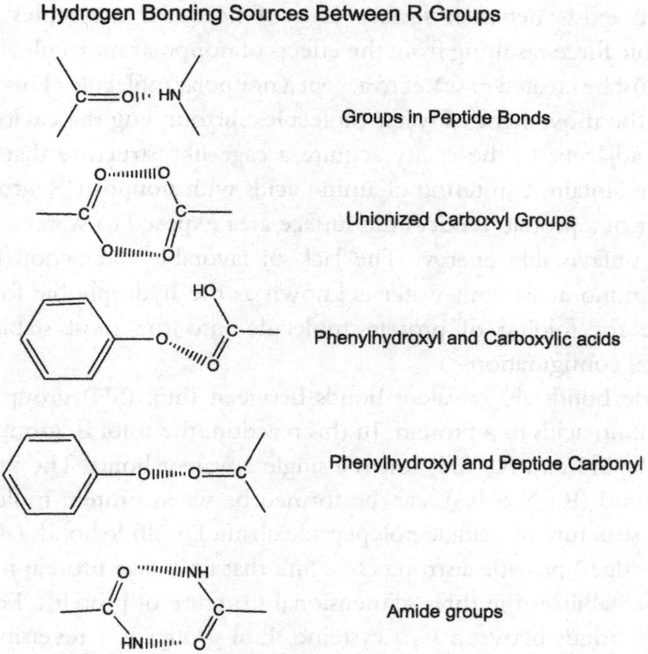

Hydrogen Bonding Sources Between R Groups

C=O··· HN   Groups in Peptide Bonds

O······O / O······O   Unionized Carboxyl Groups

HO ... Phenylhydroxyl and Carboxylic acids

O······O=C   Phenylhydroxyl and Peptide Carbonyl

O ······NH / HN······ O   Amide groups

Fig. 2.12 Hydrogen bonding between R groups.

Extensive hydrogen bonding exists between hydroxyl (OH) containing molecules such as water. Proteins contain several groups of atoms with strong dipoles. Carbonyl oxygen, amide nitrogen, and others illustrated in Fig. 2.12 provide multiple opportunities for hydrogen bonding in proteins.

Electrostatic interaction between atoms and molecules is based on the property that oppositely charged particles attract and like charged particles repel. Positive and negative charges on protein are contributed by polar amino acids. Specifically, amino acids contain a weak acid or weak base as their side chains. Acidic amino acids, such as aspartic and glutamic, contain an ionizable carboxylic acid as their R group. Basic amino acids, such as arginine, lysine, and histidine, contain an ionizable amine as their R group. Polar amino acids are almost always found on the surface of a protein molecule. At the appropriate pH, negative and positive charges contributed by polar amino R groups represent a significant force of attraction between proteins and/or between structural elements within a polypeptide.

Hydrophobic interaction: Interaction between nonpolar groups is a major force stabilizing the three-dimensional structure of proteins. A direct relationship exists between surface area of nonpolar molecules and the hydrophobic force resulting from the effects of nonpolar molecules in water. A cavity must be created in water to accept a nonpolar molecule. This results in restricting the movement of water molecules surrounding the cavity. Water molecules adjacent to the cavity acquire a cage-like structure that requires energy to maintain. Clustering of amino acids with nonpolar R groups into the interior of a protein reduces the surface area exposed to water and minimizes the unfavorable energy. The lack of favorable interaction between nonpolar amino acids with water is known as the hydrophobic force. It is responsible for folding of protein molecule into the most stable three-dimensional configuration.

Disulfide bonds are covalent bonds between thiol (SH) groups of two cysteine amino acids in a protein. In this reaction, the thiol R group of each cysteine (R_1-SH and R_2-SH) forms a single covalent bond. The product, a disulfide bond (R_1-S-S-R_2), can be formed between protein molecules or within the structure of a single polypeptide chain. Disulfide bonds (also called disulfide bridges) provide a strong cross link that links two protein molecules together or stabilizes the three-dimensional structure of proteins. Formation of disulfide bonds between two cysteine thiol groups is a reversible redox (reduction-oxidation) reaction. Oxidizing conditions favor the formation of disulfide bonds.

$$R_1 - SH + R_2 - SH \rightarrow R_1 - S - S - R_2$$

Disulfide bonded proteins are subject to exchange reactions that occur through the addition of a cysteine thiol that has lost its proton and become an ion (R_3S^-). The original disulfide (R_1-S—S-R_2) is split by the thiol ion (R_3S^-), resulting in a new disulfide product (R_1-S—S-R_3) and thiol ion (R_2S^-).

$$R_1 - S - S - R_2 + R_3S^- \rightarrow R_1 - S - S - R_3 + R_2S^-$$

So, what is responsible for the structure of a protein?

Ultimately protein structure is dependent on its sequence and composition of amino acids in the polypeptide chain. Amino acid sequence determines whether the polypeptide becomes a helix, sheet, or random coil structure. In turn, the amino acid composition and the resulting secondary structural elements determine how a polypeptide folds into a three dimensional structure we call a protein.

Denaturation

A protein is said to be denatured when there is a loss of secondary, tertiary, or quaternary structure without breaking covalent bonds. Denaturation results from a change in a protein's environment that reduces its stabilizing forces, such as hydrogen bonding and hydrophobic interaction. Proteins are often referred to as being either in native (folded) or denatured (unfolded) states. The process of unfolding a protein typically occurs as a sharp, noticeable transition rather than a gradual one. During denaturation, a protein's three-dimensional structure can undergo several rearrangements over the very brief time between folded and unfolded states. A denatured protein seldom exists as a random coil, but can contain portions of its original structure. For example, beta sheet or alpha helix may remain in the unfolded molecule. It is possible, albeit rare, for a denatured protein to refold into its native configuration once the cause of denaturation is removed. A reversion to the native state, however, is limited to monomeric proteins (proteins composed of a single polypeptide chain). Proteins with multiple subunits will not regain their native state because assembly requires additional factors present only *in vivo*. Denaturation substantially changes the properties of proteins in several ways. First, denatured proteins are typically insoluble in water. Loss of solubility for denatured proteins results from exposure of their nonpolar regions to water. Unfolded proteins aggregate through hydrophobic interaction between exposed nonpolar regions. Most large aggregates of protein molecules without sufficient interaction with water precipitate from the solution. Second, a denatured protein has higher viscosity compared to that protein in the native state. An unfolded molecule is less compact and behaves in solution like a much larger molecule. Third, denatured proteins lose their biological activity. Proteins, such as an enzyme or antibody, lose activity because the native conformation is essential. Fourth, denaturation improves a protein's digestibility. In order for a protein to be of nutritional value, it must first be broken down into individual amino acids in the digestive process. Proteins with a highly stable structure are more difficult to unfold and less digestible.

Agents of denaturation: Heating is the oldest and most common method of food preparation. Thermally induced protein denaturation makes food safe and generally increases its nutritional value by making proteins more digestible. Heating also enhances the flavor and appeal of food through Maillard reactions occurring between proteins and carbohydrates.

Why does an egg turn solid when dropped in boiling water?

The effect of heat on proteins is easily demonstrated by poaching an egg. When a liquid egg is dropped into boiling water, it almost immediately changes from translucent to a white, soft gel. The important question is how and why does this happen? The answer is that heat's most significant effect is on water. Specifically, hydrogen bonding is lost at boiling water temperatures. As described in Chapter 1, water's very high boiling temperature is due to the extensive hydrogen bonding between molecules. Water molecules transition to the gas phase only after their hydrogen bonds are broken. In proteins, hydrogen bonding stabilizes secondary structures and many interactions between elements of a protein's folded structure. Dropping an egg into boiling water demonstrates heat-induced denaturation. As the egg is heated to 100 °C, its proteins lose the stabilization provided by hydrogen bonding and unfolding begins. Heat causes increased hydrophobic interaction between nonpolar regions and new disulfide bonds are formed between protein molecules. In summary, the loss of hydrogen bonding, increased hydrophobic interaction, and disulfide bond cross linking results in formation of aggregates and precipitation of egg protein.

High pressure causes protein denaturation. A recently developed food processing technology uses high hydrostatic pressures to denature proteins and produce products equivalent in quality and safety to cooking, but without heat. In high pressure processing (HPP) the food is placed in a closed, water-filled container and extremely high pressure (30,000 to 60,000 psi) is applied while the temperature is kept low (30–40 °C). Pressures developed by this technology can be greater than those found in the deepest parts of the ocean. High pressure treatment causes proteins to unfold, inactivates enzymes, and kills microorganisms. While the mechanism of protein denaturation caused by high pressure is not completely understood, it is thought to result from the loss of hydrogen bonding and subsequent effects on protein structure. High pressure processing is currently being used on an industrial scale to process avocadoes. An advantage of this technology is inactivation of the enzyme that causes browning (polyphenol oxidase) without heat and preserves nutritional value.

pH is well known to cause protein denaturation. A majority of proteins have isoelectric points (pI) in the range of 5–6 and carry a net negative charge at neutral pH (7). Lowering the pH by addition of acid results in protein denaturation. This occurs through alteration of their intrinsic charge and disruption of ionic forces that stabilize structure. The very low pH (1–2) found in our stomach begins the digestion process by unfolding proteins. A loss of structure-stabilizing electrostatic bonds occurs when hydrogen ions displace interactions between charged amino acid R groups. For example, electrostatic bonds between positively charged amines and negatively charged carboxylic acids are lost when protons displace the interaction between them. High pH (9−10) also causes denaturation. Under alkaline conditions, serine, threonine, and tyrosine R groups become negatively charged by losing a hydrogen atom from their OH group. Negative charges formed in previously uncharged protein regions cause the molecule to expand and denature due to charge repulsion. Interestingly, proteins denatured by alkaline pH are typically more soluble because of an increase in total charge.

Effect of hydrogen bind disruptors (urea): The addition of chemical agents such as urea cause protein unfolding by disrupting the hydrogen bonds. Hydrogen bonds are responsible for stabilizing secondary structures such as alpha helix and beta sheet. They also stabilize the three-dimensional structure of proteins and the interactions between protein subunits. Although the use of urea at high concentrations is a common laboratory method to solubilize protein, it has no use in foods. Urea is used in laboratory investigations to solubilize soybean, corn, and other plant seed proteins that are otherwise insoluble. Seed storage proteins are tightly packed in clusters that are resistant to solubilization in water. Urea disrupts hydrogen bonding of water and the hydrophobic interaction of proteins. Urea solubilization is used to isolate and study the properties of these proteins.

Protein nutritional quality

All animals need protein in their diet to grow and thrive. Proteins provide energy when digested in the small intestine. About 4 calories per gram are provided by digestion of protein and carbohydrate, while about 9 calories per gram are obtained from fat. Protein breakdown (catabolism) requires energy in the form of ATP (Adenosine Triphosphate) to hydrolyze peptide bonds. The net amount of energy derived protein is therefore less than that from carbohydrate or fat. As such, protein represents the body's last reserve of energy. Dietary protein also represents a source of essential amino acids. (Table 2.1). Essential amino acids are those that humans are unable to synthesize from other compounds. It should be noted that the terms essential and indispensable are often used synonymously when referring to amino acids. Food proteins vary substantially in their content of amino acids, especially

Table 2.1 Amino acids needed in the diet.

Essential	Conditional
Lysine	Arginine
Histidine	Serine
Methionine	Cysteine
Threonine	Asparagine
Isoleucine	Glutamine
Leucine	Glycine
Phenylalanine	Proline
Valine	Tyrosine
Tryptophan	

essential amino acids. Most animal proteins are considered to be complete sources of essential amino acids, while many plant proteins are deficient in one or more. However, it is quite easy for vegetarians to eat a diet containing all the essential amino acids by combining plant food sources. For example, combinations of beans and rice or soy and corn can be used to provide a complete profile of essential amino acids.

What is a conditional amino acid?

In addition to the nine essential amino acids, others may be needed by individuals under special circumstances, such as illness and stress (Freidman, 1996; Trumbo et al., 2002). Amino acids needed by individuals experiencing illness are termed conditional amino acids (Table 2.1).

Several methods, including chemical score, protein efficiency ratio (PER), biological value (BV), and protein digestibility–corrected amino acid score (PDCAAS), are available for assessing the nutritional quality of food proteins. While each has its limitations, the PDCAAS method is more rigorous and therefore has been adopted by health agencies. The chemical score method relies on laboratory-based determination of all amino acids in a protein. In this procedure, a sample of the protein is chemically hydrolyzed and the amount of each amino acid is determined and expressed as gram of amino acid per gram of protein. The chemical score method has major limitations because chemical hydrolysis of proteins destroys some essential amino acids. The chemical method has obvious differences from normal digestion that uses enzymes from the host and bacterial sources. The protein efficiency ratio (PER) method of evaluation involves feeding test and control (usually casein) proteins to growing mice or rats. The method measures animal weight gain divided by the total grams of protein fed and enables direct comparison between test and control proteins. While PER is an improvement over the chemical score method, it is limited by substantial differences between mouse and human digestive systems. The biological value (BV) method for evaluation of protein nutritional quality is based on absorption of nitrogen by the organism. The BV method determines protein quality from a ratio of nitrogen incorporated into body tissues divided by the nitrogen absorbed from food. A major disadvantage of the BV method is that evaluations are made under strict conditions that do not reflect the everyday life of individuals and are subject to variation resulting from age, sex, and health. Presently, method of choice for evaluation of protein nutritional quality is the protein digestibility-corrected amino acid score (PDCAAS). The method has been adopted by Food and Agricultural Organization and World Health Organization (FAO/WHO) and by FDA in the United States. PDCAAS is a chemical score method based on a ratio between the first limiting amino acid in the test protein and the same amino acid in a reference protein. Specifically, PDCAAS is a ratio of the amount (mg) of the first limiting essential amino acid in 1 g of test protein divided by the amount (mg) of the same amino acid in 1 g of reference protein, corrected for fecal nitrogen. While PDCAAS is the most widely accepted method of assessing protein nutritional quality, its limitations include over estimation of requirements for elderly versus the young and does not account for anti-nutritional factors like trypsin inhibitor and phytate in plant protein sources. A comparison of protein nutritional quality in several plant and animal sources i.e., PER, BV, and PDCAAS methods is given in Table 2.2. In general, the nutritional quality of proteins from animal sources is higher than that for plant sources. Egg and milk (especially whey) protein are consistently ranked with the highest scores.

Table 2.2 Comparison of methods for evaluating protein nutritional quality.

Protein Type	PER	BV	PDCAAS
Beef	2.9	80	0.92
Black Beans	0	0	0.75
Casein	2.5	77	1.00
Egg	3.9	100	1.00
Milk	2.5	91	1.00
Whey Protein	3.2	104	1.00
Wheat Gluten	0.8	64	0.25
Peanuts	1.8	—	0.52
Soy Protein	2.2	74	1.00

Adapted from Hoffman, J.R., Falvo, M.J. 2004. Protein — which is best? J. Sport. Sci. Med., 3(3), 118—130.

Food allergy

Food allergies are serious and potentially life-threatening conditions for about 5% of adults and 8% of children (Sicherer and Sampson, 2014). Food allergy is often confused with food intolerance but, is different. Food allergies arise from an immunological reaction to proteins. Food intolerance is a reaction to a component such as lactose in milk or oligosaccharides (FODMAP) in wheat resulting in gastrointestinal symptoms (e.g., gas and diarrhea). The most common cause of food allergy occurs in 8 foods milk, eggs, fish, crustacean shellfish, peanuts, tree nuts, soybean and wheat. Allergic responses to these foods vary from swelling of the lips, skin rash and hives, to anaphylactic shock that can result in death. The type of food causing allergy varies with geographical location, and cultural influence on food selection. Food allergy typically occurs earlier in life. However, many adults develop allergy to a food previously eaten without symptom. Much is unknown about how food allergy is caused, but it is suggested to result from a loss of oral tolerance (Chehade and Mayer, 2005). Biologically, oral tolerance is a term used to describe the suppression of immune response to an antigen (protein) when the initial exposure occurs via an oral route. While some research suggests that feeding allergenic foods such as, peanuts to children at 4–11 months, may improve tolerance later in life, it is strongly cautioned that doing so is not without substantial risk. The best method for food-allergic individuals to prevent reaction, is still avoidance (Sicherer and Sampson, 2018). Unfortunately, avoidance is hard to practice because of the difficulty in knowing the complete composition of packaged and restaurant foods.

Effects of processing on proteins

Food materials are processed for a variety of reasons, not the least of which is to prevent microorganism - caused food-borne illness and spoilage. Centuries ago the practice of adding salt to meat was adopted because it preserved perishable food source and prevented the illness and death that often resulted from eating spoiled meat. Today, the reasons for improved shelf life and food safety provided by salt are better understood on a microbiological level. Added salt lowers water activity and inhibits the growth of spoilage organisms. Food processing technology converts raw, perishable commodities like meat, milk, and eggs into food products with extended shelf life. It also reduces food waste and increases the ability to feed more of the planet's population. Processing technologies include cooking, freezing, dehydrating, and fermentation, to name a few. While the goal of food processing is to create safe and nutritious foods, undesirable changes in food components can occur. Heat (thermal processing) is the method that most significantly affects the nutritional value of proteins. As in almost all chemical reactions, the higher the temperature, the faster its rate and extent of reaction. In chemical terms, heat represents an oxidizing environment. Oxidation is chemically defined as the loss of electrons from an atom or molecule. Two essential amino acids, cysteine and methionine, are the most sensitive to chemical changes caused by oxidizing environments. Heating in the range of 60—80 °C causes cysteine oxidation of the SH (thiol) group and joining of two cysteines together into one molecule called cystine, via a disulfide bond (Fig. 2.13). The change in cysteines R group can be reversed at this stage by chemical reduction. In the laboratory, converting cystine back to cysteine is accomplished by adding a chemical reducing agent. More intense thermal treatment (i.e., 110 °C) under alkaline conditions causes non-reversible changes to cysteine.

When proteins are heated in the range of 100—110 °C, in combination with pH greater than 7, cysteine undergoes substantial chemical change. First, the thiol (SH) R group is eliminated from cysteine resulting in the volatile and unpleasant compound, hydrogen sulfide (H_2S) (Fig. 2.14). Second, the rest of the amino acid becomes a compound known as dehydroalanine. Dehydroalanine formation from cysteine also results from thermal treatment under low

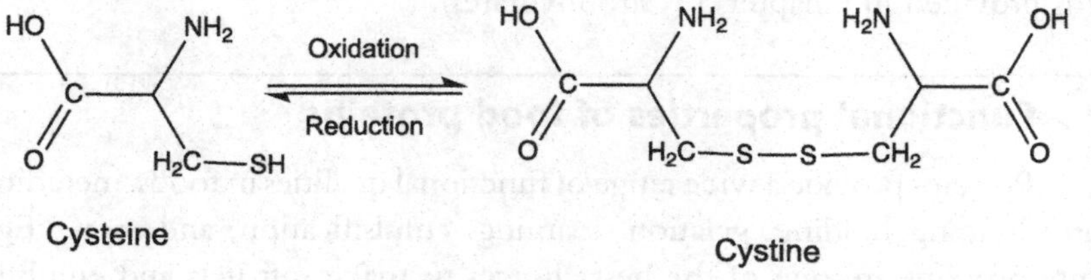

Fig. 2.13 Oxidation of the amino acid cysteine.

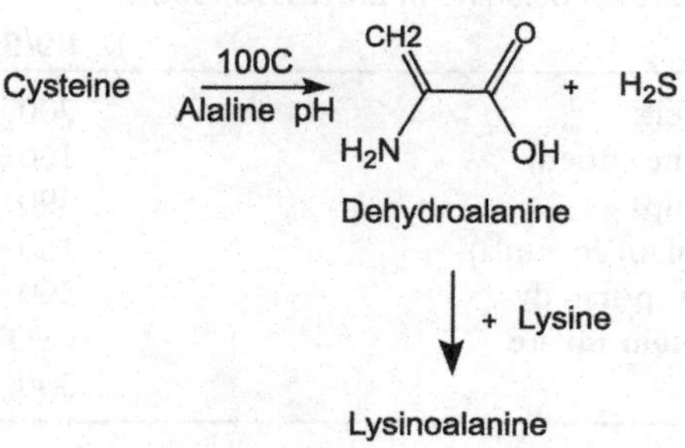

Fig. 2.14 Formation of Lysinoalanine.

moisture conditions. Specifically, heating promotes a reaction between dehy-droalanine's double bonded carbons and the amine R group of lysine. The toxic product of this reaction is lysinoalanine (LAL) (Fig. 2.14). LAL has greater toxicity in free versus protein-bound forms. Its negative effects result from binding copper and zinc ions essential to enzymes involved in catabolism, the process by which molecules are broken down for energy. LAL is notably high in heat-treated proteinaceous foods that are slightly alkaline (Table 2.3). Examples of foods containing a high level of LAL are dried egg white, evaporated milk, and pretzels. The net effect of this chemistry is destruction of two essential amino acids, cysteine and lysine, and production of undesirable products hydrogen sulfide and lysinoalanine.

Maillard reaction takes its name from the French scientist, Louis Maillard, who first described it in 1913. John Hodge, a USDA scientist, provided the first detailed description of its chemical mechanism in 1953. Maillard reaction occurs between protein and carbohydrate molecules. Specifically, the reaction occurs between the amine (NH_2) group of lysine in proteins and carbohydrates such as fructose, glucose and sucrose. Products of the reaction include brown pigments and a variety of flavors and aromas. Grilled meat, French fries, bread, coffee, and cocoa (chocolate) owe much of their flavor and color to the Maillard reaction. More recently it has been discovered that some Maillard reaction can also result in toxic products such as acrylamide (FDA Survey, 2004). A detailed description of Maillard chemistry is provided in Chapter 3 (Carbohydrates).

Functional properties of food proteins

Proteins provide a wide range of functional qualities in foods, including water binding/holding, gelation, foaming, emulsification, and more. Egg white proteins are one of the best choices to make soft gels and emulsify

Table 2.3 Lysinoalanine in processed foods.

Food	μg/g
Taco shells	200
Egg white (dried)	160—1800
Corn chips	390
Milk (infant formula)	150—640
Milk (evaporated)	590—860
Soy protein isolate	0—370
Pretzels	500

fats. Most of all, egg white proteins are superior for use in applications that require a heat stable foam like soufflés and meringues. A variety of protein ingredients isolated from animal and plant sources are commercially available for use in these applications. In general, animal proteins have better functional qualities compared to plant proteins. The ability of any protein to perform well in functional applications depends on intrinsic and extrinsic factors. The intrinsic factor is a protein's primary structure. The amino acid sequence and distribution of polar and nonpolar residues are critical to most functional properties. Extrinsic factors affecting protein functionality are temperature, pH, and salt. The following section provides an overview of intrinsic and external factors affecting protein functionality.

Hydration: The most important determinant of food protein functional quality is the ability to interact with water. Proteins with poor water compatibility do not work well in applications requiring solubility or dispersibility. A protein's affinity for water depends upon its composition of amino acids, specifically those amino acids with polar or polar non-ionized R groups. Dry protein ingredients typically contain 5%—10% water that is bound in a very thin layer to polar amino acid R groups. This class of water, referred to as bound water, is defined as water that cannot be removed by conventional drying methods. Approximately 6—7 water molecules are bound to polar R groups of amino acids (aspartic, glutamic, arginine, histidine and lysine). An additional 2 to 3 water molecules are bound to polar non-ionized amino acids (asparagine, glutamine, serine, threonine, and tyrosine). Bound water significantly aids in rehydration of proteins. The amount of water bound to a protein is a measurable quantity termed the water binding capacity, expressed as grams of water bound per gram of protein. Typical commercial protein ingredients like milk casein, whey protein isolate, and soy protein isolate have water binding capacities ranging from 0.3 to 0.9 g of water per gram of protein (Table 2.4). Sodium caseinate's water binding capacity is unusually high and is more than twice that of the other proteins in

Table 2.4 Water binding capacity of food proteins.

Protein	gWater/gProtein
Ovalbumin (egg white)	0.30
Soy Protein (soybean)	0.33
Myoglobin (Meat)	0.44
Collagen (Meat)	0.45
B-Lactoglobulin (Milk)	0.54
Na Caseinate (Milk)	0.92

Table 2.4. Milk caseins contain 6-7 polar phosphate groups on each molecule. Each phosphate group has a strong negative charge that attracts water molecules and contributes to the overall water binding capacity of caseins. Sodium caseinate is a type of protein ingredient made from milk by selectively precipitating casein protein with acid. Acid precipitated caseins are then neutralized with NaOH, creating a sodium salt form of the protein. In the dry form, called sodium caseinate, positively charged sodium ions are electrostatically bonded to each phosphate group. As a result, sodium caseinate is a more extensively hydrated, free acid form of casein. It has very good water solubility and is used as a creamer for coffee or tea. It also performs well whipped in food products like Cool Whip^{TM}.

While water binding capacity can be used to select a protein source with good hydration properties, it does have limitations. Specifically, water binding capacity of protein ingredients varies with temperature and other external factors. A more useful tool for assessing protein functionality in foods is water holding capacity. Water holding capacity is defined as the ability of foods to hold endogenous and exogenous (added) water when exposed to external forces such as gravity, centrifugation, or heating (Zayes, 1997). Water holding capacity measurements are typically performed on whole food systems. For example, differences in water holding capacity between cooked and raw meat samples are often used as an indicator of textural quality. Water holding capacity determinations are made by centrifuging a known weight of sample to separate solid materials from the expressed liquid. The solid is re-weighed following centrifugation. The difference in weight before and after centrifugation is divided by the original sample weight. This test indicates its water holding capacity. Multiplying this value by 100 gives water holding capacity as a percentage. The water holding capacity of meat is due to the nature of its proteins and the structure of the myofibril organelle. Myofibrils are composed of thick and thin filaments that slide past each other during contraction and relaxation. Myofibrils are a key factor in its water holding capacity because their open structure facilitates trapping water by capillary action. The final pH of meat and the rate at which it declines post-mortem strongly influences its water holding capacity. Final pH values between 5.7 and 6.0 are necessary for optimal water holding capacity. An example of the effect of water holding capacity in beef is demonstrated by animals having low levels of glycogen in their muscle. Glycogen is the polysaccharide store of glucose that muscle uses for energy. Under post-mortem conditions, muscle deficient in glycogen ends up with a high final pH (6.8 or higher). As a result, high pH beef is darker red than normal, has an undesirably firm texture, and an excessive water loss (drip loss). This condition, resulting from poor water holding capacity, is referred to as dark, firm, and exudative beef. This defect causes economic loss for the producer and tough steaks for the consumer.

Solubility is essential to many protein functional properties. Protein solubility is generally defined as the amount of protein that can be dissolved in solution. Proteins differ greatly in their solubility characteristics because of difference in their amino acid composition. In general, proteins with a higher content of polar amino acids are more soluble compared to those with a lower polar amino acid content because favorable water interaction is principally provided by ionizable groups. Solubility properties of most proteins can be improved by environmental factors such as pH, temperature, and presence of salts. pH has the greatest effect on protein solubility compared to the other extrinsic factors. It alters protein solubility through ionization of R groups on polar amino acids aspartic, glutamic, arginine, histidine, and lysine. As described earlier, polar amino acids contain ionizable carboxylic acids and amine groups. The pH of the environment determines whether these groups are ionized or un-ionized. A majority of food proteins, especially those of animal origin are acidic. This means that they have isoelectric points (pI) less than 7.0. In these proteins, the number of acidic amino acids (aspartic and glutamic) is greater than that of basic amino acids (histidine, lysine, and arginine). When the pH is higher than a protein's isoelectric point, its net charge is negative. Conversely, when the pH is lower than a protein's isoelectric point, its net charge is positive. As shown in Fig. 2.15, protein solubility is greater when the pH is lower or higher pH than the isoelectric point. Most proteins are more soluble when pH is greater than the pI. The greatest solubility occurs at alkaline pH. The solubility of proteins is generally increased by adjusting the pH well above its isoelectric point (Fig. 2.15). At a pH in the range of 9—10 ionization of hydroxyl (OH) groups occurs. Hydroxyls are found in the R groups of serine, threonine, and tyrosine and their ionization ($-OH \rightarrow -O^-$) creates a negative charge in regions of the molecule that were previously uncharged. Hydroxyls are found in the R groups of serine, threonine, and tyrosine and their ionization creates a negative charge in regions of the molecule that were previously uncharged.

Why do proteins precipitate at their isoelectric point?

When the pH of a solution is the same as the isoelectric point (pI) of a protein, precipitation often results. The pI of a protein is defined as the pH at which the sum of all positive and negative charges equals zero. A protein in

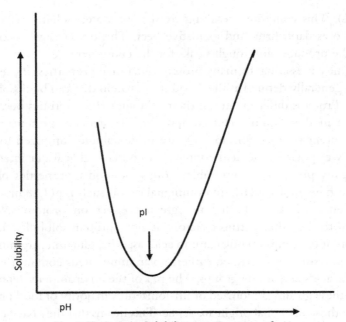

Fig. 2.15 Protein Solubility as a Function of pH.

solution at a pH equal to its pI has the least amount of ionization and therefore minimal charge. As a result, protein-protein interaction is favored and they are prone to form aggregates and precipitate from solution. Isoelectric precipitation is used in the laboratory to purify proteins. Precipitation also occurs in milk when spoilage bacteria produce lactic acid from lactose. Isoelectric precipitation of milk caseins is also an essential step in processes used to make yogurt and cheese.

Solubility and salt: When neutral salts such as sodium chloride are added to water, they dissociate completely into sodium (Na^+) and chloride (Cl^-) ions. Each ion retains its respective charge and is hydrated by several water molecules. When a charged ion is electrostatically attracted to an oppositely charged R group, ion-associated water molecules are also complexed with the protein, increasing its hydration. The net effect of added salt in the range of 0.1 M is increased protein solubility, and is sometimes referred to as "salting in". Fig. 2.16 shows the effect of added salt on the solubility of the milk protein, beta lactoglobulin at various pH levels. Solubility is lowest at the isoelectric point, about pH 4.5, when no salt is added (solid line in Fig. 2.16). However, in the presence of salt, about 0.1 M, beta lactoglobu-lin's solubility is increased for all pH values, even at its isoelectric point

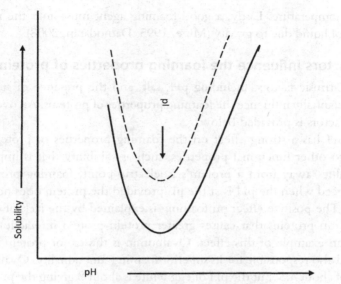

Fig. 2.16 Effect of salt on protein solubility.

(dashed line in Fig. 2.16). Conversely, if the amount of salt added to the protein solution is very high, for example 10 times more, the effect on solubility is reversed causing the protein to precipitate from solution. The effect occurs because salt ions are stronger competitors for water. The resulting loss of hydration promotes protein–protein aggregation and precipitation. This effect, known as salting-out, is also used in the laboratory to purify proteins. Precipitated proteins can be re-solubilized by dialyzing away the salt.

Foaming is an important property in many foods, including meringues, soufflés, whipped cream, and ice cream to name just a few. Foams are formed when a liquid containing a foaming agent is agitated. This causes air to become trapped in small bubbles. The ability of a protein to foams depends upon several factors. First, a good foaming agent must be able to lower the surface tension of the liquid at the interface between air and water. Molecules that lower the surface tension are called surfactants (surface-acting agent). Second, a good foaming protein unfolds in response to the agitation that forms bubbles. Unfolding rearranges the protein structure so that polar regions are oriented toward water (outside the bubble) and non-polar regions are toward air (inside the bubble). Third, proteins with good foaming ability must be able to form a very thin, elastic film surrounding air bubbles. The elastic membrane must have considerable strength. In a soufflé, for example, the light airy texture created by the foam must survive expansion caused by cooking, contraction when cooled to room temperature. Lastly, a good foaming agent must limit the natural drainage of liquid due to gravity (Mine, 1995; Damodaran, 2008).

What factors influence the foaming properties of proteins?

Several extrinsic factors, including pH, salt, and the presence of sugar or lipid, substantially influence the foaming properties of proteins. An overview of these factors is provided below.

pH: pH has a strong effect on the foaming properties of proteins. In contrast to other functional properties, such as solubility that is improved at pH values away from a protein's isoelectric point, foaming properties are enhanced when the pH is at the pI (provided the protein does not precipitate). The positive effect on foaming is explained by the lack of charge repulsion on proteins that causes greater protein–protein interaction. Egg white is an example of this effect. Ovalbumin is the major protein in egg white and also responsible for its superior foaming functionality. Ovalbumin has a pI of about 4.5, but the pH of egg white is about 8 giving the protein a net negative charge. Adding cream of tartar (an acidic salt) or vinegar to egg white substantially improves its foaming properties. This well-known culinary practice works because the added acid lowers egg white pH closer to ovalbumin's isoelectric point.

Salt: Low concentrations (approximately 0.1 M) improves the solubility and foaming properties of globular and albumin type proteins. The terms globular and albumin protein come from an older and more general protein classification system based on solubility. In this system, albumins are proteins soluble in water and globulins are soluble in dilute salt solutions.

Added salt improves the foam capacity and stability of proteins such as egg white and soy protein. It is presumed that improvement in foaming is due to neutralization of charges on proteins by sodium and chloride ions. Addition of magnesium Mg^{+2} or calcium Ca^{+2} containing salts substantially improves foam capacity and stability by providing ionic (electrostatic) linkages between protein molecules. In contrast, addition of salt to whey proteins such as beta lactoglobulin reduces its foam capacity and stability (Zhu and Damodaran, 1994a). Increased protein stability to unfolding provided by salt may explain this effect on beta lactoglobulin.

Sugar: The combination of sugar (sucrose) and egg white protein in foods such as meringues has a substantial effect on the foaming properties. Specifically, added sugar reduces the foam capacity of egg white. Sugar also greatly improves stability of the foam. Reduction in foam capacity with added sugar is likely a result of sucrose's ability to stabilize protein structure and prevent unfolding. The increase in foam stability provided by sucrose results from increased liquid viscosity that slows drainage of water from the foam. The important message for making high volume foams with egg white containing sugar is clear: whip egg white first and add sugar later.

Lipid: As anyone who has tried to whip egg whites knows, even a small bit of yolk completely destroys the ability to make it foam.

I just don't crack eggs very well. Will that little bit of yolk in my egg whites really matter when I am making meringues?

It is common practice to crack eggs one at a time and discard any with even the smallest drop of yolk in it. The presence of yolk in egg white effectively inhibits any foaming functionality. Whipping egg white contaminated with yolk makes a foam that collapses as soon as the whipping stops. This happens because egg yolk lipids (principally the phospholipid lecithin) bind to proteins, preventing the formation of complexes required to stabilize the interfacial films of gas bubbles.

Protein denaturation: Foaming properties of protein ingredients, such as whey and soy protein isolates, can be a challenge to use in foaming applications. As described above, proteins that function well in foaming such as egg white, are easily unfolded (denatured) enabling them to act as surfactants. However, soy and whey proteins are more resistant (stable) to unfolding and their foaming properties are poor compared to egg white. A moderate heat treatment (e.g., 70 °C) of whey and soy protein causes partial denaturation making them to easier to form stable foams (Zhu and Damodaran, 1994b).

Is a copper bowl better for whipping egg white?

Copper bowls are better than glass for making good quality egg white foams because of copper's chemistry. As egg proteins are unfolded during whipping, the sulfur containing amino acid cysteine is exposed to the copper surface, causing an oxidation reaction. Disulfide bonds formed in this reaction create links (disulfide bonds) between egg white proteins that provide increased stability. (Kitts and Weiler, 2003).

Emulsification: An emulsion is a dispersion of one phase as small droplets in another phase. The emulsification properties of proteins are relied upon to make processed foods such as meats and cheeses, salad dressings, and frozen desserts. There are two basic types of food emulsions. The first, an oil-in-water emulsion, occurs when lipid (oil or fat) is dispersed in an aqueous phase.

Salad dressing, sauces, and soups are examples of oil-in-water type emulsions. The second, a water-in-oil emulsion, occurs when water is dispersed in lipid (oil or fat). There are fewer examples of water-in-oil emulsions, but butter and margarines are prominent ones. Proteins with good emulsification properties share a common characteristic of being good emulsifiers. Proteins that are good emulsifying agents are amphiphilic. They contain both polar and nonpolar amino acids with the ratio of the two types slanted toward nonpolar. Additionally, proteins with good emulsification properties must unfold easily in response to mixing and re-orient amino acids R groups according to their compatibility with water and lipid phases. In emulsified systems, polar and nonpolar protein regions are oriented by the mixing process to face water and lipid fractions, respectively. Milk, for example, benefits from the emulsifying properties of its casein protein. Before the days when homogenization was used in processing milk, its fat content would typically separate and float on top of the liquid. The presence of a thick fatty layer on top of a milk container is a defect called creaming. Today, the process of homogenization eliminates separation of milk fat by applying high pressure to force the liquid through a small aperture. This process generates a shearing action that causes casein proteins to unfold. Shearing action also causes fats to be liberated from a milk vesicle called the milk fat globule. Liberated fat molecules quickly bind to hydrophobic regions of unfolded casein molecules. The result is a uniformly dispersed, stable complex of fat and protein in which separation no longer occurs. Caseins, with their high degree of molecular flexibility and amphiphilic properties, are uniquely suited to this function.

Gelation: Gels are an important functional property of proteins often experienced in foods. Examples of protein gels include boiled egg, processed meat and cheese, Jell-O™, and tofu. Gels in these foods provide a desirable soft texture because they contain a high water content (e.g., 90%) that does not become liquid and separate. Protein gels can be thought of as acting like a sponge. They hold a great deal of water without leaking unless acted upon by some external force. On a microscopic scale, the open, sponge-like character of a protein gel is created by a three-dimensional network of denatured molecules. Gel structures are stabilized by protein-protein interactions that include hydrophobic interaction between nonpolar regions, hydrogen bonds, and disulfide bonds. Proteins must be unfolded to form a gel, but some structural elements such as beta sheet or alpha helix may remain. Food protein sources such as egg, milk, or soy are composed of mixtures of protein with varying stabilities. For this reason, some protein sources form gels more readily than others. Water in a gelled structure is held by capillary action within the three-dimensional network. Water is held by its molecular interactions with the protein. Polar amino acids are hydrated with 6—7 water molecules per charged group. Water is also held through hydrogen bonding to polar non-ionized amino acids groups. Such hydrogen bonding is responsible for water binding to the carbonyl oxygen and amide nitrogen in peptide bonds of the unfolded molecule (Zayas, 1997).

How are protein gels made?

Heat: Heat treatment in the most common way to make a protein gel. Heating egg white, for example, results in a solution to gel transition almost as soon as it hits boiling water. The high concentration of protein in egg white, approximately 10%, is a contributing factor to gel formation. Ovalbumin is the predominant protein of egg white and principally responsible for its gelation properties. Like most proteins with good gelation properties, ovalbumin has a high proportion of nonpolar amino acids. Ovalbumin unfolds at 80—85 °C and forms gel networks principally through hydrophobic interactions and disulfide bonds. Cross-linking is a disulfide exchange reaction favored by heat and the alkaline (pH 8—9) environment of egg white. A model disulfide exchange reaction between proteins is illustrated in the equations below. Initially, a disulfide bond is present between the two proteins, P_1 and P_2. The reaction begins with ionization of an additional protein (P_3SH) containing a free thiol group (SH). The alkaline environment is responsible for loss of the thiol's hydrogen to hydroxyl ion (OH^-). The product of this reaction is the negatively charged, reactive sulfur species called the thiolate anion (P_3S^-). Subsequently, thiolate anion (P_3S^-) splits the disulfide bond between proteins P_1 and P_2 creating a new pair of disulfide-linked proteins (P_1-S-S-P_3) and a new thiolate anion (P_2S^-). The repeated cycle of exchange reactions results in multiple cross-links between protein molecules and formation of a stable gel.

$$P_1 - S - S - P_2 (\text{disulfide linked proteins } P_1 \text{ and } P_2)$$

$$P_3SH + OH^- \rightarrow P_3S^- (\text{thiolate anion})$$

$$P_1 - S - S - P_2 + P_3S^- \rightarrow P_1 - S - S - P_3 + P_2S$$

$$- (\text{new pair of disulfide linked proteins} + \text{thiolate anion})$$

Protein gel properties and environmental affects: The appearance of protein gels can be either an opaque coagulum or a translucent gel. Egg white, for example, forms a soft, opaque gel (known as a coagulum) as a result of thermally-induced reactions. The optical property of opaqueness results from light scattering of coagulated proteins. In general, proteins with a high proportion of nonpolar amino acids form opaque gels. These gels are principally stabilized through hydrophobic interaction and disulfide bonds and are non-reversible. In contrast, proteins with a small proportion of nonpolar amino acids tend to form soluble complexes upon denaturation principally through hydrogen bonding. Proteins of this type remain soluble during heating and form reversible translucent gels only upon cooling. Collagen (gelatin) is an example of a protein that forms translucent gels with reversible solution to gel transitions. Translucent gels also have greater water holding capacity and less syneresis because of their great capacity for hydrogen bonding.

Salt: Salt is known to alter the properties of protein gels formed. Low concentrations of sodium chloride (0.1 M) can affect the gelation behavior of proteins. Binding salt ions to charged R groups reduces charge-repulsion between proteins, an effect that promotes hydrophobic interaction and gel formation. In contrast, added salt can have the opposite effect, increasing protein hydration and solubility. Some proteins form gels as a direct result of adding salt. Tofu, for example, is made by adding calcium sulfate salt to extracted soy proteins. Soft soy proteins gels are initially stabilized by electrostatic cross-linking of soy proteins with divalent cation calcium (Ca^{+2}). Subsequent heating (75 °C) creates the final gelled product known as tofu.

pH: The pH of the solution influences the type of gel formed and its strength. When the pH is at or near the isoelectric point of the protein, coagulum type gels are most often formed. In general, gel strength is greater when the pH is at the protein's isoelectric point because electrostatic repulsion is at a minimum, promoting protein-protein interaction.

Protease treatment: Cheese is the best example of protease treatment resulting in a gel. The proteolytic enzyme chymosin is added to milk in the cheese-making process. Chymosin action hydrolyzes only one peptide bond in the milk protein, kappa casein and creates two fragments. The shorter of the two fragments is very hydrophilic and is released into the soluble fraction. However, the larger and more hydrophobic fragment initiates a change in micellar structure of milk caseins causing them to aggregate and precipitate from the liquid aggregation and precipitation of the protein from milk. The gel-like coagulum of casein proteins is collected and represents the first step in making cheese.

Transglutaminase: Treatment of proteins with the enzyme transglutaminase catalyzes the cross-linking of protein molecules by forming covalent bonds between glutamine and lysine R groups. The result is a very strong, irreversible type of gel with elastic properties. Applications of this enzyme are principally used to bind different types of meat (e.g., beef, pork, and chicken) together in processed products. It is also used in artisanal breads to provide a hard crust.

Enzymes in food

Enzymes are important to many aspects of food. When a tomato ripens on the vine, its flavor results from two amino acids (aspartic and glutamic acid) derived from enzymatic break-down of its proteins. Similarly, snow crab protein break-down begins soon after harvest and creates free amino acids. Most notably, glycine, alanine, arginine, and glutamine produced by protease enzymes are key components of crustacean seafood flavor. Fermentation is a centuries old process driven by microbial enzymes resulting in unique foods with extended keeping qualities.

For example, yeast enzymes ferment carbohydrates in grapes and grains into wine and beer. Soy sauce is made by fermenting soy beans and wheat using the mold, *Aspergilis oryzae*. Glutamic acid, liberated from protein, combines with salt in the mixture to give soy sauce its unique, umami flavor (Lioe et al., 2010). Enzymes occurring naturally in food are termed endogenous and are responsible for reactions that affect almost every aspect of food quality, including color, flavor, and texture. Enzymes that are added directly or indirectly to foods are termed exogenous. Fermentation is perhaps the most widely used form of indirect enzyme addition. In such addition, isolated enzymes are used like chemical reagents for industrial scale processes. Several enzymes are immobilized in a large reactor to convert starch into the sweetener known as high fructose corn syrup.

What is an enzyme? Simply put, an enzyme is a protein that catalyzes (speeds up) chemical reactions. A catalyst is defined as a substance that increases the rate of a chemical reaction without being altered itself. In the reaction, a substrate molecule binds to the enzyme's active site and creates the enzyme-substrate complex (Fig. 2.17). As a result of forming this complex, several changes occur that increase the rate at which substrate is converted to product. First, the enzyme's amino acid R groups are brought in close proximity to the substrate molecule. This change effectively increases the concentration of reactants. Second, binding the substrate to the enzyme's active site places strain on bonds within the substrate molecule and lowers its stability. Third, the enzyme's active site creates a micro-environment that facilitates the

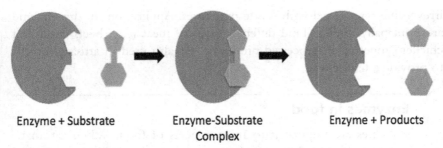

Enzyme + Substrate   Enzyme-Substrate   Enzyme + Products
                                  Complex

Fig. 2.17 Enzyme-substrate reaction. *http://blogs.scientificamerican.com/lab-rat/ speeding-up-reactions-biological-vs-chemical-catalysts/.*

reaction. For example, an active site containing acidic amino acid R groups can donate protons and promote hydrolysis of bonds in the substrate molecule.

Enzymes catalyzed reactions vary substantially in their substrate specificity. Enzyme specificity is classified into three general categories termed; bond, group, and absolute. The lowest level of specificity is the type of bond acted upon. Bond-specific enzymes act on substrates containing one type of chemical bond. Lipase, for example, is an enzyme that acts on almost any type of ester bond in lipids and liberates fatty acids. The next level of specificity is for a group of atoms within a molecule. Group-specific enzymes act on substrates containing a specific bond and adjacent atoms necessary for its binding to the enzyme's active site. Trypsin, for example, is a digestive enzyme that hydrolyzes peptide bonds in proteins. However, trypsin specifically only cleaves peptide bonds following lysine or arginine amino acids. The R group atoms of these amino acids are necessary for binding to the enzyme's active site. The highest level of enzyme specificity is for a single substrate containing a unique configuration. Stereo-specific enzymes like L-amino oxidase act only on the L isomer of an amino acid and will not act on its mirror image, the D isomer. Biologically, enzymes are essential to the synthesis and catabolism of all cellular components, including carbohydrates, lipids, nucleic acids, and other proteins. Enzymes can catalyze a single reaction or work concertedly in a metabolic pathway. Glycolysis, for example, is a metabolic process in which sugar molecules are metabolized through a series of enzyme reactions to produce energy in the form of adenosine triphosphate (ATP). An enzyme's activity is influenced by environmental factors such as pH and temperature. Typically, the range of pH and/or temperature at which an enzyme functions with maximal activity is small. It is not surprising that most enzymes function at or near normal biological pH and temperature, namely pH 7 and 37 °C. However, there are exceptions of note. Enzymes have been found in thermophilic microorganisms that thrive at high temperatures, 80—120 °C. There is considerable interest in these enzymes for use in industrial scale processes such as ethanol production from cellulosic sources. One of the most widely used thermostable enzymes is Taq-polymerase, an enzyme essential for the polymerase chain reaction (PCR). This enzyme can amplify tiny amounts of DNA found in a drop of saliva so that sequence analyses can be performed.

Enzymatic browning
Why do my apple slices always turn brown?

Enzymatic browning is responsible for fresh apples, bananas, potatoes, and other foods turning brown after being cut. Browning of these foods and others is caused by several related enzymes performing the same function, namely oxidation of phenolic substrates and subsequent formation of brown pigment. Enzymatic browning is different from Maillard browning in that the latter is strictly chemical reaction between sugars and proteins and requires heat. Brown color produced by the enzymatic process results from oxidation of phenolic substrates (Fig. 2.18) affecting a variety of foods.

The enzymes responsible for browning are referred to by several names including phenolase, polyphenol oxidase, tyrosinase, and catecholase. Enzymatic browning proceeds in a two-step reaction that requires oxygen and copper ions. A reaction scheme using polyphenol oxidase as the model enzyme and tyrosine as the substrate is shown in Fig. 2.18.

The first and rate limiting step of the enzyme reaction involves the addition of a hydroxyl (OH) groups to the phenol ring. The second step of the enzyme reaction involves oxidation of hydroxyl groups to carbonyls ($C=O$). In chemical terms, oxidation of a phenol hydroxyl groups to carbonyls converts the molecule to a colorless compound known as a quinone. The rest of the process is a chemical cascade in which quinones undergo condensation reactions to form large melanin polymers. Melanin is a light-absorbing compound reflecting brown colors (Parkin, 2008). Some foods depend on enzymatic browning for the color normally associated with them. Examples include raisins, unroasted coffee and cocoa beans, and tea (Table 2.5). However, brown color is not desirable in foods such as lettuce and fruits because of its association with spoilage. The control of

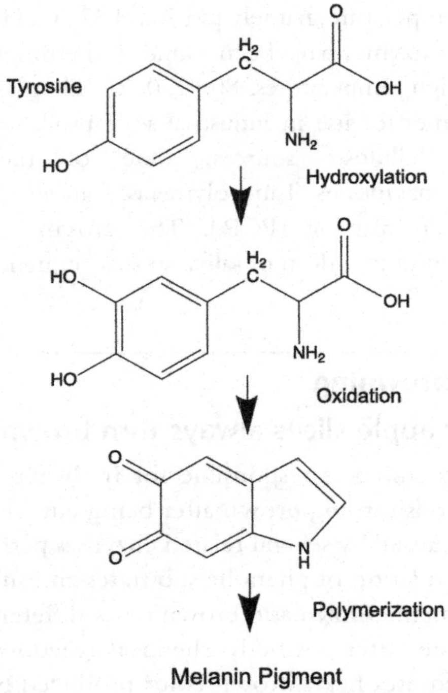

Fig. 2.18 Enzymatic browning reaction. *Source file = Enzymatic browning.eps.*

Table 2.5 Foods with enzymatic browning activity.

Apples and apple cider
Pears
Bananas
Raisins and Prunes
Potatoes
Lettuce
Coffee and cocoa beans
Tea
Mushrooms
Shrimp and other Crustaceans

the enzyme's activity is therefore an important aspect of food quality. Most enzymes that cause browning are moderately heat stable and the temperature required to inactivate them (approximately 60 °C) also renders lettuce and most fruits unacceptable. Fortunately, there are other ways to control enzymatic browning. Oxygen is a necessary component of the reaction and replacing it in packaging with nitrogen and/or CO_2 is an effective way to control the reaction. This approach combined with lower temperature is typically used in packaged salad mixes. Enzymatic browning in apples and other fresh fruits can be controlled by the adding chelating agents and/or acids. A chelating agent is a compound that tightly binds the metal ions essential to polyphenoloxidase enzymatic activity. EDTA is a food grade compound with excellent chelating activity and tightly binds copper ions essential for polyphenoloxidase activity. Combining a chelating agent with an acid is an effective strategy for controlling enzymatic browning. Added acid limits enzyme activity by lowering the pH to levels below its optimum range. Ascorbic acid is a very effective inhibitor of enzymatic browning. Ascorbic acid lowers pH with minimal addition. Additionally, it binds oxygen and chelates metal ions, especially copper. Ascorbic acid is a strong antioxidant that inhibits the conversion of phenols to quinones. While the above options work well in commercial applications, a simpler approach might be desired at home. Adding fresh orange juice to cut apples is an easy way to control enzymatic browning because orange juice is acidic and contains ascorbic acid (vitamin C) and citric acid.

Summary

Proteins are unique in their ability to provide the desirable attributes to food (i.e., flavor, texture) and are essential to nutrition and health. Proteins are noted for their functional properties such as the ability to foam (e.g., ovalbumin), and form gels (e.g., casein, collagen, ovalbumin). Proteins as enzymes create color and enable the process of fermentation that preserves food and makes products such as beer, wine, and cheese. Proteins undergo chemical reaction (Maillard) in response to heating with carbohydrates. These reactions are responsible for producing the desirable colors and flavors of foods such as bread, coffee, and chocolate. Nutritionally, a varied protein diet containing all essential amino acids is required for optimum growth and maintaining health. As world population grows, it is important to expand our sources of protein beyond traditional commodities to meet that need. Greater knowledge of proteins from non-traditional sources (described in Chapter 9) will contribute to that goal.

Glossary

ATP (Adenosine Triphosphate) Organic compound the produces energy and ADP (adenosine diphosphate) plus inorganic phosphate during a hydrolysis reaction

Amphiphilic Property of a molecule with both hydrophilic (water loving) and hydrophilic (water hating) parts

Albumin proteins Generalized group of proteins have good solubility if water. The term albumin is based on an older system of protein classification

Chiral molecule A chiral molecule is one that is indistinguishable from its mirror image but can't be superimposed upon it. L and D isomers of amino acids are examples

Conditional amino acid Amino acids required in the diet of individuals who are rapidly growing, ill, or under stress

Essential amino acid Amino acids that can't be synthesized by us and therefore required in the diet

Enzyme A protein that functions as biological catalyst. They speed up the rate of a reaction without being transformed.

Enzymatic browning A browning reaction in fresh foods resulting from oxidation of phenolic substrates

Globulin Proteins that are soluble in water and dilute salt solution. The term globulin refers to an older system of protein classification.

Isomer Molecules with the same chemical formula but differ in structure

Oligopeptide A polypeptide composed of approximately 20–40 amino acids (e.g., 20–40)

Peptide A polypeptide consisting of between 2 and 20 amino acids

pI/Isoelectric point The pH at which a molecule's positive and negative charges sum to zero

Peptide Bond A covalent bond formed between two amino acids

Polypeptide A linear chain of amino acids joined together through a covalent (peptide) bond

Salting in Adding salt at a low concentration to a protein solution can increase its solubility.

Salting out Adding salt at high concentration to a protein solution can decrease its solubility, promote aggregation, and cause precipitation

Subunit A single polypeptide chain having its own folded conformation

van der Waals Force Weak attractive or repulsive forces that are driven by electrical interactions between atoms or molecules.

Water binding capacity A measure of water bound to a protein, expressed as grams of water bound per gram of dry protein

References

Brown, J.H., 2010. How sequence directs bending in tropomyosin and other two-stranded alpha-helical coils. Protein Sci. 19, 1366–1375.

Chehade, M., Mayer, L., 2005. Oral tolerance and its relation to food hypersensitivity. J. Allergy Clin. Immunol. 115, 3–12.

Damodaran, S., 2008. Amino Acids, Peptides and Proteins. Fennema's Food Chemistry, fourth ed. CRC Press, Boca Rotan FL.

Freidman, M., 1996. Nutritional value of protein from different sources: a review. J. Agric. Food Chem 44, 16–29.

Hoffman, J.R., Falvo, M.J., 2004. Protein — which is best? J. Sport. Sci. Med. 3 (3), 118—130.

Kamtekar, S., Schiffer, J.M., Xiong, H., Babik, J.M., Hecht, M.H., 1993. Protein design by binary patterning of polar and nonpolar amino acids. Science 262 (5140), 1680—1685.

Kitts, D.D., Weiler, K., 2003. Bioactive proteins and peptides from food sources. Applications of bioprocesses used in isolation and recovery. Curr. Pharmaceut. Des. 9 (16), 1309—1323.

Lioe, H.N., Selmat, J., Yasuda, M., 2010. Soy sauce and its umami taste: a link form the past to current situation. J. Food Sci. 75, R71—R76.

Luke, H.B., Thumfort, P.P., Hecht, M.H., 2006. De novo proteins from binary-patterned combination libraries. Methods Mol. Biol. 340, 53—69.

Mine, Y., 1995. Recent advances in the understanding of egg white protein functionality. Trends Food Sci. Technol. 6 (7), 225—232.

Parkin, K.L., 2008. Enzymes. Fennema's Food Chemistry, fourth ed. CRC Press, Boca Rotan FL.

Sicherer, S.H., Sampson, H.A., 2014. Food Allergy, epidemiology, pathogenesis, diagnosis and treatment. J. Allergy Clin. Immunol. 133, 291—307.

Sicherer, S.H., Sampson, H.A., 2018. Food Allergy: a review and update on epidemiology, pathogenesis, diagnosis and treatment. J. Allergy Clin. Immunol. 141, 41—58.

The Science Behind Whipping Egg White in Copper Bowls. The Kitchn.com http://www.thekitchn.com/the-science-behind-whipping-egg-whites-in-copper-bowls-221943.

Trumbo, P., Schlicker, S., Yates, A.A., Poos, M., 2002. Food and Nutrition Board of the Institute of Medicine, the National Academies. Dietary reference intakes for energy, carbohydrate, fiber, fat, fatty acids, cholesterol, protein and amino acids. J. Am. Diet. Assoc. 102, 1621—1630.

Velisek, J., 2014. The Chemistry of Food. Wiley and Sons, Oxford, UK.

Zayes, J.F., 1997a. Water holding capacity of proteins. In: Functionality of Proteins in Food. Springer-Verlag, Berlin, pp. 76—133.

Zayes, J.F., 1997b. Gelling properties of proteins. In: Functionality of Proteins in Food. Springer-Verlag, Berlin, pp. 310—366.

Zhu, H., Damadoran, S., 1994a. Effects of calcium and magnesium ions on aggregates of whey protein isolate and its relation to foaming properties. J. Agric. Food Chem. 42, 856—862.

Zhu, H., Damadoran, S., 1994b. Heat-induced conformational change in whey protein isolate and its relation to foaming properties. J. Agric. Food Chem. 42, 846—855.

Further reading

Brown, A.C., 2011. Understanding Food: Principles and Preparation. Wadsworth Pub, Belmont, CA.

Hillman, H., 2003. The New Kitchen Science: A Guide to Knowing the Hows and Whys for Fun and Success in the Kitchen. Houghton Mifflin, Boston.

McGee, H., 2004. On Food and Cooking: The Science and Lore of the Kitchen. Scribner, New York, N.Y.

Myhrolvd, N., 2012. Modernist Cuisine at Home. The Cooking Lab, Bellevue WA.

Schaafsma, G., 2012. Advantages and limitations of the protein digestibility-corrected amino acid score (PDCAAS) as a method for evaluating protein quality in human diets. Br. J. Nutr. 108 (Suppl. 2), S333—S336.

Sicherer, S.H., Sampson, H.A., 2018. Food Allergy: a review and update on epidemiology, pathogenesis, diagnosis and treatment. J. Allergy Clin. Immunol. 141, 41–58.
Zayes, J.F., 1997. Functionality of Proteins in Food. Springer-Verlag, Berlin, pp. 76–133.

Review questions

1. Describe the structure of an amino acid, including its various groups.
2. Define the terms polar, polar non-ionized, and nonpolar as it refers to the R groups of amino acids.
3. Define the term chirality and its importance to amino acids.
4. Which amino acid R groups are affected by changes in pH?
5. Describe the elements of protein structure (primary, secondary, tertiary, quaternary).
6. What are the chemical forces that stabilize protein structure?
7. Define the term "hydrogen bonding" and give an example of its importance to protein secondary structure.
8. What is a disulfide bond and how does it stabilize protein structure?
9. Why does pH affect the solubility of proteins?
10. Define the term isoelectric point (pI) as it refers to a protein.
11. Why do proteins precipitate when the pH is adjusted to their isoelectric point?
12. Why does added salt improve the solubility of proteins?
13. Define the term denaturation in regard to proteins.
14. Why does heat cause proteins to denature (e.g., egg in boiling water)?
15. Why does denaturation improve protein digestibility?
16. What is the difference between conditional and essential amino acids?
17. Describe the PDCAAS method for assessing the nutritional quality of proteins.
18. What factors influence the foaming properties of proteins?
19. Why is egg yolk detrimental to foaming egg whites?
20. What is cream of tartar and why does it improve egg white foams?
21. Why does beating egg whites in a copper bowl improve foam stability?
22. What is a surfactant?

23. Give an example of a protein gel.
24. Why does egg white "gel" when heated?
25. What is an enzyme?
26. Give three examples of enzymes in foods.
27. What is enzymatic browning and how can it be controlled?

CHAPTER THREE

Carbohydrates

Learning Objectives

This chapter will help you describe or explain:

- Structure of common monosaccharides, oligosaccharides, and polysaccharides
- How Maillard reaction contributes color and flavor in food
- Undesirable products of Maillard reaction
- Functional properties carbohydrates in food
- Nutritional and health benefits of oligosaccharides, polysaccharides, and dietary fiber
- The structure of starch and how it affects digestion and glycemic index
- Hydrocolloids and their functional properties in food

Introduction

Biologically, carbohydrates represent a major source of energy for plants and animals. Plants make simple sugars (monosaccharides) through the process of photosynthesis and subsequently store them in the polymer

Introduction to the Chemistry of Food
ISBN: 978-0-12-809434-1
https://doi.org/10.1016/B978-0-12-809434-1.00003-7

known as starch. When needed, starch is broken down into individual glucose units and used for energy. Plants also convert some of their photo-synthetically derived glucose to make the closely related polymer called cellulose. Cellulose is a fibrous material providing structural integrity to plant cell walls and tissues. However, cellulose is non-edible because it is only broken down by microbial enzymes. Nutritionally, carbohydrates provide us with a large part of our daily calorie requirement. It is important to note that glucose is our brain's primary source of energy. Recent discoveries have shown that carbohydrates, as prebiotics, are energy sources supporting the growth of beneficial bacteria in our gut's microbiome. Carbohydrates in the form of fiber also contribute to our health as the source of short chain fatty acids and fiber that aid in controlling the level of cholesterol in our body. In food, carbohydrates provide sweetness, contribute to color and fla-vor via Maillard chemistry. They are responsible for functional properties needed to make bread, pasta, and numerous other products that are part of our daily lives.

Questions that will help you explore information in this chapter are:
• Why does my grandmother have lactose intolerance?
• Can I use a FODMAP to get me to California?
• What is a reducing sugar and what does the term reducing mean?
• Why doesn't pudding survive a freeze-thaw?

Carbohydrates forms

Carbohydrates consisting of monosaccharides, disaccharides, oligosac-charides, and polysaccharides are important to food. Monosaccharides are known as simple sugars. They bind water, taste sweet, and are principal reactants in the Maillard reaction. This chemistry is responsible for creating flavor and color in cooked foods. Sucrose is a disaccharide containing two monosaccharides (glucose and fructose) linked by a covalent bond. Sucrose is sweet, but most other disaccharides (lactose and maltose) are not. Longer carbohydrate polymers consisting of 2—10 monosaccharides units are termed oligosaccharides. This class of carbohydrate has little or no sweetness, but some are nutritionally important as dietary fiber and prebiotics. While we lack the ability to digest many oligosaccharides, they are fermented by bacteria in the large intestine and produce health promoting components such as short chain fatty acids. Polysaccharides, such as starch and cellulose, are much larger molecules composed of numerous glucose units joined by covalent bonds. Starches exist in linear (amylose) and branched chain (amylopectin) forms varying in digestibility and functional properties. The structure of amylose is compact, slowing the digestion process and subsequent release of glucose into blood. Conversely, amylopectin has a more open structure, facilitating its digestion and quick release of glucose into blood. The composition of amylose and amylopectin in starch varies with plant source. Wheat, corn, and potato starches differ in their ratio of amylose to amylopectin. They also behave differently in digestibility and functionality. Cellulose is a glucose polysaccharide, but its structure is substantially different from starch because of the chemical nature of the bond linking glucose units. Glucose polysaccharides in cellulose are tightly complexed linear polymers impervious to water hydration and digestive enzymes.

Carbohydrate structure and nomenclature

Monosaccharides are the simplest form of carbohydrate. They are typically composed of 3—6 carbon atoms joined by single (glyosidic) bonds. Monosaccharides are polyols, meaning that most carbon atoms in a monosaccharide have an attached hydroxyl (OH) group. The polyol nature of monosaccharides gives them high water solubility. It is possible to make aqueous solutions of glucose in water containing 60%—70% solids. Most monosaccharides also contain a carbonyl group in the form of an aldehyde or ketone and are termed aldose or ketose sugars, respectively. Carbonyl groups participate in chemical reactions involving the gain or loss of electrons, called reduction or oxidation (redox), respectively. Carbohydrate nomenclature is complex because both common and systematic names are used. Common names such as glucose, fructose, sucrose, and lactose are used in non-technical communications when referring to these carbohydrates. Systematic naming of carbohydrates provides information about carbohydrate structure. Monosaccharides are described using the stem name indicating the number of carbon atoms combined with a suffix indicating whether the molecule contains an aldehyde or ketone group. For example, a six-carbon monosaccharide containing an aldehyde is named by combining the stem "hex" with the suffix "ose" to give the name "hexose". By analogy, a six-carbon monosaccharide containing a ketone is named by combining the stem "hex" with the suffix "ulose" to give the name "hexulose".

Fisher Structure: Systematic nomenclature also numbers the carbon atoms in carbohydrate molecules. For example, Fig. 3.1 shows the structure

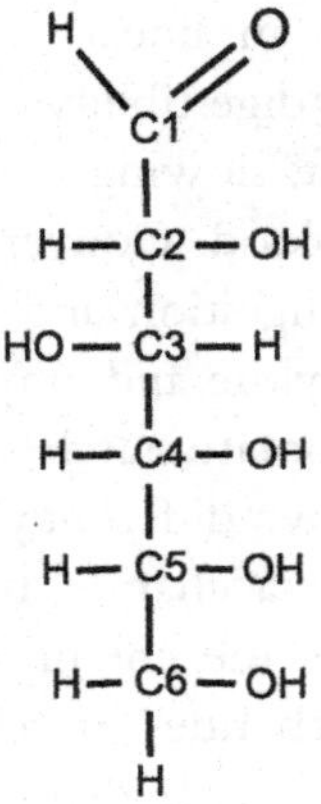

D Glucose
Fig. 3.1 Fisher structure of D glucose.

of glucose, a hexose monosaccharide, using the Fisher projection system. By convention in the Fisher system, monosaccharide atoms are numbered from top to bottom with the carbonyl group placed at the top (C-1). Carbons 2 through 6 are arranged with OH groups pointing either to the right or left of the carbon backbone. Carbon atoms can make four bonds to other atoms. A carbon atom is said to be asymmetrical when each of its four bonds are made to a different atom. A close look at the glucose structure in Fig. 3.1 shows carbons numbered 2 through 5 are asymmetrical. This means that each carbon atom is bonded to four different atoms. An asymmetrical configuration is important because it results in optical activity, a physical property meaning it is able to rotate a beam of polarized light in one of two directions, either right handed (clockwise) or left handed (counter-clockwise). Molecules with asymmetrical centers are chiral, which adds another level of complexity to carbohydrate nomenclature. Specifically, monosaccharides are designated as D or L based on the chirality of the molecule. In the Fisher model, when the OH of the highest numbered asymmetrical carbon is positioned to the right of the carbon backbone, it is in the D form. Conversely, if the OH group is on the left of that carbon, it is in the L form. Therefore, the molecule in Fig. 3.1 is named D-glucose. D and L forms of glucose are enantiomers, a word of Greek origin meaning opposite. The property of chirality of molecules is biologically important. For example, glucose and other monosaccharides are ultimately substrates for enzymes that selectively catalyze a reaction with one enantiomer or the other, but not both.

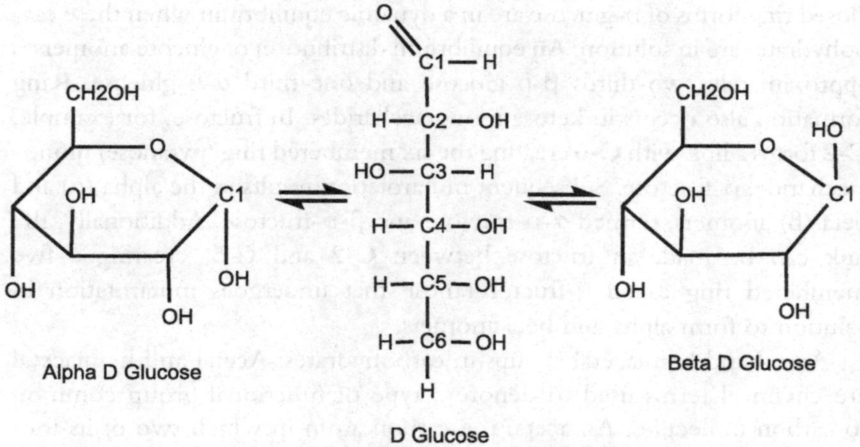

Fig. 3.2 Alpha and beta D glucose.

Haworth Structure: While the Fisher projection system conveniently represents monosaccharide structures as straight chains, the model does not adequately represent the structural complexity. Fig. 3.2 shows the numbering of carbon atoms for both Fisher and Haworth conventions using glucose as the example. In the Haworth convention, numbering of carbon atoms proceeds in a clockwise order for the ring structure. Hydroxyl groups pointing to the right of the carbon backbone in the Fisher projection are placed pointing down in Haworth system. Similarly, OH groups to the left of the carbon backbone in the Fisher projection are placed pointing up in Haworth structures.

Monosaccharides exist both as open chain and closed structures resulting from the process of mutarotation (Fig. 3.2). Mutarotation is a property of carbohydrate molecules that results from the ability to interconvert from open chain to closed ring structures. For example, glucose dissolved in water exists in equilibrium between open chain and two ring forms. A link between C-1 and C-5 atoms creates a six-membered ring named a glucopyranose. Alternatively, a ring can be formed between C-1 and C-4 atoms creating a five membered ring called a glucofuranose. In either case, forming a ring structure introduces an additional chiral center at C-1, making it the anomeric carbon. Orientation of the hydroxyl group at C-1 can be either down or up in the Haworth convention, corresponding to a designation of alpha (α) or beta (β), respectively. Mutarotation of D-glucose results in two cyclic anomers named α-D-glucose and β-D-glucose. Open chain and closed ring forms of D-glucose are in a dynamic equilibrium when these carbohydrates are in solution. An equilibrium distribution of glucose anomers is approximately two-thirds β-D-glucose and one-third α-D-glucose. Ring formation also occurs in ketose monosaccharides. In fructose, for example, C-2 forms a link with C-6 creating the six membered ring (pyranose) monosaccharide, D-fructose. Subsequent mutarotation results in the alpha (α) and beta (β) anomers, termed α-D-fructose and β-D-fructose. Additionally, the link can be made in fructose between C-2 and C-5, creating a five membered ring called D-fructofuranose that undergoes mutarotation in solution to form alpha and beta anomers.

Acetal and hemiacetal groups in carbohydrates: Acetal and hemiacetal are chemical terms used to denote a type of functional group common to carbon molecules. An acetal is a carbon atom in which two of its four bonds are linked to other carbons. The other two bonds of the carbon atom are linked to oxygen atoms that share a bond with other carbon atoms (Fig. 3.3). A hemiacetal differs from an acetal in that one of the linked oxygen atoms shares its other bond with a hydrogen instead of another carbon, a hydroxyl (OH) group. Acetal and hemiacetal groups are especially important to the chemistry of carbohydrates. Hemiacetal groups are found in monosaccharides at the site of ring closure (C-1). A hemiacetal is formed when the hydroxyl on a carbon, for example C-5 attacks the carbonyl group on C-1. Hemiacetal formation is reversible and this property is essential to the mutarotation process. Acetal groups are also formed at the site of ring closure (C-1) in monosaccharides. A notable feature of acetal formation is that they are not reversible. Once an acetal is formed, a monosaccharide can not open its ring and participate in mutarotation. Various groups or molecules can be coupled to monosaccharides via acetal formation and are termed glycosides. Glycosidic bonds are covalent

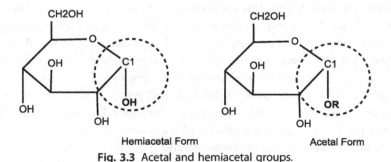

Fig. 3.3 Acetal and hemiacetal groups.

chemical bonds used to link monosaccharides in polymers such as starch. A glycosidic bond is also used to link glucose and fructose and create the disaccharide known as sucrose. Chemically, a carbohydrate must be in the hemiacetal form to participate in a redox reaction (important to the Maillard reaction described below). Presence of hemiacetal at C-1 enables the carbohydrate to open the ring and form the aldehyde. However, if the carbohydrate is in the acetal form, the ring can not open. A consequence of acetal formation is the loss of chemical reducing activity. Carbohydrate molecules with their anomeric carbon in the hemiacetal form are able to donate electrons to other molecules and are termed reducing sugars.

What is the easiest way to remember the definition of redox reactions?

The essential definitions of redox reactions include two key concepts:
- Oxidation is the loss of electrons by an atom, ion, or molecule.
- Reduction is the gain of electrons by an atom, ion, or molecule.

The chemistry of reducing sugars is essential to the Maillard reaction and is responsible for both desirable and undesirable outcomes in foods.

Monosaccharides: Glucose is a hexose sugar existing predominantly in D form (Fig. 3.4). Glucose, also known as dextrose, is only mildly sweet compared to table sugar (sucrose). It is not widely used for sweetening in food. Glucose occurs in combination with other sugars and contributes to the sweet taste of fruits, such as grapes and cherries.

Fructose is the major source of sweetness in fruits. The combination of fructose and glucose is responsible for the intense sweetness of honey. Fructose was discovered in 1847 by Augustin-Pierre Dubrunfaut and the name, fructose, was subsequently coined by William Miller in 1857. Sweetness is a major advantage for fructose, which is about twice as sweet as sucrose. Structurally, fructose is a ketose sugar that exists primarily as D-fructose. Its ring structure is stabilized by the molecule's internal hydrogen bonding. Fructose is produced as a dry solid from beet sugar,

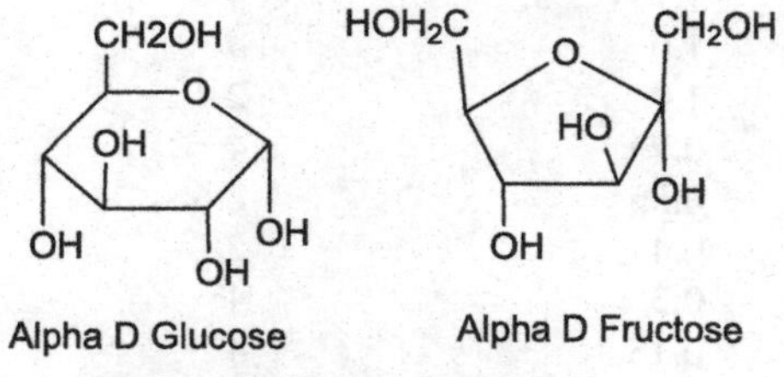

Fig. 3.4 Glucose and fructose.

sugar cane, honey, or corn starch. Corn starch represents the largest source of fructose. Using an enzymatic process, corn starch is hydrolyzed into glucose followed by its conversion to fructose. The final product is a 50/50 mixture of glucose and fructose called high fructose corn syrup (HFCS). HFCS is 50% sweeter than sucrose and widely used in food, especially soft drinks and candy. Fructose is also found in fruits and vegetables (Table 3.1). The sweetness of fructose is perceived more quickly and intensely than sucrose or glucose. Chemically, fructose is a stronger reducing sugar and browns more quickly in Maillard reaction. It is added to baked goods in the form of invert sugar or high fructose corn syrup to create more flavor and brown color. Nutritionally, fructose has a lower glycemic index than sugar (19 for fructose vs. 65 for sucrose), but has several negative aspects. Fructose and glucose are metabolized differently, and are preferentially made into 2 carbon (acetate) units that are the basic building blocks of fatty acids. Fructose over consumption contributes to non-alcoholic fatty liver disease, gout, and obesity. Over consumption of fructose is more likely to cause gastric distress and diarrhea when it enters and is fermented in the large intestine. Consumption of apple and pear juice by infants is a concern because fructose content of these juices can promote diarrhea.

Why is honey so sweet and milk is not?

Honey is a highly concentrated carbohydrate food containing about 20% water, 30% glucose and 40% fructose as monosaccharides. It is well known for its intense sweetness. Lactose is a disaccharide composed of glucose and galactose. It is the major component of cow's milk making up about 5% of its composition. Even if honey and lactose sugars are tasted on an equal gram for gram basis, honey would be 2 or 3 times sweeter. The full explanation for

Table 3.1 Sweetness and caloric value of polyols.

Sugar	Relative sweetness	Calories/g	Glycemic index
Fructose	1.7	4	23
Sucrose	1.0	4	65
Xylitol	1.0	2.4	12
Glucose	0.75	4	100
Glucatol/Sorbitol	0.55	2.6	4
Lactitol	0.4	2	3
Galactose	0.3	4	23
Lactose	0.15	4	45

the difference in sweetness perception is provided in Chapter 6 (Flavors) but the simple answer is that molecular structure dictates the response of taste bud receptors on your tongue.

Disaccharides: Sucrose is a disaccharide composed of glucose and fructose (Fig. 3.5). Sucrose refined from cane or beet sugars is used in many food products to add sweetness and body and to provide an overall pleasurable taste sensation. The link between monosaccharide units is termed a glycosidic bond and is used in making carbohydrate polymers (disaccharides, oligosaccharides, and polysaccharides). The glycosidic bond ($\alpha 1 \rightarrow 2$) between glucose and fructose involves both anomeric carbons, making sucrose a

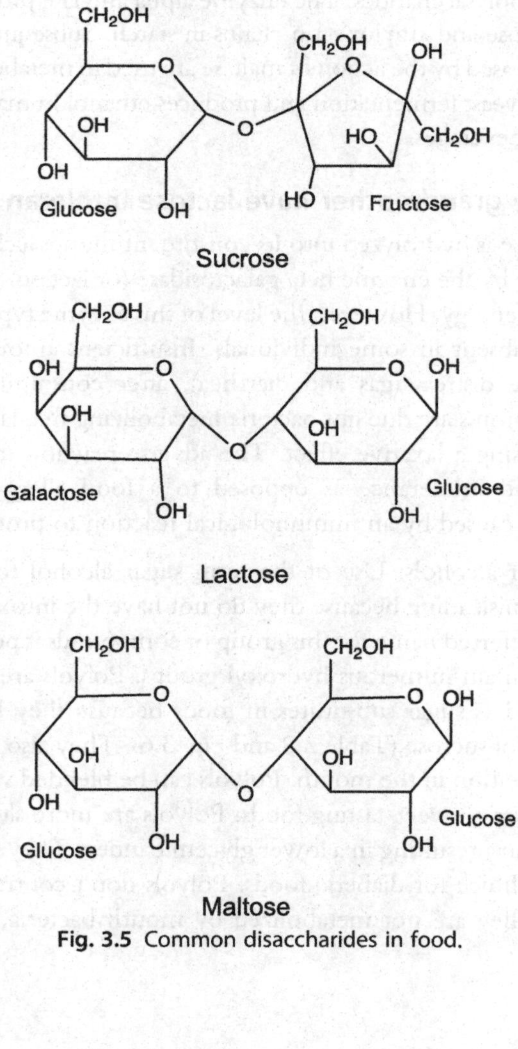

Fig. 3.5 Common disaccharides in food.

non-reducing sugar. However, the bond between the sugars is broken by heat. This results in release of two reducing sugars, glucose and fructose. Lactose represents the largest component of cow's milk. Lactose is composed of galactose and glucose linked by an β1 → 4 glycosidic bond. Lactose is a reducing sugar because the anomeric carbon (C-1) of glucose component is free to form an aldehyde. Lactose is beneficial in breast milk by promoting the growth of *Lactobacillus bifidus* bacteria in the infant's gut. *L. bifidus* bacteria help in the prevention of infection. However, it can cause a substantial problem known as lactose intolerance, resulting in gas and diarrhea. Maltose is a disaccharide composed of two α D-glucose units connected via an α1 → 4 glycosidic bond. The major source of maltose comes from enzymatic degradation of starch polysaccharides. The enzyme alpha amylase produces maltose units from amylose and amylopectin chains in starch. Subsequently, the two glucose units released by the action of maltase are used in metabolism. Maltose is a substrate for yeast fermentation and produces ethanol in making beer and other alcoholic beverages.

Why does my grandmother have lactose intolerance?

Normally, lactose is hydrolyzed into its constituent monosaccharides, galactose and glucose by the enzyme beta galactosidase (or lactase) in the gut and metabolized for energy. However, the level of this enzyme typically declines with age or is absent in some individuals. Insufficient amounts of lactase results in gastric distress (gas and diarrhea), after consuming milk. The unpleasant symptoms are due gut bacteria metabolizing free lactose producing gas and causing a laxative effect. The adverse reaction to lactose is an example of food intolerance as opposed to a food allergy. The latter, (food allergy) is caused by an immunological reaction to protein.

Polyols–sugar alcohols: Use of the term sugar alcohol for these compounds is a bit misleading because they do not have the intoxicating effects of ethanol. A preferred name for this group of compounds is polyol based on the fact they contain numerous hydroxyl groups. Polyols are sweet-tasting compounds used as sugar substitutes in foods because they have less than half the calories of sucrose (Table 3.2 and Fig. 3.6). They also create a desirable cooling sensation in the mouth. Polyols can be blended with sucrose to create lower calorie, sweet-tasting foods. Polyols are more slowly absorbed than regular sugars, resulting in a lower glycemic index. They are, therefore, the sweetener choice for diabetic foods. Polyols don't contribute to tooth decay because they are not metabolized by mouth bacteria. The number

Table 3.2 Polyol food applications.

	Parent sugar	Calories/g	Sweetness	Food applications
Sorbitol	Glucose Monosaccharide	2.6	60%	Sweetener, humectant, low calorie ice cream
Maltitol	Maltose Glucose Disaccharide	2.4	70%−80%	Chewing gum, hard candy
Lactitol	Lactose Glucose & Galactose Disaccharide	2.0	35%	Baking, low calorie chocolate and ice cream
Xylitol	Xylose Monosaccharide	0.4	100%	Beverage, sweetener in ice cream and yogurt, mouth wash
Erythritol	Erythrose Monosaccharide	0.2	60%−70%	Chewing gum, baked good, and beverages
Sucrose	Disaccharide Glucose & Fructose	4.0	100%	

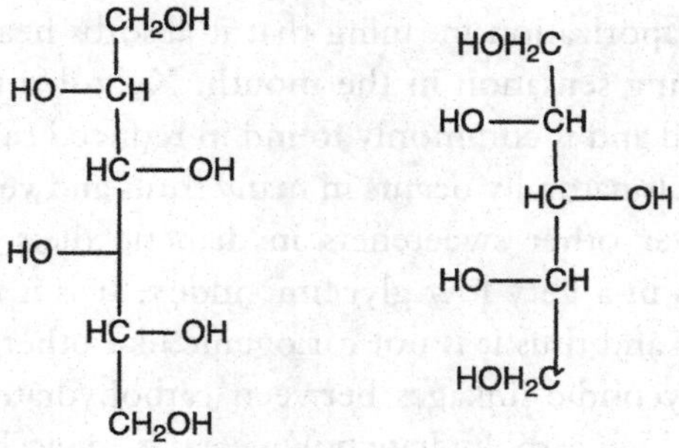

Sorbitol (Glucitol)   **Xylitol**

Fig. 3.6 Glucitol and Xylitol.

of hydroxyl groups on a polyol aids in reducing the water activity (a_w) level in food and can be used as humectants. A major disadvantage of polyols is their tendency to cause gas and diarrhea. Polyols are poorly absorbed in the small intestine and passed to large intestine where they are fermented by gut bacteria, producing gas and causing a laxative effect. Examples of polyols commonly used in foods include sorbitol, maltitol, lactitol, xylitol, and erythritol (Table 3.2).

Sorbitol, also known as glucitol, is 50%—70% as sweet as sucrose. It is used as a sweetener in candy, baked goods, and frozen desserts. Sorbitol naturally occurs in fruits such as peaches, apricots, prunes, and raisins. The disaccharide maltitol is 60%—80% as sweet as sucrose. Maltitol has greater potential to cause gas, bloating, and diarrhea compared to other polyols. In cooking, maltitol is one of the few polyols that can participate in Maillard browning reactions. High temperature causes hydrolysis of the glycosidic bond in maltitol and release of glucose to participate in browning reactions. Lactitol is about 30%—40% as sweet as sucrose, but it is poorly absorbed in the gut. While only 2% is absorbed in the small intestine, the remainder is fermented in the large intestine into short chain fatty acids and lactic acid. Lactitol has beneficial properties as a prebiotic, promoting the growth of desirable gut bacteria. However, it can also have as a substantial laxative effect. Erythritol is about 60%—70% as sweet as sucrose. It is a major exception to the rule of limited absorption. Approximately 90% of erythritol is absorbed through the small intestine into the blood. Erythritol provides very few calories because most is excreted in the urine. Erythritol occurs naturally in pears, watermelon, and grapes and can also be derived from the fermentation of starch. Xylitol is about as sweet as sucrose and is used in chewing gums and candies. It also has a negative heat of vaporization meaning that it absorbs heat as it volatilizes and creates a cooling sensation in the mouth. Xylitol is the most widely used polyol in food and is commonly found in reduced calorie mint chocolates and candies. It naturally occurs in many fruits and vegetables. Xylitol has advantages over other sweeteners in diabetic diets because its low digestibility results in a very low glycemic index. It is is not metabolized by mouth bacteria and thus it is not cariogenic like other sweeteners.

Glycosides: Glycosidic linkages between carbohydrate units represent the chemistry by which carbohydrate polymers (i.e., disaccharides, oligosaccharides, and polysaccharides) are formed. Glycosidic links are also made between carbohydrates and hydroxyl (OH)-containing R groups of amino acids, such as serine and threonine. These amino acids are the sites of carbohydrate attachment to proteins in the biological process known as glycosylation. Proteins tagged with carbohydrates via glycosylation have important biological roles. For example, blood group typing as A, B, or O is based on glycosylation of red blood cell proteins. Some viruses and flu, for example, have glycosylated proteins on their protective envelop, shielding them from detection by the immune system. Glycosylation of plant and animal proteins represents a system for targeting proteins to storage sites in seeds and cells. Glycosylation also occurs in flavonoid molecules. Glucose is the most common monosaccharide attachment. Glycosylated flavonoids provide plants with protection against fungal infection and contribute to the color of their flowers. In foods, flavonoids are responsible for the color of blueberries and grapes as described in Chapter 8. Glycosylation of flavonoids makes them more water soluble and enhances their contribution to the color of wine. Flavonoids compounds are phytonutrients that have health promoting properties, such as antioxidant activity.

Oligosaccharides: While there is no hard rule, carbohydrate polymers containing between two and 10 monosaccharide units are termed oligosaccharides. Several examples of oligosaccharides naturally occuring in food include stachyose, raffinose, fructose oligosaccharides (FOS), and galactose oligosaccharides (GOS) (Fig. 3.7). Raffinose and stachyose are among the smallest oligosaccharides composed of three and four monosaccharide units, respectively. Raffinose is a tri-saccharide composed of galactose, glucose, and fructose. Specifically, the raffinose molecule contains α-D-galactose, α-D-glucose, and β-D-fructose. Stachyose is very similar in structure to raffinose, except that it contains an additional D-galactose that makes it a tetra-saccharide (Fig. 3.7). Although neither of these oligosaccharides are digestible by humans, they are fermented in the large intestine. These oligosaccharides cause gas (flatulence) after consuming foods that contain them. Green beans, peas, soybeans, and other legumes are sources of these oligosaccharides. Maltotriose is a trisaccharide composed of three α-D-glucose units derived from the enzymatic digestion of starch. Substantial amounts of Maltotriose are produced during the process of mashing to make "wort", an essential

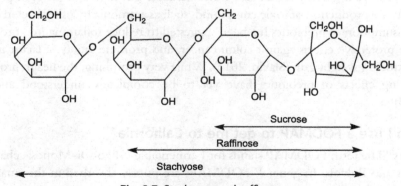

Fig. 3.7 Stachyose and raffinose.

step in brewing beer. Yeast subsequently ferment maltotriose into ethanol and carbon dioxide components of beer. Fructose oligosaccharides (FOS) are polymers of fructose varying in size from 2 to 10 units. They have a mildly sweet taste and low caloric value because they are not absorbed in the gut. FOS naturally occur in foods such as agave, asparagus, onion, garlic, leek, and banana. Inulin is a type of FOS composed of a longer fructose chain (about 60 monosaccharide units) and naturally occurs in the roots of the chicory plant. Inulin's major benefit is as a source of dietary fiber and promotion of gut health through support of beneficial bacteria growth and mineral absorption (Sabater-Molina et al., 2009; Kalyani et al., 2010). Galactose oligosaccharides (GOS) are composed almost entirely of galactose. They are resistant to digestion in the upper GI tract, but, fermented in the large intestine. GOS are mildly laxative and not prone to cause gas or diarrhea. GOS and FOS have been identified as prebiotics with beneficial effects on the growth of *L. bifidus* and the potential to stimulate the immune system (Fanaro et al., 2007).

Oligosaccharides as prebiotics: Prebiotics are defined as components (mostly carbohydrates) that promote the growth of beneficial microorganisms in the gut. Carbohydrates fitting that definition are typically oligosaccharides with low- or non-digestible properties. Characteristics of an effective prebiotic include resistance to digestion in the upper GI tract, fermentation by microflora of the large intestine, and stimulation of growth and/or activity of bacteria associated with health. Organisms of the genera bifidobacteria and lactobacilli are responsible for the fermentation of prebiotics, and therefore termed probiotics. Fermentation of prebiotics results in several beneficial effects. For example, bifidobacteria ferment oligosaccharides into short chain fatty acids (SCFA), such as butyrate and propionate, that lower intestinal pH. SCFA serve as energy sources for other desirable gut flora and lowered pH limits the production of toxic compounds such as ammonia by other bacteria. Consumption of prebiotics has been suggested to reduce inflammation, provide protective effects against colon cancer, and promote satiety, a factor in controlling weight gain (Slavin, 2013). While very promising, the health promoting effects of prebiotics have yet to be completely understood and verified.

Can I use a FODMAP to get me to California?

Sorry! The term FODMAP stands for Fermentable Oligo-Di-Monosaccharides and Polyols. In general, FODMAPs are poorly absorbed in the small intestine, but rapidly fermented by bacteria in the colon. This results in gas, bloating, diarrhea, and cramps attributed to symptoms of irritable bowel syndrome (IBS).

FODMAPs and food intolerance: Examples of FODMAPs include fructose, lactose, polyols, and some FOS and GOS (Table 3.3). Foods containing these carbohydrates can cause IBS-like symptoms. Wheat gluten, its major protein fraction, is also associated with causing IBS-like symptoms. Many have selected gluten-free diets to avoid the unpleasant symptoms caused by eating bread and pasta. However, recent studies suggest that IBS-like reactions resulting from eating gluten-containing foods are more likely caused by carbohydrates. Specifically, fructose oligosaccharides are suggested to be the major cause of IBS-like symptoms, previously attributed to gluten protein. Controlled studies using FODMAP restricted diets significantly reduced the occurrence of IBS-like symptoms (Shepard and Gibson, 2006; Fedewa and Rao, 2014).

Maillard reaction: Discovery of Maillard chemistry is attributed to the French physician and scientist, Louis Camille Maillard in 1912. He studied the reaction that occurred when mixtures of amino acids and sugars are heated. His work established that this chemistry was responsible for desirable flavors and brown pigments. Details of the reaction mechanism were unknown until John Hodge, a USDA scientist, published "Chemistry of browning reactions in model systems" in 1953. His work described the compounds produced in the early stage of reaction and those responsible for its flavors. Maillard chemistry is used in the creation of synthetically derived food flavors. In food, the essential reaction occurs between protein and carbohydrate, in a heated system. For example, Maillard chemistry is primarily responsible for color, flavor, and aroma of baked bread. During cooking, sugars and starches react with wheat proteins to form hundreds of compounds contributing to the flavor and aroma of bread. The surface

Table 3.3 FODMAPs in foods.

	Type	Food source
Fermentable oligosaccharides	FOS, Fructans, GOS, Galactans	Wheat, Spelt, Rye, barley, Legumes
Disaccharides	Lactose	Milk
Monosaccharides	Fructose	Fruits, Sweeteners, HFCS
Polyols	Sorbitol, Maltitol, Lactitol, Xylitol, Erythritol	Fruits and sweeteners

of bread experiences higher temperatures during baking, causing the desirable brown crust. Application of egg white wash to the loaf's surface further intensifies the color. Maillard chemistry is responsible for transforming bland commodities like raw meat into flavors and aromas of cooked foods (e.g., burgers and steaks). High temperature cooking methods, such as grilling, enhance meat flavors by creating a more intense Maillard reaction. Maillard chemistry is ubiquitous in foods such as cookies, toasted marshmallows, roasted coffee, cocoa beans (chocolate), peanuts, and more. However, Maillard chemistry also has adverse effects in food. It causes substantial destruction of essential amino acids and vitamins.

Reducing sugars: Reduction-oxidation chemistry is essential to Maillard reactions involving carbohydrates. Reduction-oxidation (redox) reactions are those in which atoms, ions, or molecules lose or gain electrons. Redox reactions are coupled. This means that oxidation (loss of an electron) in one molecule is linked to reduction (gain of an electron) in another. Biologically, redox chemistry is essential to all living things, producing energy derived from the metabolism of glucose.

Mechanism of Maillard reaction: Maillard is a series of chemical reactions between an amine (NH_2) group and a carbonyl group. Amines (NH_2) are supplied to the reaction by amino acids, peptides, or proteins. Carbonyl to the reaction by reducing carbohydrates containing aldehydes or ketones. The reaction begins with chemical condensation of the protein's amine group with C-1 of glucose. This results in glucosamine and a water molecule as initial products. Subsequent reactions result in products such as hydroxymethyl furfural that are responsible for brown color. Flavors and aromas are produced by reactions between amino acids and α-dicarbonyl, in a process known as Strecker degradation (Fig. 3.8). Aldehyde products of Strecker degradation are a major source of flavors and aromas in cooked food. Methional, for example, is an aldehyde generated from a Strecker reaction between α-dicarbonyl and the amino acid methionine. Methional has the unmistakable aroma of baked potatoes. Strecker reaction is responsible for pyrrole and pyrazine compounds having the characteristic flavors of honey, bread, crackers, coffee, chocolate, peanuts,

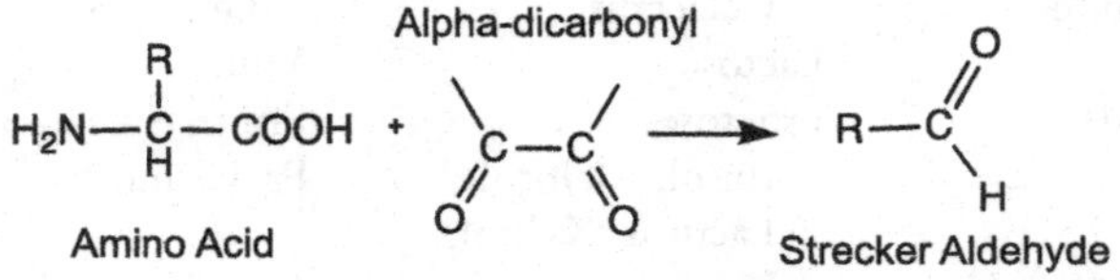

Fig. 3.8 Strecker degradation reaction.

potato chips, and grilled meat. Maillard chemistry is commercially used in the creation of synthetic flavors by heating mixtures of amino acids and sugars.

Factors affecting Maillard reaction: Environmental conditions influence the extent of Maillard reaction and the flavor and color of food resulting from this chemistry. It is logical that higher temperature and/or time of the thermal treatment increases food flavor and color. For example, high temperature cooking methods, such as frying or grilling, produce greater flavor and color than baking or boiling. The extent of Maillard reaction is also greatly influenced by type of carbohydrate. Monosaccharides have the fastest rate of reaction with some variation between types of sugars. Glucose, for example, browns more extensively compared to an equivalent amount of lactose. Most polyol monosaccharides do not brown when heated because they do not contain aldehyde or ketone reducing groups. Disaccharides, such as lactose, sucrose, and maltose, are slower reactants compared to monosaccharides. Sucrose, while not a reducing sugar itself, browns during cooking following breakdown into its constituent monosaccharide sugars (glucose, and fructose). Oligosaccharides and polysaccharides represent the slowest group of Maillard reactants because almost all potential sources of reducing groups (carbonyls) are tied up in glycosidic linkages between constituent monosaccharides. Polysaccharides participate to some degree in Maillard browning. The rate and extent of reaction is increased following enzymatic hydrolysis of their glycosidic bonds. Each site of glycosidic bond hydrolysis creates a new carbonyl reducing group and increases the extent of browning. For example, process such as worting and fermentation used in making beer causes hydrolysis the starch and provides more reducing groups to participate in Maillard reaction. The extent of Maillard reaction is well known to be influenced by the pH of the environment. Slightly acidic conditions (pH 5 to 6) slow the reaction, resulting in less flavor and color. Neutral pH greatly improves the reaction rate and is typical for flavor and aroma components produced in baking bread. At alkaline pH (8−9), the extent of browning is much greater and unique flavors are produced. Pretzels, for example, are made in a two-step process involving boiling in baking soda solution (alkaline pH), followed by baking. The combined treatment gives pretzels their unique flavor and brown color. Moisture content also affects Maillard chemistry. In general, high moisture level inhibits Maillard reaction. Boiling meat and potatoes, for example, results in mild flavors and little or no color. However, if meat and potatoes are first pan-seared at high temperature, darker colors and more intense flavors are created.

Undesirable effects of Maillard chemistry: While Maillard is essential to desirable effects in cooked foods, its chemistry is also responsible for undesirable consequences. The nutritional quality of food is adversely affected when heat is used in processing operations, such as drying milk or egg, to make shelf-stable products. Essential amino acids, such as cysteine and lysine, are destroyed when they participate in Maillard reactions. Maillard chemistry is also responsible for the destruction of B vitamins (vitamin B_6 pyridoxamine) and folate in heated foods. The loss of these nutrients is accelerated in cooking at high temperature. Knowledge of Maillard's potential to cause heat-induced loss of essential nutrients was a major factor that prompted food companies to add vitamins, post processing, to breakfast cereals. While many of the adverse consequences are rectified by adding back important nutrients, some products of the reaction have toxic or genotoxic properties. Most concerning of these products resulting from Maillard chemistry is acrylamide. Acrylamide, a neurotoxin and carcinogen, came to international attention in 1997 when it was linked to an incident that killed animals and sicken construction workers in Sweden. Swedish farmers noticed that cows suddenly became paralyzed and died. Large numbers of dead fish were found floating in aquafarm pools. Construction workers using a sealant material to plug leaks in tunnels and levies, experienced numbness. The ultimate cause of these unfortunate outcomes was the acrylamide component of sealant material, leaching into the water. The incident brought attention to the wider prevalence of acrylamide in food. It was found that cooking also produced acrylamide as a product of Maillard chemistry. The principal reactant responsible for acrylamide in food is the amino acid, asparagine. Acrylamide is produced by the reaction between asparagine and reducing carbohydrates. A condensation reaction between asparagine and the carbohydrate's carbonyl group is the first step. Subsequently, a cascade of reactions results in acrylamide as a significant product (Fig. 3.9). The amount of acrylamide formed in food is influenced by environmental factors, notably pH, temperature, and moisture level. Acrylamide formation is favored at alkaline pH (8—9) and high temperature. A threshold of 120 °C is needed for the reaction to produce significant amounts acrylamide. The highest level of acrylamide is formed in low moisture foods cooked at high temperatures (e.g., frying). Conversely, high moisture cooking methods (e.g., boiling) produce little or no acrylamide. The potential for acrylamide to occur in food has resulted in considerable effort

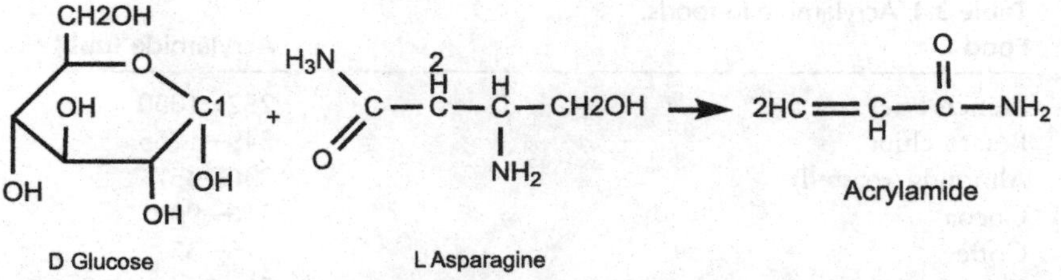

Fig. 3.9 Acrylamide formation. *Adapted from FDA Data Survey on Acrylamide in food. 2002–4.*

to reduce or prevent its production. Asparagine is the target of methods designed to reduce its level in starchy foods that are cooked at high temperature. For example, the acrylamide level in French fries and potato chips is reduced by substituting conventional potatoes with low asparagine varieties. Genetic engineering is being used to obtain potatoes with lower asparagine content. It is also possible to lower the level of asparagine in potatoes by treating them with the enzyme asparaginase. Asparaginase causes removal of an amide group in asparagine, converting it to aspartic acid, a slow reactant in the acrylamide formation pathway.

Which foods contain the greatest amount of acrylamide?: The answer may come as an unpleasant surprise. Values for each product listed in Table 3.4 are given as a range because processing variations can alter the amount of acrylamide produced. Coffee and chocolate, for example, vary in acrylamide levels due to differences in heating methods. Higher temperature and longer roasting times are required to produce darker beans. However, these conditions result in higher levels of acrylamide in dark coffee or chocolate. The reader is referred to the US Food and Drug Administration web site (fda.gov) for more specific findings on acrylamide in food. Black olives may seem out of place in this table because they are not heated to high temperature. However, black olives are treated with alkali (lye) that increases potential for acrylamide formation. Limiting consumption of food cooked by high heat methods (i.e., frying) represents a simple way to reduce acrylamide exposure (Linebeck et al., 2013).

Polysaccharides: Polysaccharides are polymers of monosaccharide units joined together through glycosidic bonds. Examples of polysaccharides include starch, glycogen, chitin, cellulose, pectins, and gums. Polysaccharides composed of a single type of monosaccharide are termed homopolysaccharides. Conversely, those composed of different monosaccharides are termed heteropolysaccharides. Polysaccharides serve major biological functions.

Table 3.4 Acrylamide in foods.

Food	Acrylamide (ppb)
French Fries	252–1030
Potato chips	249–1265
Almonds (roasted)	236–457
Cocoa	316–909
Coffee	175–351
Breakfast cereals	71–266
Cookies and crackers	61–944
Black olives	41–130
Breads	34–364
Peanuts butter	64–125

Starch and glycogen represent energy stores for both plants and animals. Cellulose, pectin, and chitin are examples of structural polysaccharides that constitute a major component of plant cell walls and the exoskeleton of invertebrate animals.

Starch is a polysaccharide of glucose and a major energy source for growth and development in plants and animals. It is located in the endosperm of seeds and the roots of tubers. Commodity starches (i.e., those used as food ingredients) are derived from cereals such as wheat, corn, rice, oats, and tubers like potatoes and cassava. Additionally, beans such as lentils, fava beans, black beans, and kidney beans represent a diverse group of plant foods that contain substantial amounts of starch. Starch from wheat, corn, and potato is used to make a countless number of food products. Wheat starch, for example, is essential to making bread and pasta. Starches are relied upon to provide texture in extruded food products like breakfast cereals and snack foods. All starch occurs in varying size and shape granules, depending on their source. Differences in starch granules are evident in images obtained from the scanning electron microscope (Fig. 3.10). SEM micrographs illustrate the differences in starch granule morphology and the distribution of their sizes. Corn, wheat, and potato, starch granule size increases in that order. On average, the diameter of corn, wheat, and potato starch is approximately 15, 25, and 40 µm, respectively. Starch granules are dense and poorly soluble in water, but hydrate when soaked in cold water. Crystalline regions in starch granules have an optical property called birefringence observable when viewed under the light microscope. Native starch granules refract polarized light and crate a luminous feature called the Maltese cross.

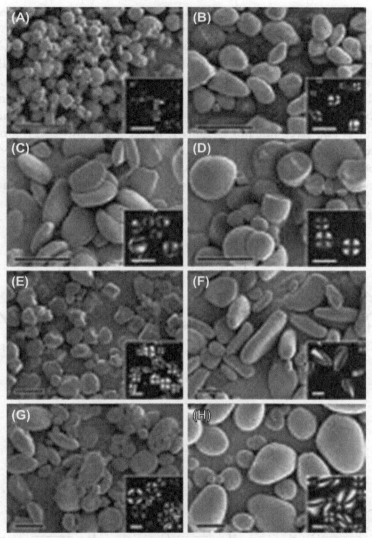

Fig. 3.10 Sem micrograph of starch granules. *Bertoft, E. 2010.*

This feature that can be seen in the lower right-hand side of each panel in Fig. 3.10.

Starch structure: The hydroxyl groups of monosaccharides in starch stabilize its structure through hydrogen bonding. Starch molecules are tightly packed in granules and explains why they are slow to hydrate.

Starch polysaccharides are composed of glucose units in two types of polymers (amylose and amylopectin) numbering between 50 and several thousand molecules. The linear glucose polymer is termed amylose, and the branched chain polymer is termed amylopectin (Fig. 3.11). Both amylose and amylopectin occur in starch and their respective ratio varies with the source. A type of corn starch termed "waxy" is composed almost entirely of amylopectin. Amylose is a mostly linear chain of α-D-glucose molecules. The glycosidic link of glucose molecules in amylose, is formed between C-1 and C-4 and described as an α1-4 linkage. The "α" designation indicates direction of bond at the C-1 atom. Amylose is mostly a linear polymer, but it contains a few α1-6 branches, about 1 in every 100 glucose units, on average. Amylose chains exist as a left-handed double helix, as

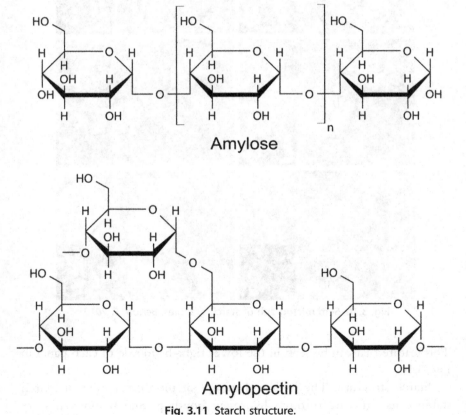

Fig. 3.11 Starch structure.

opposed to a random coil. Each turn of the helix contains approximately 6 glucose units. The interior of amylose chains is somewhat hydrophobic and can bind components with similar properties, such as, flavor compounds, and lipids. Notably, amylose binds iodine and results in a deep blue color that is the basis of the well-known starch test. Amylose chain lengths vary among starches, ranging between several hundred and a thousand glucose units. In contrast, amylopectin has a branched structure containing both α1-4 linked glucose polymers with α1-6 links serving as branch points linking other chains. Branch points in the polymer give it a bush-like structure (Fig. 3.12). In starch granules, amylopectin is predominantly composed of several structures, termed A, B, and C chains, that differ in branching and length. Only the C chain contains reducing ends in the polymer. Amylopectin molecules are assembled as clusters in the granule

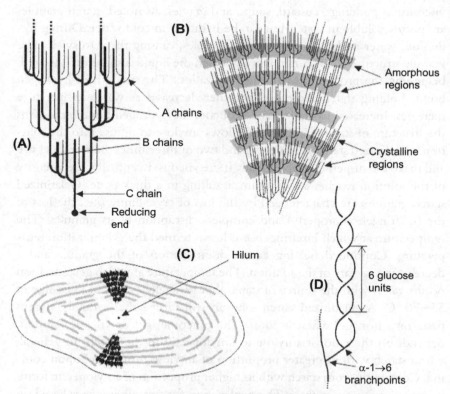

Fig. 3.12 Amylopectin + Starch Granule Structure. *Coultate, TP., 2009.*

structure. Amylopectin molecules are arranged with overlapping A and B chains organized in radial fashion with the reducing end (C chain) toward the granule's hilum (Fig. 3.12). Overlapping clusters of A and B chains are responsible for alternating light and dark concentric rings observed when starch granules are viewed in the light microscope (Fig. 3.12). Light areas in the granule correspond to crystalline regions of the amylopectin molecule. Conversely, dark areas in the granule correspond to amorphous regions of amylopectin that characteristically contain most of the branch points. Amylopectin is very similar in structure to glycogen, the storage polysaccharide in muscle and liver tissues. However, glycogen in muscle is completely metabolized during the postmortem process when muscle is converted to meat.

Food applications: Starch is responsible for a wide range of functional properties in foods. Starches provide the texture of bread and pasta and viscosity in puddings, custard, soups, and gravies. As noted, starch granules are poorly soluble in water, but can be hydrated in cold water. During hydration, water molecules diffuse into granules, causing them to swell. The granule structure changes dramatically when the liquid suspension is heated, beginning the process known as gelatinization. The strength of hydrogen bonds holding starch molecules together decreases as water temperature increases. Increased hydration and reduction of hydrogen bonding loosens the structure of starch granules and allows amylose to diffuse into the solution. Gelatinized starch granules expand two or three times their original size and make the suspension substantially more viscous. Eventually, the viscosity of the solution reaches a maximum, resulting in a thick paste. Gelatinized starch granules are characterized by the loss of crystallinity (i.e., the loss of the birefringence property) and complete disruption of its granules. The temperature at which birefringence is lost is termed the gelatinization temperature. Continued heating causes disintegration of the granule and a decrease in viscosity of the solution. The temperature at which gelatinization occurs varies with the source of starch, but generally occurs in the range of 55–70 °C. A gelatinized starch solution will form either a highly viscous paste or a firm gel when it cools. The outcome, gel or viscous solution, depends on the ratio of amylose to amylopectin in the starch. In general, wheat starch with its greater proportion of amylose forms a gel upon cooling. Conversely, corn starch with its higher proportion of amylopectin forms viscous paste upon cooling. The explanation for this difference is based on the chemistry of hydrogen bonding and the structure amylose versus amylo-pectin molecules. The strength of hydrogen bonds holding polysaccharide molecules together weaken as the temperature increases. At 100 °C hydrogen bonding is almost nonexistent and there is little structure-stabilizing force. When the temperature drops, hydrogen bonds are reformed. Amylose molecules are able to gel upon cooling because their structural flexibility enables formation of hydrogen bonds that link many polysaccharide molecules. In contrast, a solution of gelatinized high amylo-pectin starch forms relatively few intermolecular hydrogen bonds because its branched structure limits the amount of hydrogen bonded links that can form between molecules.

Why doesn't pudding survive a freeze-thaw?

Pudding is a sweetened dessert food typically made by cooking a mixture of sugar, flavorings, and a high amylose starch. This mixture is heated and forms a gel after being allowed to cool. However, if the pudding is frozen, it col-lapses into a mixture of goo and water then thawed. Want to know why this happens? See the next section for explanation about retrogradation.

Retrogradation: Pudding is a gelled form of starch containing an open structure that entraps large amounts of water. The expansion of water during freezing damages the starch gel and causes water to separate and leave a shrunken gooey mess. Retrogradation is the cause of this undesirable change. Retrogradation is caused by reformation of hydrogen bonds between starch molecules, resulting in exclusion of water molecules. High amylose starch is particularly prone to retrogradation, thus high amylopectin starch is preferred for freeze-thaw applications. Stale bread is another example of retrogradation. Bread becomes stale a few days after baking but, its stiff texture is not the result of drying out. While some water is lost through evaporation, but, it retains a majority of its moisture for 2−3 days. This is demonstrated when stale bread is toasted. Heating makes it soft again by weakening hydrogen bonding forces and redistributing its water content. Commercial breads resist staling for days and longer, through the use of emulsifiers to slow the staling process. An alternative to the use of additives to prevent staling, is to replace 1/4 cup of water with two egg whites in the recipe. Retrogradation is inhibited because the egg protein forms a complex with carbohydrate that prevents recrystallization.

What is instant starch?

Instant starch is used in applications where cooking for extended periods of time is not desirable or possible. Products like instant pudding, gravy, and sauces contain instantized starch. Oatmeal, for example, is a breakfast food typically requiring cooking time that may be a luxury on a rushed morning. Instant oatmeal is an alternative to traditional oatmeal. It can be prepared in the microwave in less time than it takes to text someone. The convenience factor in these food products is provided by starch that has been instantized. Instant starch is made in the factory by heating the starch ingredient (like oats for oatmeal) to the gelatinization point followed by quickly cooling and drying it. High viscosity is formed in the product as soon as water is added. Heat is not necessary for instant pudding and simply adding cold milk produces a highly viscous product. Instant oatmeal benefits from a brief heating to finish the gelatinization process.

Extruded food products: Extrusion is a type of food processing that employs a combination of heat (120–150 °C) and high pressure (10–20 Bar or 145 to 290 psi) to continuously cook food materials. A food extruder consists of a heated cylinder approximately 4 feet in length, containing a rotating auger feeding the food matrix toward a narrow opening at the front end. A die with holes of approximately 1 cm or less, placed at the front, creates extruded products whose shape are varied by the die. Extrusion processing is used to make products like macaroni, breakfast cereals, pet food, textured vegetable protein, and snack foods. Popular corn snack food products (e.g., cheese puffs) are made by extrusion processing. In this process, corn starch hydrated to about 15% moisture is fed into the preheated extruder where the auger compresses and moves it along the barrel. The pressure generated by compression is sufficient to increase the boiling point of water above the process temperature. As the material exits through the small die, the resulting pressure drop causes water in the material to become steam and expand the product. A knife cuts the rope-like extruded material into short segments as it exits. Salt, colorants, and flavorings are added to create the finished product. Extrusion is a widely used processing technology that also impacts the nutritional quality of food. Extrusion processing of blended soy and corn, for example, can be used to create plant-based food products with a higher level of protein and better balance of essential amino acids. This processing can reduce the level of antinutritional factors, such as phytate and trypsin inhibitor, typically present in legume protein sources. There are some negative aspects associated with extrusion processing, such as the loss of vitamins (ascorbic acid, thiamine, and folate) and essential amino acids (lysine, methionine). Extrusion processing increases starch digestibility, compared to other forms of cooking and may increase its glycemic index.

Starch digestibility: The enzymatically driven process of converting starch into energy occurs in the small intestine. Two enzymes important to this process are alpha amylase and maltase. Alpha amylase hydrolyzes $\alpha 1 \rightarrow 4$ glycosidic bonds in the polysaccharide, producing the disaccharide, maltose. Maltase subsequently hydrolyzes the glycosidic bond in maltose, yielding two glucose molecules. The overall structure of starch affects the degree to which the polysaccharide is broken down into glucose. For example, alpha amylase does not hydrolyze $\alpha 1 \rightarrow 6$ chain branch points in amylopectin structures. The hydrolysis activity of alpha amylase is slower in amylose compared to amylopectin because it has a tight, alpha helical structure with extensive hydrogen bonding between chains. Amylopectin structure, in contrast, is less compact, enabling greater access of the enzyme to hydrolysis sites. Structure is therefore an important factor affecting the digestibility of starch from various plant sources. For example, starch derived from legume sources typically contains a higher proportion of amylose and is less digestible than corn or wheat starch. Conversely, corn and wheat starches are higher in amylopectin content and more readily digested. Starch digestibility is improved by cooking. Raw potato starch, for example, is poorly digested because its structure is more crystalline. Boiling potatoes gelatinizes the starch and increases its digestibility. Beans are a similar example in that digestibility is lowest before cooking. Bean starch hydrates slowly because of its tight structure. For this reason, beans should be soaked overnight before being cooked. Boiling raw beans without a soaking step can convert its starch into a hard, fibrous material with little digestibility.

Fiber: Dietary fiber refers to polysaccharides that are not readily broken down into glucose and metabolized for energy. Dietary fiber is derived from food such as vegetables, beans, peas, lentils, and some grains. There are two major categories of dietary fiber: soluble fiber and insoluble fiber. Soluble fiber is composed of polysaccharides that form a viscous liquid in the small intestine. These polysaccharides are subsequently fermented by bacteria in the colon into short chain fatty acids (SCF) such as acetic, propionic, and butyric acids. SCF are taken up and utilized by intestinal epithelial cells as a major source of energy. Butyric acid has been shown to provide health benefits to the host in reducing the risk of colon cancer (Russo et al., 2018). Insoluble fiber is composed of polysaccharides, such as cellulose, that are indigestible by us or the bacteria in our gut. This form of dietary fiber serves as a bulking agent that reduces constipation, binds excess cholesterol, and facilitates its elimination. Resistant starch (RS), a term coined by Hans Englyst in 1992, refers to starch that is poorly digested in the small intestine. It is subsequently passed on to the colon where it is fermented by bacteria and produces substantial amounts of butyric acid and other SCF. Resistant starch is not a unique form of starch. It is found in several sources including legumes, raw potatoes, and green bananas. Legume starch is particularly resistant to digestion due to linkages with fibrous cell walls. Starch in foods like potatoes and rice that are cooked and cooled becomes more resistant to digestion due the process of retrogradation. Resistant starch can also be made by chemical cross-linking of its chains.

Glycemic index (GI): GI is a measure of the blood glucose-raising potential associated with carbohydrates in food, compared to that of pure glucose. GI values are designated as low (less than 55), moderate (56–69), and high (greater than 70). Eating a high GI food causes a sharp increase in blood glucose level, followed by a rapid decline. In contrast, eating low GI foods results in a lower rise in blood glucose level that declines gradually. The spike in blood glucose level stimulates beta cells in the pancreas to secrete extra insulin, resulting in a rapid decline of glucose level in blood. Chronic over-stimulation of the pancreas can result in loss of its ability to produce insulin, leading to the disease known as type 2 diabetes. While GI values provide indication of which foods are likely to result in higher or lower blood glucose levels, it does not account for the serving size. The concept of glycemic load (GL) is used to provide evaluation of potential impact on blood glucose (GI) and the amount eaten. GL is determined by multiplying a food's GI, by the grams of carbohydrate in a serving of the food. That number divided by 100, gives the GL value. The GL values are designated low (less than 10), moderate (11–19), and high (greater than 20). Comparison of GI and GL values in a food illustrates the difference. For example, GI value for watermelon and a donut are nearly the same (71 and 76). However, the GL for these foods are 8 and 17, when adjusted for the amount of carbohydrate contained and the serving size.

Cellulose is the most abundant organic molecule on the planet. However, the structure of this polysaccharide limits its use as a food. Cellulose does not hydrate in water and only modified versions of it have applications in food. Chemically hydrolyzed cellulose powders, for example, are used to provide high viscosity and desirable mouth feel in diet drinks and smoothies. Cellulose and starch are both glucose polysaccharides, but their structure is very different due to the bond that joins glucose units in the respective polymers. Cellulose is composed of glucose molecules, linked by $\beta1 \rightarrow 4$ glycosidic bonds (Fig. 3.13). The corresponding bond in starch links glucose units via $\alpha1 \rightarrow 4$ glycosidic bonds. As a result, cellulose polysaccharides are long, linear chains that tightly associate by hydrogen bonding. Cellulose chains are so tightly joined that water molecules are unable to penetrate and hydrate its structure. Cellulose is indigestible to humans and most animals because we lack the enzyme that hydrolyzes the $\beta1 \rightarrow 4$ bond. Ruminant animals are the exception and can use cellulose as an energy source. Bacteria in the foregut of ruminants provide enzymes necessary to break down cellulose into glucose.

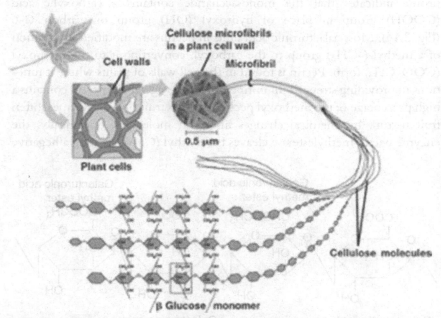

Fig. 3.13 Cellulose Structure. *Image Source: A Level Biology, Carbohydrate polymers https://alevelbiology.co.uk/notes/carbohydrate-polymers/.*

Food applications of cellulose: Cellulose in its native form is not useable as a food ingredient, but it can be modified to provide functional properties. Microcrystalline cellulose is made by extensive chemical hydrolysis of cellulose, resulting in shorter polymer segments that are more readily hydrated. Microcrystalline cellulose is soluble in dilute acid and provides beverages (e.g., smoothies) with high viscosity without adding calories. This cellulose is also used to provide high viscosity in salad dressing and whipped toppings. Another form of modified cellulose, carboxymethyl cellulose (CMC), is made by oxidation of the extensively hydrolyzed form. CMC has a wide range of applications in food and cosmetics as a result of the introduction of carboxyl (COOH) groups in the polymer. The very polar nature of this group provides good interaction with water (especially in the ionized form). In foods, CMC is used as a protein stabilizer, keeping milk casein proteins in stable suspension over a pH range of 3—6. It is used in processed cheese spread and as a dispersing agent in dry powder drink mixes.

Pectin is a polysaccharide composed of galacturonic acid. The suffix uronic indicates that the monosaccharide contains a carboxylic acid (COOH) group in place of hydroxyl (OH) group on carbon C-6 (Fig. 3.14). Most galacturonic acid units in pectins are modified by addition of a methyl (-CH$_3$) group to the carboxyl, converting it to the methoxyl (COO-CH$_3$) form. Pectin is found in the cell walls of plants where it functions in providing strength. In unripe fruits and vegetables, pectin contains a high percentage of the methoxyl pectin form. As fruit ripens, enzymes soften fruit texture by chemical changes in pectin molecules. Specifically, the enzyme pectin methylesterase cleaves the methyl (CH$_3$), creating a negative

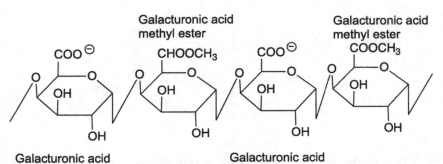

Fig. 3.14 Pectin.

charge. Another enzyme, polygalacturonase, cleaves glycosidic bonds in pectin, weakening the polymer. Foods, such as jam and jelly, use pectin to make the product form a gel. A gel is a three-dimensional network of molecules connected by non-covalent forces (hydrogen bonds) in which a large amount of water is held by capillary action. Low methoxyl pectin makes poor gels because its ionized carboxyl groups contribute a large negative charge preventing formation of hydrogen bonds between molecules. High methoxyl pectin makes strong gels because of its lower level of charge. Commercially, pectins are graded as high methoxyl (HM) or low methoxyl (LM) based on the percentage of its carboxyl groups containing methyl groups. The percentage of high methoxyl groups in HM and LM pectins is approximately 70% and 50%, respectively. High methoxyl pectin is the form most often used in home canning. While LM pectin is generally a poor gel former, it can be used to make a firm gel if calcium is added. Calcium is a divalent cation (two positive charges) that forms electrostatic cross links between negatively charged carboxyl groups in adjacent pectin molecules. The pectin structure formed with calcium resembles an egg crate (Fig. 3.14) Sugar content and pH are critical factors in making strong gels with pectin. Sugar (sucrose) helps to make a strong gel because it competes with pectin in binding water molecules. High levels of sugar (50%—65%) result in greater hydrogen bonding between pectin molecules and increases gel strength. Low pH also contributes hydrogen bonding between pectins. When pectin carboxyl groups are fully protonated (approximately pH 3.0), its carboxyl groups are protonated and do not carry a negative charge. The resultant drop in charge on pectin molecules promotes intermolecular hydrogen bond formation and gel strength.

Food applications of pectin: Pectin is the traditional ingredient from which jellies, jams, and soft gel candies are made. It is also used as a thickening agent and stabilizer in acidic foods like yogurt. Acidic milk drinks are made possible by the addition of pectin. Charged carboxyl groups in pectin enable it to bind proteins via electrostatic interaction. The complex of protein and polysaccharide is advantageous in processed cheese products by preventing precipitation of milk caseins under acidic conditions. Similarly, pectin is used in processed meats to increase water retention in low fat hot dogs.

Hemicellulose is a heteropolysaccharide typically found in combination with cellulose in the cell walls of plants. While the names hemicellulose and

cellulose are similar, they are completely different polysaccharides. Hemicellulose is a relatively small polymer containing both linear and branched segments. Hemicelluloses are composed of several types of monosaccharide
units including D-xylose, L-arabinose, and D-glucuronic acid. Its glucuronic
acid content provides a modest degree of solubility. The best characterized
hemicellulose is xylan, derived from cereal grains. Xylan is composed of a
D-xylopyranose (a 5-carbon sugar in a 6 membered ring) backbone that is
linearly linked $\beta 1 \rightarrow 4$. The backbone is occasionally branched $\beta 1 \rightarrow 2$ to
another xylopyranose chain. Additional substitutions occur to the backbone
with L-arabinose and D-glucuronic acid.

Food applications of hemicellulose; Hemicellulose and other complex
carbohydrates have little or no caloric value, but provides health benefits
as dietary fiber. Hemicellulose and other complex carbohydrates, like βglu-
can and inulin, are indigestible in the small intestine but, are fermented by
bacteria in the large intestine. Hemicellulose is a heterogenous carbohydrate
composed of xylose, mannose, galactose, and glucose. Gut microorganisms
in turn produce beneficial short chain fatty acids, such as butyric acid.
Butyric acid and other short chain fatty acids have been suggested as preventative to bowel disorders and cancers of the colon. Hemicellulose is also
noted for its ability to bind cholesterol in the gut, a property that contributes
to lowering serum cholesterol levels. As a food ingredient, hemicellulose is
principally used in baked foods to improve loaf volume and dough mixing
properties.

Hydrocolloids: The terms hydrocolloid and gum are often used
interchangeably, but hydrocolloid is preferred. Hydrocolloids are polysaccharides derived from a variety plant and microbial sources. They are used
in food applications to provide functional properties, such as thickening,
gelling, stabilizing emulsions, preventing ice crystal formation, and binding
flavors. Hydrocolloids provide these functionalities without contributing
calories because they are indigestible and used at low concentration,
typically less than 1%. Examples of hydrocolloids described in this section
include plant exudates, seed gums, seaweed gums, microbial gums, and
starch and cellulose derived polysaccharides. A summary of hydrocolloid
properties and their uses can be found in the review by Saha and
Battacharya (2010). Hydrocolloids have become very popular for making
novel foods such as cola caviar, cheese fries, yogurt eggs, and chocolate spaghetti. New methods for using hydrocolloid ingredients to create novel

foods is part of the movement termed molecular gastronomy. For those who might be interested in exploring food hydrocolloid concepts, a collection of recipes can be found in the 2014 publication called *Texture*. This is a hydrocolloid recipe collection edited by Martin Lersch (Lersch 2014).

Seed gums

Guar gum: Seeds of the guar plant is the principal source of this polysaccharide gum. Guar (*Cyamposis tetragonoloba*) is a member of legume plant family and its functional polysaccharide is found in the endosperm of its seeds. Guar gum is a polymer composed of two monosaccharides, mannose and galactose, and is referred to as a galactomannan. Its structure consists of two D-mannose units linked $\beta 1 \rightarrow 4$ as the backbone of the polymer with a D-galactose unit linked $\alpha 1 \rightarrow 6$ to every second mannose unit. The galactose-mannose unit is repeated throughout the guar gum polymer. The overall ratio of mannose to galactose units in guar gum is 2:1.

Food applications of guar gum: Guar gum hydrates readily in cold water to form high viscosity solutions at a low concentration, but it does not form gels unless other polysaccharides are added. Guar gum can be used to thicken salad dressing and sauces. Reduced calorie versions of dairy products like ice cream and yogurt maintain their creamy texture through added guar gum. Ice cream and other frozen desserts are less likely to develop large ice crystals when guar gum is included. Guar gum is also used to provide a smooth texture to plant-milk products such as soy, almond, and coconut milk. Processed meat products benefit from addition of guar gum because of its ability to act as a binder. Combinations of guar gum and xanthan gum have a synergistic effect on solution viscosity in food systems. Nutritionally, it has desirable properties as insoluble dietary fiber. Guar gum absorbs water in the gut, producing a mild laxative effect. It contributes to lowering the increase of blood glucose level following a high starch meal. It binds to starch in the intestinal lumen, slowing starch digestion and glucose release.

Locust bean gum: Locust bean gum is derived from seeds of the carob tree. Like guar gum, it is composed of mannose and galactose units and classified as a galactomannan. However, there are important differences between the two polysaccharide polymers. Mannose and galactose in locust bean gum share the same type of glycosidic links as found in guar gum, but their ratio is different. The ratio is four to one (mannose to galactose) in locust bean gum. Additionally, locust bean gum contains long stretches of mannose units enabling formation of hydrogen bonds between molecules.

Food application of locust bean gum: Locust bean gum hydrates in water to form viscous solutions, but like guar gum does not gel. Gels can be formed with locust bean gum by adding xanthan or carrageenan. Locust bean gum is principally used in dairy products such as cheese spread and frozen novelties. Locust bean gum is sweet tasting and is roasted to produce a cocoa-like material called carob powder. It is used as a replacement for cocoa powder in candies and cookies, but does not have the same taste or texture.

Plant exudate gums

Gums derived as exudates from several types of plant sources are used as additives in food systems. Most notable of these exudates are gum tragacanth and gum arabic. Both of these polysaccharides are quite polar due to their content of ionizable carboxyl groups found on glucuronic and galacturonic units. Segments of these polysaccharides also contain non-polar regions, making them good emulsifiers. Dual character contributed by polar and non-polar groups enables emulsification of lipids, preventing separation in water suspensions. Salad dressing is an example of this type of stabilized emulsion.

Gum tragacanth is derived from the sap of an *Astragalus* plant called goat's thorn, a type of legume typically found in the Middle East. Tragacanth sap is harvested from the plant's roots. After drying, it is ground into a powder. When rehydrated, this crude form of the gum can be made into a thick paste or used in leather burnishing to produce a smooth shiny edge in tanned leather. Purified forms of tragacanth are used in a number of food applications, including thickening agent and emulsifiers. Gum tragacanth is a large and highly branched heterogeneous polysaccharide composed of several monosaccharides including, D-galacturonic acid, D-galactose, L-fucose, D-xylose, L-arabinose, and L-rhamnose.

Food application of tragacanth: Tragacanth is highly hydrophilic due to its high galacturonic acid content. Tragacanth is known for its ability to stabilize emulsions at low pH. As little as 0.25% tragacanth forms a stable oil-in-water emulsion. Tragacanth is added to cake and cookie icings to prevent drying and cracking. Tragacanth has been shown to have potential use as a prebiotic. Its hydrolysis in the gut provides oligosaccharides that promote the growth of beneficial bacteria, *Bifidobacterium longum* (Ahmadi Gavlighi et al., 2013).

Gum arabic is an exudate of the acacia tree typically found in tropical climates. It has been used for hundreds of years in numerous non-food applications, such as a binder for paint pigments, postage stamp adhesive, and in lithography. Today, gum arabic is also widely used in the foods industry. Gum arabic is a mixture of polysaccharide and glycoprotein. The polysaccharide component is large in size and its composition is heterogeneous. It is made up of two fractions. The major fraction, approximately 70%, contains D-galactose, L-arabinose, D-glucuronic acid, L-rhamnose, and 4-O-methyl-D-glucuronic acid. There is little or no protein in the major fraction. The minor fraction, approximately 30%, is made up of high molecular weight molecules that are covalently linked to various proteins. Gum arabic's content of polar polysaccharides (i.e. glucuronic acid) provides excellent interaction with water and thus a high solubility. Solutions containing as much as 50% solids can be made without major increase in viscosity. This property results from the high content of negatively charged groups (ionized carboxylic acid groups) that effectively inhibits hydrogen bonding between molecules. For this reason, gum arabic does not form gels.

Food applications: Gum arabic is a good emulsifying agent, keeping polar and non-polar components from separating in food systems. For example, it is used in whipped cream to make a stable dispersion in this high fat food. Gum drops, marshmallows, and edible glitter are also notable foods made using gum arabic. Guar arabic prevents crystallization of sucrose in soft drinks. Commercial syrups from which soda pop is made often contain 50% sucrose and/or fructose. High concentrations of sugar in this product can cause formation of crystals that precipitate from solution. Addition of 0.1% gum arabic prevents sugar crystal formation. Gum arabic is also an excellent flavor binding agent. Flavors in food systems can be problematic because they readily volatize. Mixing the flavor agent with this gum binds it and prevents loss. The best-known application of gum arabic is as flavor binder in the original Kool aid™ soft drink. Like many polysaccharide gums, gum arabic is not digestible in the small intestine and only minimally fermented in the large intestine. It contributes little in the way calories. It may be beneficial by inhibiting the absorption of cholesterol from the gut and as a source of dietary fiber. However, it is typically used at very low levels (i.e., less than 1% in foods) and thus its health benefits are also low.

Sea weed gums

Agar is one of oldest polysaccharide hydrocolloids still in use today. It was discovered in Japan during the 17th century where it was principally used to make gelled food materials, most likely as a means of preservation. Fruits and vegetables have long been made into jellies using agar extracted from seaweed. Agar has interesting properties as hydrocolloid. It is insoluble in room temperature (20 °C) water, but dissolves when boiled. A translucent gel will form when the heated solution is cooled down to 35 °C. The gel can subsequently be liquified again by reheating to it 85 °C. Agar has the advantageous properties of mechanical strength and reversible translucent gel formation that enable its use as a substrate for culturing, enumerating, and identifying microorganisms. Agar is made from red seaweed (*Rhodophycea*). Once harvested from the ocean and washed, a crude form of agar is extracted by boiling it in a very dilute solution of sulfuric or acetic acid. The mixture is filtered while hot to remove insolubles, allowing it to solidify on cooling. Finer grades of agar are made by incorporating bleaching steps and repeated cycles of melting and solidification before it is dried and ground to a powder. The structure of agar is a linear polysaccharide containing repeating galactose disaccharide units. Specifically, the repeat consists of D-galactose linked $\beta 1 \rightarrow 3$ to the anhydro form of L-galactose. Two types of polysaccharide exist in red seaweed, agarose and agaropectin. The major difference between these forms is the presence of negatively charged sulfate groups (SO_4^{-2}) in agaropectin. Approximately 3—10% of agaropectin is sulfated. This gives it a strong negative charge and prevents it from gelling. Thus, it has little functional use. In contrast, agarose contains little or no sulfate and readily forms reversible gels. Agarose in water forms hydrogen bonded crosslinks and stabilizes the three-dimensional gel network. Purified agarose (minus any agaropectin) is used to make the final agar product used in foods and microbiological media.

Food applications: A range of food applications exist for agar. In baked foods, agar is used as a stabilizer for pie fillings, meringues, and soufflés.

Agar thickens pie fillings and provides a lighter texture to whipped egg foods. It also provides desirable properties in icings used on everything from cakes to doughnuts. Incorporation of agar into the mixture of fat, sugar, and flavorings is typically used in icing to prevent cracking and/or becoming gooey in high humidity. A Jell-O™-like product with firmer texture is made from flavorings, sugar, and agar. Cubed bits of flavored agar are added to fruit in making dessert dishes. Agar is added to frozen desserts and processed cheese to provide desirable textures.

Carrageenan is a polar polysaccharide derived from a red seaweed called Irish Moss (*Chondrus crispus*). Carrageenan is common in Ireland where it is known by its Gaelic name, *carraigin*. Carrageenan's structure consists of a long linear polymer, predominantly composed of D-galactose units linked either $\alpha 1 \rightarrow 3$ or $\beta 1 \rightarrow 4$. Some galactose residues have anhydro links that represent a bridge between carbons 3 and 6. An important feature of carrageenan's structure that affects its functional properties, is the presence of sulfate groups. Sulfate (SO_4^{-2}) is strongly negatively charged at any pH above 2 and the percent of sulfate in carrageenan determines its functionality either as a gelling or viscosity enhancing agent. Sulfates are attached to the hydroxyl groups of C-2, C-6, by an ester bond (Figure 17). The total amount of bound sulfate can be much as 40%. Crude carrageenan is extracted from red seaweed using a very alkaline solution of potassium hydroxide at high temperature. The extract is filtered to remove cellulose and other insolubles. Carrageenan extract is concentrated before being fractionated into types based on its degree of sulfate content. Carrageenans listed in the order of increasing sulfate content are kappa (κ), iota (ι), lambda (λ), and mu (μ). Kappa and iota carrageenan have the lowest content of sulfate groups and are used to form gels, especially with small amounts of K^+ or Ca^{+2} ions that electrostatically bind and neutralize the charge on sulfate groups.

Food applications: Proteins are able to complex with kappa carrageenan and form gels. However, lambda and mu carrageenan have the highest content of sulfate that results in a strong negative charge and prevents gel formation. They are highly soluble and form highly viscous solutions at low concentration. Carrageenan has a wide range of food applications, especially in dairy foods. Dairy based desserts like ice cream, pudding, and milk shakes require very low level of carrageenan, approximately

0.02%, to have a substantially higher viscosity and creamy texture. Milk caseins form complexes with carrageenan through electrostatic interactions between the positively charged protein and negatively charged carrageenan. Anyone who tried making chocolate milk with cocoa powder knows it does not suspend well. However, mixing carrageenan with cocoa powder before adding to milk results in a homogenous and stable suspension. Plant milks such as soy, almond, and coconut also use carrageenan to provide a stable emulsion and desirable viscosity. Without a hydrocolloid such as carrageenan, extracts of almond pastes, for example, would not mix well with water. Carrageenan's ability to bind proteins is also used in making in processed meat. Sausages and hams benefit from carrageenan addition by absorbing water, improving texture, and increasing the bind between meat particles. Adding carrageenan at 0.1% can increase the water retention in ham by 20%.

Alginate is a polar polysaccharide derived from brown seaweed (*Phaeophyceae*) and constitutes 18−40% of the kelp plant. The polysaccharide is a linear polymer composed of D-mannuronic acid and L-glucuronic acid. Overall, alginate is a very polar and acidic polysaccharide. The process of making alginate begins with the extraction of the polysaccharide using sodium carbonate, an alkaline treatment, from washed brown seaweed. Insoluble materials are removed from the extract and soluble fraction, now called alginic acid, is precipitated by addition of calcium chloride. The process of washing, and solubilization is repeated, but precipitation is achieved by adding sodium carbonate. This converts the carboxylic acid groups to the sodium salt, sodium alginate. Sodium alginate has good solubility in water and makes highly viscous solutions. The viscosity of alginate solution is affected by pH. A low pH solution results in greater viscosity than a high pH solution. The explanation for this effect results from protonation of its carboxyl groups removing negative charges and eliminating repulsive forces between chains.

Food applications: Sodium alginate is best known for its gel forming ability when calcium ions are added. Calcium is a divalent cation ion (positively charged) that forms electrostatic links with negatively charged carboxylic acid group in the alginate polymer. The gelling effect of calcium on alginate is easily demonstrated by adding a few drops of calcium chloride to an alginate solution. Each droplet immediately becomes a small spherical gel. This bit of chemistry is the basis for making "cola caviar". A cola soft drink containing about 1% alginate is added dropwise into a solution of calcium chloride, resulting small pearls of gelled cola that can be scooped out and used as a topping for ice cream. High viscosity beverages, like smoothies, are readily made using sodium alginate. Restructured products, like pimentos-stuffed olives, onion rings, and fruit strips are made using alginate. In these products, diced food (e.g., onion, olives) and flavorings are mixed with alginate and extruded to form the desired shape. Alginates are used to make cheese cake that can be frozen and thawed without loss of texture. Processed meat represents one of the largest uses for alginates in food. Alginate added to meat batter makes it possible to reduce fat level by 50% in hot dogs and other sausage products.

Microbial gums

Xanthan is a polysaccharide derived from a bacteria *Xanthomonas campestris*. In nature, the organism is responsible for a condition called black rot in green leafy vegetables like broccoli. The organism produces a sticky substance that is found to have interesting and useful properties. Xanthan is composed of several monosaccharides, specifically glucose, mannose, and glucuronic acid. Glucose molecules linked $\beta 1 \rightarrow 4$ constitute xanthan's backbone. Every other glucose in the backbone has a branch point containing a glucuronic acid and two mannose units. Additional modifications of xanthan can be found on mannose residues which can be substituted with pyruvate or acetyl groups. The composition of polar monosaccharides in xanthan provide it with good water solubility. Xanthan is produced in large scale by fermentation. It is isolated by precipitation using isopropyl alcohol, dried, and ground into a powder for use in food systems.

Food applications: Xanthan is a large and polar polysaccharide that forms highly viscous solutions at very low concentration, typically less than 0.5%. The viscosity of a xanthan solution is not affected by a broad range of pH. An interesting property of a xanthan thickened solution is its ability to become gel-like when allowed to sit undisturbed. However, its gel-like property disappears, reverting to a viscous solution, upon stirring. Solutions with reversible sol to gel behavior are termed thixotropic. Transformation from gel to sol occurs because mechanical action disrupts weak intermolecular attractions in a process called shear thinning. Thixotropic materials are useful in food products like ketchup and salad dressing. A ketchup bottle can be turned upside down without flowing out of the bottle. A quick shake or two makes it flow easily. Xanthan has excellent functional properties as a stabilizer and emulsifier in foods containing fats or oils. Xanthan is used to keep large food particles in suspension. The viscosity of xanthan changes very little with temperature. For example, barbeque sauce containing xanthan has nearly the same viscosity whether it is cold or hot. A fat-free egg substitute can be made using egg white and xanthan. In this application, xanthan stabilizes egg white proteins and creates the desirable texture of fat. Xanthan does not form a gel on its own, but mixing with locust bean gum results in a synergistic interaction that gels. One of the more recent popular application of xanthan is in gluten-free products. Xanthan provides a desirable texture to these products and is able entrap gas in raised gluten-free breads.

Gellan is a water-soluble polysaccharide derived from the bacterium *Sphingomonas elodea*. This bacteria, isolated from aquatic plants, is used for the commercial scale production of gellan. Gellan is a linear polysaccharide containing four monosaccharides in a repeating unit. Each repeat is composed of two D-glucose units and one unit each of L-rhamnose and D-glucuronic acid. This large polymer is polar and water soluble due to its glucuronic acid content. In water, the gellan polymer forms double helices that contribute to the solution's viscosity. A typical gellan polysaccharide contains over 500,000 monosaccharide units. In its native form, glucose units are substituted to a high degree with glycerol. This modification can be removed by treatment with alkali which hydrolyzes the bond between glucose and glycerol. This process is used commercially to produce two forms of the polysaccharide, termed high- and low-acyl gellan. Either form gels at low concentration (typically less than 0.3%), but with some differences. High-acyl gellan gels at fairly warm temperature (70 °C), whereas low-acyl gellan gels in the range of 30–50 °C. Low-acyl gellan forms very high viscosity solutions at low concentration and gels if calcium is added.

Food applications: Gellan use involves its ability to thicken, emulsify, or stabilize a food system. For example, gellan is used to keep poorly soluble soy proteins from precipitating out in soy milk. Gellan is a vegan alternative to gelatin for making gummy candies.

> ## Summary

Carbohydrates represent the predominant source of energy for plants and animals. In food, monosaccharide and disaccharide carbohydrates (fructose and sucrose) are responsible for the attribute of sweetness. Oligosaccharides once thought to be valuable only as fiber, are now known to support the growth of beneficial gut bacteria. Polysaccharides, such as starch, are essential to making bread, a food that has supported life for thousands of years. Hydrocolloids are a type of polysaccharide that contribute textural properties (e.g., gelling and thickening) to food and facilitate the creation of novel foods such as, cola caviar and chocolate spaghetti in the cuisine of molecular gastronomy. Carbohydrate chemistry is responsible for generating the desirable flavors of chocolate, coffee, and grilled meat. Unfortunately, it is also responsible for undesirable substances such as acrylamide. In the fermentation process, carbohydrates are essential to making wine, cheese, and beer. Knowledge of carbohydrate components and their chemistry contributes to the pleasurable aspects of food and provides better understanding of the relationship between diet and health.

Glossary

Aldose A carbohydrate molecule containing an aldehyde group

Anomeric Carbon A chiral center created in the closed ring structure of monosaccharides. In Haworth structures, the OH group attached to the anomeric carbon can be either down or up, corresponding to its designation as alpha or beta, respectively

Anomer Stereo isomers of the closed ring form of carbohydrate molecules. Alpha and beta isomers are formed at the anomeric carbon (C-1 of an aldose or C-2 of a ketose) as a result of mutarotation

Birefringence Optical property of rotating plane polarized light, found in ungelatinized starch granules

Chirality Molecules that are not superimposable upon its mirror image, but are otherwise identical

Dextrose Another name for glucose, and dextrose equivalents represent the degree of hydrolysis in starch determined by measuring reducing sugar level

Dietary fiber Carbohydrates (mostly oligosaccharides) that are not broken down by human digestive enzymes

Enantiomer Chiral forms of the same molecule

FODMAP A group of carbohydrate compounds that contribute to symptoms of irritable bowel syndrome (gas, bloating and diarrhea).

Food intolerance An adverse reaction to a food component that does not involve a response by the immune system

Furanose A monosaccharide in a five membered ring configuration

Gelatinization Breaking of intermolecular hydrogen bonds in starch granules allowing greater interaction with water and eventual dissolution of the granule

Gelatinization temperature The temperature at which birefringence (Maltese cross) disappears in starch granules. It also corresponds to the temperature of maximum paste viscosity.

Glycation The covalent modification of a protein or lipid by a carbohydrate molecule, usually a mono- or oligosaccharide.

Glycemic index A ranking of polysaccharides that measures how much blood sugar level is raised.

Glycoside/Glycosidic link A covalent bond between the anomeric carbon of one carbohydrate molecule and a second molecule's hydroxyl group.

Glycoprotein A protein to which a carbohydrate molecule is covalently attached to a serine or threonine side chain

Hilum Beginning points at which starch granules synthesized in concentric (growth) rings Invert sugar

Invert Sugar A product resulting from splitting the sucrose into its constituent parts, glucose and fructose. Invert sugar is sweeter than sucrose

Ketose A carbohydrate molecule containing a ketone group.

Mutarotation A spontaneous and reversible change in the structure of a monosaccharide resulting in a change of its optical properties, specifically the rotation of polarized light

Prebiotics Non-digestible oligosaccharides that promote the growth of beneficial bacteria in the large intestine.

Probiotics Microorganisms consumed orally that provide a health benefit to the host

Pyranose A monosaccharide in a six membered ring configuration

Resistant starch (RS) is defined as any starch that is not digested in the small intestine and subsequently passed on to the large intestine.

Retrogradation A process of reformation of hydrogen bonds between linear regions of starch molecules

Soluble fiber Polysaccharides that attract water and forms a gel in the intestines.

Syneresis The loss of water molecules occurring when gelatinized starch undergoes retrogradation

References

Coultate, T., 2009. The Chemistry of its Components, fifth ed. RSC Publications, Cambridge, UK.

Fanaro, S., Boehm, G., Garssen, J., Knoi, J., Moscia, F., Stahl, B., Vigi, V., 2007. Galacto-oligosaccharides and long-chain fructo-oligosaccharides as prebiotics in infant formulas: a review. Acta Paediatr. S449, 22–26.

Fedewa, A., Rao, S.C., 2014. Dietary fructose intolerance, fructan intolerance and FODMAPs. Curr. Gastroentrol. Rep. 16, 370–378.

Hodge, J.E., 1953. J. Agric Food Chem. 928-9xx.

Kalyani, N.K., Kharb, K., Suman, K., Thompkinson, D.K., 2010. Inulin dietary fiber with functional and health attributes—a review. Food Rev. Int. 26 (2), 189–203.

Lersch, M., 2014. Texture: A Hydrocolloid Recipe Collection, vol. 30.

Linebeck, D.R., Coughlin, J.R., Stadler, R.H., 2013. Acrylamide in foods: a review of the science and future considerations. Annu. Rev. Food Sci. Technol. 3, 15–35.

Russo, G.L., Pietra, V.D., Mercurio, C., Zappia, V., 2018. Protective effects of butyric acid in colon cancer. Adv. Exp. Med. Biol. 472, 131—147.

Sabater-Molina, M., Larque, E., Torella, F., Zamora, S., 2009. Dietary fructo-oligosachharides and potential benefits on health. J. Physiol. Biochem. 65, 315—328.

Saha, D., Batttacharya, S., 2010. Hydrocolloids as thickening and gelling agents in food: a critical review. J. Food Sci. Technol. 47, 587—597.

Sajilata, M.G., Singhal, R.S., Kulkarni, R.K., 2006. Resistant starch-A review. Comp. Rev. Food Sci. Safety. 5, 1—17.

Shepard, S., Gibson, P.R., 2006. Fructose malabsorption and symptoms of irritable bowel syndrome: guidelines for effective dietary management. J. Am. Diet. Assoc. 106, 1631—1639.

Slavin, J., 2013. Fiber and prebiotics: mechanisms and health benefits. Nutrients 5, 1417—1435.

Somoza, V., Foglano, V., 2013. 100 years of maillard reaction: why our food turns brown. J. Agric. Food Chem. 61, 10197.

Further reading

Prebiotics

Sabater-Molina, M., Larque, E., Torella, F., Zamora, S., 2009. Dietary fructo-oligosachharides and potential benefits on health. J. Physiol. Biochem. 65, 315—328.

Shepard, S., Gibson, P.R., 2006. Fructose malabsorption and symptoms of irritable bowel syndrome: guidelines for effective dietary management. J. Am. Diet. Assoc. 106, 1631—1639.

Slavin, J., 2013. Fiber and prebiotics: mechanisms and health benefits. Nutrients 5, 1417—1435.

FODMAPS

Fedewa, A., Rao, S.C., 2014. Dietary fructose intolerance, fructan intolerance and FODMAPs. Curr. Gastroentrol. Rep. 16, 370—378.

Maillard Chemistry

Somoza, V., Fogliano, V., 2013. 100 years of the Maillard reaction: why our food turns brown. Somoza and Fogliano. J. Agric. Food Chem. 61, 10197.

Linebeck, D.R., Coughlin, J.R., Stadler, R.H., 2013. Acrylamide in Foods: A Review of the Science and Future Considerations. Annu. Rev. Food Sci. Technol. vol. 3, 15—35.

Nursten, H., 2005. The Maillard Reaction Chemistry, Biochemistry and Implications. Royal Society of Chemistry, London, UK.

Starch

Perez, S., Bertoft, E., 2010. The molecular structures of starch components and their contribution to the architecture of starch granules: a comprehensive review. Starch 62, 389—420.

Singh, J., Dartois, A., Kaur, L., 2010. Starch Digestibility in food matrix: a review. Trends Food Sci. Technol. 21, 168—180.

Sajilata, M.G., Singhal, R.S., Kulkarni, R.K., 2006. Resistant starch-A review, Comp. Rev. Food Sci. Safety. 5, 1—17.

Hydrocolloids

Banerjee, S., Battacharya, S., 2012. Food Gels: gelling process and new applications. Crit. Rev. Food Sci. Nutr. 52, 334—346.
Saha, D., Batttacharya, S., 2010. Hydrocolloids as thickening and gelling agents in food: a critical review. J. Food Sci. Technol. 47, 587—597. Texture, a Hydrocolloid Recipe Collection edited by Martin Lersch.

Review questions

1. Give one example each of aldose and ketose sugars
2. Describe the structure of glucose and sucrose
3. What disaccharide is found in milk?
4. What is a prebiotic? Give an example.
5. What is a FODMAP? Give an example and explain why it can be undesirable.
6. What is a polyol?
7. Why are polyols added to food?
8. What does the Maillard reaction do for food?
9. What are the major reactants of Maillard reaction?
10. Define the term "reducing sugar".
11. Name three foods that benefit from Maillard reaction.
12. What toxic product is formed as result of Maillard reaction?
13. Describe the structure of a starch granule.
14. Describe the structure cellulose in comparison to amylose.
15. Define the terms gelatinization and retrogradation in regard to starch.
16. What is dietary fiber?
17. If its digestion in the gut is limited, why is dietary fiber beneficial?
18. What are SCFA (short chain fatty acids)?
19. What is soluble fiber?
20. What form of starch (amylose or amylopectin) is most resistant to retrogradation?
21. Define the term glycemic index.
22. How is the glycemic index of starch affected by extrusion processing?
23. What is the composition of pectin?

24. Why is low pH and high sugar content important to making firm gels (jelly) with pectin?

25. How is "instant" starch made?

26. What foods is it found in?

27. What is gum arabic composed of? What foods is it used in?

28. What is alginate composed of? Give an example of a novel food gel made using alginate.

CHAPTER FOUR

Lipids

Learning objectives

This chapter will help you describe or explain:

- Lipid structure of fatty acids, acylglycerols, and sterols
- Importance of fat crystalline structure to food quality
- Chemistry of hydrogenation
- How trans-fat occurs in food
- Lipid oxidation chemistry
- Antioxidants and their role in controlling lipid oxidation
- Lipids that are biologically important to health and well-being

Introduction

The term lipid represents a large class of compounds contributing to the sensory attributes and the nutritional quality of food. The word lipid is generally used to denote compounds with low solubility in water and

Introduction to the Chemistry of Food
ISBN: 978-0-12-809434-1
https://doi.org/10.1016/B978-0-12-809434-1.00004-9
127

conversely, high solubility in organic solvents. Food lipids are commonly referred to as fats or oils. The difference between these terms is a somewhat arbitrary distinction. Fats are differentiated from oils based on their physical form at room temperature. Lipids that are solid at room temperature are termed fats. Conversely, those that are liquid at room temperature are termed oils. The structure of lipids (acylglycerols) in fat is important to foods such as chocolate. Cocoa butter, the fat in chocolate, is melted and cooled in a controlled process called tempering. It provides the most desirable eating properties. Tempering creates the high gloss surface and pleasurable mouth feel we expect in good quality chocolate. Nutritionally, lipids are a rich source of energy proving more calories per gram than any other food component. Lipids are biologically important as steroid hormones, vitamins (A, D, E, and K), essential fatty acids, natural antioxidants, and fats. Lipid components of food such as omega-3 (ω3) fatty acids, carotenoids, and conjugated linoleic acid are noted for their health promoting properties. Chemical reactions in food, such as oxidation, affect nutritional quality by destroying fat-soluble vitamins, essential fatty acids, and the antioxidant activity provided by some fat-soluble vitamins. The chemical treatment of food lipids, (i.e., hydrogenation of oils) is used to make margarine and shortening that provide better functional properties but, also creates trans-fat, having adverse effects on health.

This chapter contains questions that will help you explore and better understand food lipids.

What is hydrolytic rancidity and is that sour milk safe?

What is lipid oxidation?

What is hydrogenation and how does it make margarine different from butter?

What is shortening and is it bad for me?

What are LDL and HDL?

What is tempering and how does it make chocolate better?

What is an essential fatty acid?

What is an antioxidant?

Lipid structure and nomenclature

Fatty acids consist of a linear chain of carbons varying in length from 2 to 24 atoms. Each fatty acid has methyl (CH_3) and carboxylic acid groups (COOH) located at either end of the chain. The number of carbon atoms in a fatty acid is most often an even number because they are synthesized from 2 carbon units. Propionic acid contains 3 carbons and is an exception. The carboxylic acid group is polar and interacts well with water. However, the chain region of fatty acids is composed only of carbon and hydrogen atoms and is very nonpolar (hydrophobic). Fatty acids, such as acetic, propionic, and butyric, are quite soluble in water because of their polar carboxylic acid group and short chain lengths (2, 3, and 4 carbons). The solubility of fatty acids decreases with increasing chain length because its hydrophobic contribution exceeds the positive effect of the polar carboxylic acid group. Chain length, defined as short, medium, and long, is one way of classifying fatty acids. Short chain fatty (SCF) acids, containing between 2 and 6 carbons, are the most volatile and make a substantial contribution to food flavor. Acetic acid is instantly recognizable as the flavor and taste of vinegar. Propionic acid is a major component of Swiss cheese flavor and its antimicrobial activity inhibits the growth of mold and some bacteria. Propionic acid is often used as a preservative for animal feed and human food. Butyric acid has a pungent odor instantly recognizable as the taste of sour milk. It is also produced by bacterial fermentation of complex carbohydrates (fiber) in the gut. Butyric acid has positive effects on the microflora population contributing to the control of irritable bowel syndrome (IBS) and inhibiting colon cancer (Zalesky et al., 2013). It is worth noting that butter contains 3%−4% butyric acid. Additional benefits of short chain fatty acids include the ability to cross cell membranes (i.e., the blood-brain barrier) and serve as a major source of metabolic energy. Fatty acids containing between 6 and 12 carbons are termed medium chain fatty acids (MCF). For example, caproic, capric, and lauric fatty acids contain 6, 8, and 12 carbons, respectively. These fatty acids compose about 10%−15% of cow's milk lipids. Coconut and palm oils are also sources of medium chain fatty acids. Medium chain fatty acids are nutritionally beneficial because their metabolism contributes to weight reduction by promoting the metabolic state called ketosis. Ketosis promotes utilization of the body's fat reserve, instead of glucose. Ketosis occurs when the diet is low in carbohydrates (sometimes referred to as the fat burning mode). A ketogenic diet is beneficial in reducing the level of acylglycerides, LDL cholesterol, and blood glucose while increasing the level of HDL cholesterol. The overall effect of MCF may be a lower risk of cardiovascular disease. Long chain fatty (LCF) acids are composed of 14−24 carbon atoms linked by single or double bonds. These fatty acids can be saturated or unsaturated. Saturated fatty acids (SFA) do not contain double bonds in their chain of carbon atoms. The presence of at least one double bond in a fatty acid makes it unsaturated. Those containing

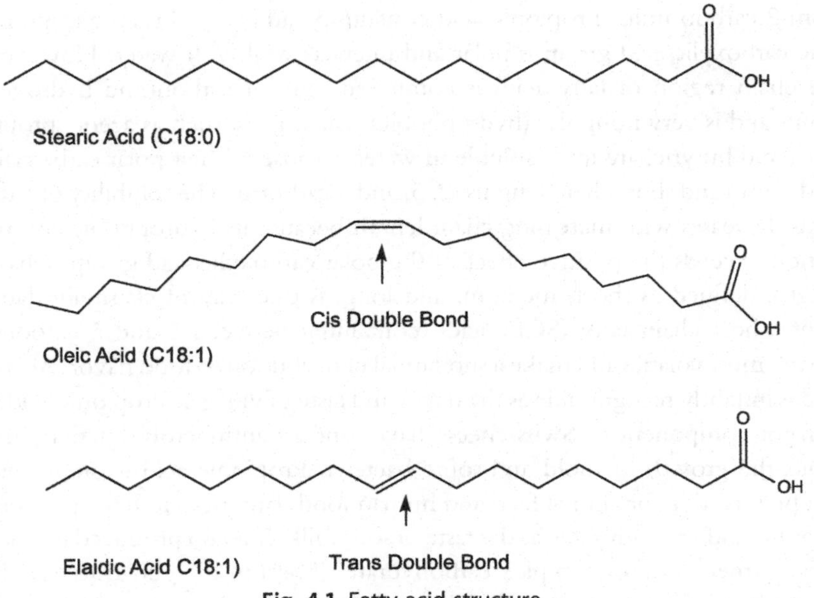

Fig. 4.1 Fatty acid structure.

two or more double bonds are termed polyunsaturated fatty acids (PUFA). The presence of double bonds in fatty acids causes a change in their structure (Fig. 4.1). However, a double bond, such as in oleic acid, represents an inflexible point, preventing rotation and limiting movement of atoms in the chain. Oleic acid's double bond is in the cis configuration and creates a V-like shape. Conversely, the trans double bond of elaidic acid's has a linear shape, closely resembling that of stearic acid. The structural difference between fatty acids with cis or trans double bonds affects their physical properties. For example, fatty acids with cis double bonds have a lower melting point compared to those with the trans configuration. The cis configuration of fatty acid double bonds is predominant in most plant and animal lipids. Trans variants, such as elaidic acid, are produced by processing (hydrogenation) and linked to cardiovascular disease. The process of hydrogenation is discussed later in this chapter.

Naming fatty acids

Unfortunately, there are multiple systems for naming fatty acids. Options for naming include, common, systematic, and numerical abbreviations. Common names, like palmitic or oleic, are widely used in non-technical communications. In contrast, systematic and numerical abbreviations provide detailed information regarding the molecular structure of fatty acids. Common names are the most widely used method for identifying fatty acids in foods (Table 4.1). A common name often indicates the source of the fatty acid. Palmitic and oleic acids, for example, are derived from palm trees and olives, respectively. Common names are useful to define the composition of fatty acids in acylglycerols. For example, stearic, oleic, linoleic and linolenic are common names for long chain fatty acids. A systematic name provides the reader with additional information about the molecule's structure and is the preferred method for use in technical and scientific publications. Systematic nomenclature begins with the number of carbon atoms in the fatty acid. The number of atoms in the molecule is specified using Greek prefixes

Table 4.1 Systematic and common names of fatty acids.

Saturated fatty acids	Common name	Numerical abbreviation/ (ω)
Saturated fatty acids		
Acetanoic	Acetic acid	C2:0
Butanoic acid	Butyric acid	C4:0
Hexanoic acid	Caproic acid	C6:0
Octanoic acid	Caprylic acid	C8:0
Decanoic acid	Capric acid	C10:0
Dodecanoic acid	Lauric acid	C12:0
Tetradecanoic acid	Myrstic acid	C14:0
Hexadecanoic acid	Palmitic acid	C16:0
Octadecanoic acid	Stearic acid	C18:0
Unsaturated fatty acids		
cis 9–Octadecenoic acid	Oleic acid	C18:1Δ9/(ω9)
trans 9–Octadecenoic acid	Elaidic	C18:1Δ9/(ω9)
All cis 9,12–Octadecadienoic acid	Linoleic acid	C18:2Δ9,12/(ω3)
All cis 9,12,15–Octadecatrienoic acid	Alpha Linolenic acid (ALA)	C18:3Δ9,12,15/(ω6)
All cis 5,8,11,14–Eicosatetraenoic acid	Arachidonic acid	C20:4Δ5,8,11,14/(ω3)
All cis–5,8,11,14,17–Eicosapentaenoic acid	Eicosapentaenoic acid (EPA)	C20:5Δ8,11,14,17/(ω3)
All cis–4,7,10,13–16–,19–Docosahexaenoic acid	Docosahexaenoic acid (DHA)	C22:6Δ4,7,10,13,16, 19/(ω3)

(e.g., hexa, octa, deca, and dodeca for 6, 8, 10, and 12 atoms respectively). For example, the systematic name for stearic acid (18-carbons) is obtained by combining the prefix "octadec" for 18 carbons, with the suffix ("anoic") indicating this fatty acid is saturated. Thus, the systematic name for stearic acid is octadecanoic. Multiple double bonds in fatty acids are specified by changing the suffix to dienoic (2) or trienoic (3), as required. Linoleic acid is another 18-carbon fatty acid, but contains two double bonds. The systematic name for this fatty acid is octadecadienoic acid. The position of double bonds in a fatty acid is specified by their location from the carboxylic acid end of the molecule. Using linolenic again as the example, the double bonds are located at carbons 9 and 12, and the resulting systematic name is 9, 12-octadecadienoic acid.

Systematic nomenclature can also define the configuration of double bonds. If both double bonds in linolenic acid are in the cis configuration, its name becomes all cis-9,12-octadecadienoic acid. Numbering carbons from the carboxylic end conforms to the convention established by the International Union of Pure and Applied Chemistry (IUPAC, 1997). Table 4.1 contains a summary of common and systematic nomenclature for several saturated and unsaturated fatty acids. Numerical abbreviations are used to provide a convenient way of identifying fatty acids structure via a shorthand notation based on their number of carbon atoms. These abbreviations use the chemical symbol for carbon "C", followed by a colon and the number of double bonds. The abbreviation for linolenic acid is written using as C18: $3\Delta^{9,12,15}$. The Δ symbol followed by numbers indicates positions of double bonds.

Omega (ω) nomenclature: The omega (ω) system is an alternative method for naming fatty acids. This method numbers carbon atoms from the methyl end of the fatty acid molecule. The Greek letter "ω" plus a numerical value indicates the position of the double bond. For example, linoleic acid contains 18 carbons with double bonds at atoms 9 and 12. The numerical designation of this fatty acid is C18:$2\Delta^{9,12}$. α linoleic acid is termed an ω6 fatty acid using the omega system (Fig. 4.2). Similarly, oleic

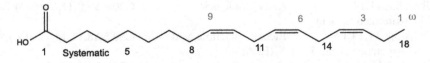

Alpha Linolenic acid (ALA)

Fig. 4.2 Fatty acids numbering (systematic and omega methods).

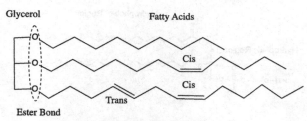

Fig. 4.3 Acylglycerol structure.

acid and linolenic acid are examples of ω9 and ω3 fatty acids, respectively. The omega system is advantageous, compared to more technical methods, in nutritional information to lay communities. For example, recommendations that diets higher in omega 3 fatty acids (fish, nuts, flaxseed, and canola) and lower in omega 6 fatty acids (corn and soybean oils) are important to health.

Acylglycerols are composed of a glycerol molecule to which 1, 2, or 3 fatty acids are linked by ester bonds that are correspondingly named as mono-, di-, and tri-acylglycerols, respectively (Fig. 4.3). Acylglycerols make up 90% or more of the lipids in food. The ester bond holding fatty acids and glycerol together is generally stable, but, can be chemically or enzymatically broken by a hydrolysis reaction. Lipases are a family of enzymes that hydrolyze ester bonds and release fatty acids into the medium. Milk acylglycerols contain about 4% short chain fatty acid i.e., butyric acid. Once hydrolyzed from the acylglycerol, butyric acid volatizes and becomes the familiar odor of sour milk. In contrast, the release of short chain fatty acids makes a desirable contribution to the flavor of cheese. The variety of acylglycerol composition found in plant and animal foods is created by differences in their fatty acid composition. Both the type of fatty acid and its position on glycerol, are important to its nutritional and functional properties. In foods, the functional properties of acylglycerols is strongly influenced by the fatty acid chain length, the degree of saturation (number of double bonds), and their configuration (cis or trans).

Polar lipids: Lipids are broadly defined as aliphatic molecules, soluble only in organic solvents and often synonymous with the word fat. However, polar lipids are amphipathic, meaning they possess both polar and nonpolar properties. In general, polar lipids contribute functional and nutritional benefits to food. Examples of polar lipids include phospholipids, sterols, and phytosterols. Phospholipids are biologically important constituents of cellular membranes. They are distinct from other acylglycerols having a phosphate group esterified to the glycerol backbone (Fig. 4.4). The phosphate group

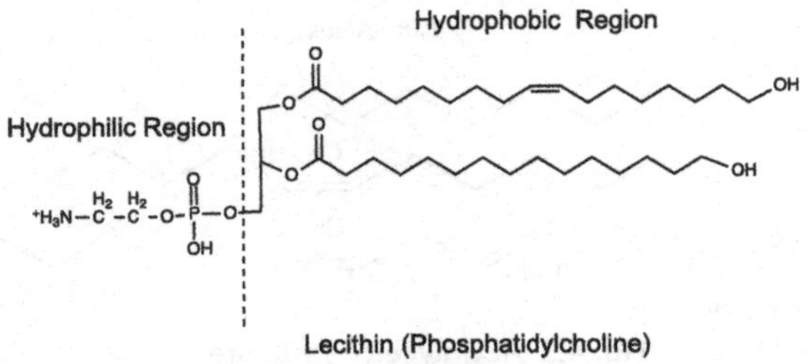

Fig. 4.4 Phospholipid structure.

carries a strong negative charge at any pH encountered in food and provides hydrophilic properties to the molecule. Fatty acids of varying chain length and degree of saturation are esterified to the other glycerol hydroxyl groups providing hydrophobic properties. The combination of hydrophilic and hydrophobic groups in the same molecule makes them good surfactants. Phospholipid molecules contain additional groups, including choline, ethanolamine, serine, or inositol, esterified to phosphate. The corresponding names for these phospholipid derivatives are phosphatidylcholine (PC), phosphatidylethanolamine (PE), phosphatidylserine (PS), and phosphatidylinositol (PI), respectively. Phospholipids, without additional groups esterified to phosphate, are termed phosphatidic acid. Biologically, phosphatidylcholine (PC) is the most nutritionally important form. PC is a component of cell membranes that supplies the essential nutrient, choline, required for the synthesis of the neurotransmitter acetylcholine. This phospholipid functions in muscle contraction and memory.

Lecithin is the common name of the phospholipid mixture found in various plant and animal tissues. It is composed of choline, acylglycerols, free fatty acids, and cholesterol. Lecithin was initially isolated by the French chemist Theodore Gobley in 1848. Gobley is credited with coining the name of this fraction as "lecithin". He also was first to chemically characterize its major phospholipid, phosphatidylcholine. Purified lecithin is widely used as a health supplement and food ingredient. As a supplement, lecithin may improve cognitive function in individuals with dementia, due to its choline content. In foods, lecithin is recognized for its ability to make stable emulsions. Lecithin's food uses include spreadable margarines to soften the texture of bakery products. Lecithin improves the flow characteristics of liquid chocolate coatings. When added to cocoa powder, it greatly improves solubility and the ability to create stable suspensions of cocoa powder in milk. Lecithin is also useful as a lubricant. It prevents food from sticking to the pan during cooking and is the principal ingredient of non-stick cooking sprays.

Sterols are a group of structurally related lipid compounds occurring in both plants and animals. Plant derived sterols are commonly known as phytosterols. Animal forms are known as zoosterols. The basic structure of a sterol consists of three, six-membered rings and one four-membered ring that are fused into one molecule. A lone hydroxyl group attached to one of the 6-membered rings adds polar character to this nonpolar molecule (Fig. 4.5). This addition enables sterols to act as surfactants. The hydroxyl group can also serve as a point of fatty acid attachment by forming an ester link with its carboxyl group. Cholesterol is the major form of sterol found in animal fat. Cholesterol is biologically important as the precursor for the synthesis

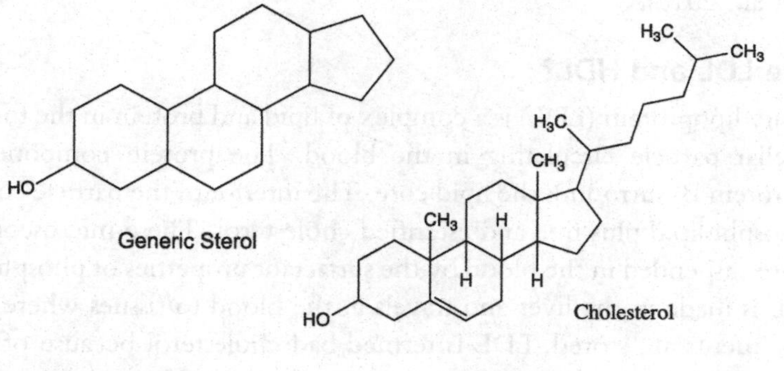

Fig. 4.5 Sterol structures.

of vitamin D. It is also the precursor of bile salt (sodium taurocholate). The function of taurocholate is emulsification of dietary fats in the gut and facilitation of their absorption. Animal foods such as meat, milk, and eggs contain substantial levels of cholesterol. Phytosterols include plant sterols and stanols found in many oils, including corn, canola, sesame, and olive. Phytosterols are also found in wheat germ, peanuts, and almonds. Stanols differ from sterols in that their structure does not contain double bonds. The hydroxyl group of stanols is esterified with a fatty acid that aids their function (Fig. 4.5). β-sitosterol and stigmasterol are two examples of phytosterol. Phytosterols are notable for their ability to compete with cholesterol for uptake in the gut, lowering cholesterol levels in the blood. Phytosterols ability to lower blood cholesterol level has been suggested as beneficial in reducing the risk of coronary heart disease, but this link has not been proven (Genser et al., 2012).

What are LDL and HDL?

Low density lipoprotein (LDL) is a complex of lipid and protein in the form of a micellar particle circulating in the blood. The protein component (apolipoprotein B) surrounds the lipid core. The interior of the particle consists of phospholipid plus free and esterified cholesterol. These microscopic particles are suspended in the blood by the surfactant properties of phospholipid. LDL is made in the liver and travels in the blood to tissues where its lipid constituents are stored. LDL is termed bad cholesterol because of its tendency to cling to arterial walls. Accumulation of LDL in the arteries forces the heart to work harder in pumping blood. Over time, the added pumping stress increases the risk of heart failure. High Density Lipoprotein (HDL) is a similar complex of lipid and protein. HDL's function reverses the effect of LDL. It removes lipid molecules from cells and tissues and transports them back to the liver where they are broken down and eliminated. The ratio of LDL to HDL is most important because high LDL level relative to that of HDL is associated with risk of coronary heart disease (Marventano et al., 2015).

Functional properties of lipids in food

Melting Point and Crystallization: Acylglycerols found in ice cream, butter, milk, meat, cheese, and chocolate, are responsible for the desirable sensory qualities of these foods. Fat in chocolate (i.e., cocoa butter) contributes flavor and a sharp melting point approximately equal to normal human body temperature. Species-specific flavors of meat (e.g., chicken, beef, and pork) are primarily a property of its fat (acylglycerol). The flavors of herbs and spices (e.g., rosemary, oregano, basil, and garlic) are derived, in part, from lipids. Fat is an important part of the diet. The desire for food containing fat is hard wired into our behavior. It is thought that seeking energy-dense foods is an evolutionary pre-disposition to survival. Lipid, as a class of nutrient, contains more than twice the calories per gram of any other food component. The calories per gram of protein, carbohydrate, and lipid are 4, 4, and 9, respectively. Eating foods high in fat also provides a greater feeling of satiety than those high in protein or carbohydrate.

The solid form of acylglycerols (fat) predominates when the temperature is below its melting point. The chemistry behind the change from solid to liquid and vice versa is explained by van der Waal forces. These are weak attractive forces between molecules and atoms. The solid form of acylglycerol predominates when the sum of these attractive forces is sufficient to hold molecules in a semi-crystalline structure known as fat. As the temperature is increased, van der Waal force decreases and the solid becomes liquid (oil). The temperature at which the energy of the system is greater than the van der Waals forces holding acylglycerol molecules together defines the melting point of the lipid. The liquid (oil) or solid (fat) states of an acylglycerol are reversibly interconverted by changing the temperature above or below its melting point.

The structure and composition of fatty acids in acylglycerols are major factors determining its melting point. Specifically, fatty acid chain length, the presence of double bonds, and their cis or trans configuration, affect melting point the most. Acylglycerols composed of short chain fatty acids have a lower melting point because the amount of van der Waals interaction is lower compared to acylglycerols with longer chain fatty acids. The presence of a double bond in one or more of the constituent fatty acids in an acylglycerol creates structural differences limiting the extent of van der Waals attraction. Acylglycerols composed of unsaturated fatty acids have a lower melting point that their saturated counterpoint. The cis or trans configuration of the double bond is also important. Fatty acids with double bonds in the trans configuration pack more closely compared to those in the cis configuration, resulting in a higher melting point. Crystallization is the opposite of melting. It occurs as a liquid acylglycerol is cooled below its melting point. As the temperature decreases, fatty acids re-associate via van der Waal attractions and eventually become a solid. Recrystallization is important to the functional properties of acylglycerols because they can exist in several forms termed polymorphs. Environmental factors, such as rate of cooling and presence of other crystals, influence the predominant form (α, β, or β') in the resultant fat. Rapid cooling of acylglycerols, for example, results in the α crystalline form. Alpha crystals are small, smooth textured, and amorphous in shape. Their density and stability is low compared to the other forms. Therefore its melting point is also lowest (Table 4.2). The melting point α, β, and β' crystalline forms increases in that order. It should be noted that a solid appearing fat is a mixture of crystalline forms and liquid. Liquid acylglycerols within the mass give fat a quality referred to as plasticity.

Table 4.2 Characteristics of acylglycerol crystal polymorphs.

	α	β'	β
Appearance	Translucent	In between	Opaque
Crystal form	Amorphous	Orthorhombic/needle-like	Triclinic
Size	Smallest	intermediate	Largest
Texture	Smooth	Fine grained	Grainy
Melting point/Stability	Lowest	Intermediate	Highest

Chocolate derives much of its desirable sensory properties from its fat content. The fat in chocolate provides a smooth, glossy surface and melting point which is nearly the same as normal body temperature. The principal ingredients of chocolate are cocoa butter, cocoa powder, sugar, and flavorings. The flavor of various types of chocolate (i.e., dark, milk, and white) are principally due to its cocoa powder content. Dark, milk, and white chocolate contains 50%–85%, 10%–15%, and zero% cocoa powder, respectively. In addition to dark brown color, cocoa powder contains a significant level of flavonoids responsible for a bitter flavor that increases with its cocoa powder content. Milk chocolate contains only 10%–15% cocoa powder and an approximately equal amount of dry milk solids, creating a milder flavor and softer texture. White chocolate flavor is principally due to its sugar, cocoa butter (or other fat), and flavorings.

Chocolate is made from the cocoa tree which is indigenous to equatorial regions of the world (i.e., Central and South America, West Africa and Southeast Asia). Manufacture of chocolate is an extensive process that begins with harvest of cocoa seed from tree pods (ADM Cocoa manual, 1999). The seeds and a carbohydrate rich mucus from the pods undergo natural fermentation on the ground. This is followed by air drying and cleaning of extraneous materials. Cocoa beans are alkalized using potassium carbonate to neutralize some of the acid produced by fermentation. Alkalizing increases cocoa bean pH and serves to enhance flavor development in the subsequent roasting phase. The alkalization process known as, "Dutching" also improves the solubility of cocoa powder and improves flavor by neutralizing natural acids. Cocoa beans are roasted at 95—145 °C to intensify development of chocolate color and flavor. A greater thermal process enhances darker color and intensifies flavors. These changes result from Maillard reaction between carbohydrates and proteins (described in Chapter 3). Roasted cocoa beans are composed of approximately 55% fat, 14% protein, and 14% —20% carbohydrate as starch and fiber. Other minor constituents include minerals, alkaloids (theobromine), and flavonoids (phenolic antioxidants). Roasted beans then are cracked to separate the shell from the kernel portion of the bean (nibs). The refining process begins with grinding of the nibs. Heat generated by grinding converts the bean's fat into liquid, allowing it separation from the solid fraction. The liquid fraction contains most of the cocoa butter (lipid). The pressed and dried solid fraction is the cocoa powder. Cocoa butter's lipid composition is primarily responsible for chocolate's desirable properties. Cocoa butter is nearly 100% lipid of which most (99%) is in the form of acylglycerols. Less than 1% of the lipid is present as free fatty acids. The relative proportion of saturated and unsaturated in cocoa butter is unusual. Two-thirds of the fatty acids are saturated and about one-third are unsaturated. Additionally, fatty acids are uniquely arranged within cocoa butter acylglycerols. Three triacylglycerol forms predominate: Palmitic-Oleic-Stearic (POSt), Stearic-Oleic-Stearic (StOSt), and Palmitic-Oleic-Palmitic (POP). These give cocoa butter and chocolate its sharp melting point (34—36 °C) (Becket, 2000). When chocolate melts, there is an accompanying endothermic (heat absorbing) process. The heat absorbed by the melting chocolate creates a pleasant cooling sensation in the mouth, adding to the sensory experience.

What is tempering and how does it make my chocolate better?

The melting point of cocoa butter can vary from 16° to 36 °C, depending on its crystalline form. Cocoa butter exists in 6 polymorphic forms, (Roman numerals I-VI). These corresponds to the preferred Greek letter system (γ, α, β'_2, β'_1, β_2, and β) (Beckett, 2000). The most desirable polymorph is the β (or VI) form. Chocolate containing cocoa butter in low stability forms (γ or α) is least desirable, very soft, and dull surfaced. In contrast, chocolate is most desirable in the β form because the melting point is close to body temperature and the texture is firm with a definite snap and glossy surface. Additionally, the endothermic effect is greatest and more desirable when melting chocolate in the β form. Fortunately, the crystalline form of cocoa butter fat in chocolate is easily changed through a process called tempering. In this process, melted chocolate is held at a temperature just below its melting point while solid chocolate (β form) is added to "seed" the liquid. The temperature is allowed to cool slowly which promotes formation of the highly desired β form. In contrast, rapidly cooled chocolate will adopt the least desirable α form. A common defect in chocolate occurs when partially melted chocolate is re-cooled. This produces a dull gray appearance termed chocolate bloom.

Interesterification is a chemically based process used to improve the functional performance of fats by blending (rearranging) the fatty acids within acylglycerols. In this process, fatty acids are released from their original positions in two or more types of acylglycerols and reattached to glycerol in random order. Acylglycerols are selected for their composition of fatty acids (e.g., chain length, degree of saturation) in creating hybrids with desired functional properties. The technology is used in making a variety of soft spread margarines, cheese spreads, and salad dressings. A major advantage of interesterification is that little or no trans fats are created.

Emulsification is a process by which two immiscible liquids are combined into a stable suspension (emulsion). Fats and oils are not compatible with water because of their hydrophobic nature and quickly separate, even after vigorous blending. However, if a small amount of detergent is added to the oil and water mixture a stable suspension is formed. So, how does a detergent keep the oil and water phases from separating? The answer is that detergents are surface-active agents (surfactants) that change the water and lipid phases. First, addition of a surfactant lowers the surface tension of water by reducing the attraction between water molecules. Second, surfactants form micelles in water with polar regions facing the aqueous environment and non-polar regions facing the interior (Fig. 4.6). Adding a surfactant to the oil and water mixture enables formation of a stable suspension called an emulsion.

Food emulsions: Homogenized milk, processed meats, mayonnaise, and butter are common examples of food products based on emulsions. The fat in fresh milk separates to the surface after sitting for just a few hours. This problem is referred to as creaming in the dairy industry. Fat separation is

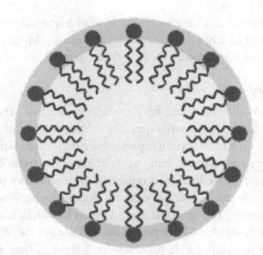

Fig. 4.6 Oil in water micelle. *Image source: https://en.wikipedia.org/wiki/File:A_lipid_micelle.png.*

prevented by the process of homogenization which forces milk through tiny orifices under high pressure. The physical shearing action caused by homogenization unfolds milk proteins and creates micelles. Milk proteins (described in Chapter 9) are good surfactants because they contain both polar and non-polar groups. Homogenization uses milk proteins to form micelles that trap fat and create a stable oil-in-water emulsion from which fat can no longer separate. Processed meats rely upon constituent proteins to make products such as hot dogs, sausages, and ham. These products are made by blending meat, salt, flavorings, and fat in an operation designed to maximize muscle protein (myosin) extraction. Salt and vigorous mixing extracts and unfolds myosin protein that acts as a surfactant. Myosin (described in Chapter 9) has a large compliment of polar and non-polar amino acids and thus is a good surfactant. Micelles made from meat proteins hold substantial amounts of fat as oil-in-water emulsions. Much of the emulsified fat in processed meat is retained even after cooking. Mayonnaise is made by blending egg yolks with large amounts of oil. Egg yolk provides the surfactant in the form of the polar phospholipid, called lecithin. Because oil is the predominant phase and water is the minor component, the mixture is an example of water-in-oil type emulsion. Butter is also an example of an emulsified food system in which fat is the predominant component and water is the minor one. Butter is made from cream (15%–30% fat) by churning or other means of agitation. The mixing action promotes formation of micelles using milk proteins as a surfactant. However, the amount of protein available in butter is relatively small, resulting in the coalescence of fat into large particles during churning. Butter typically contains a minimum of 80% fat and about 10% moisture.

Lipid chemistry

Lipid molecules of all classes (i.e., fatty acids, acylglycerols, sterols) undergo some degree of chemical reaction, especially in the presence of heat. Hydrolysis, hydrogenation, and oxidation represent the major types of chemical reactions occurring to food lipids. Fatty acids liberated by hydrolysis reactions contribute substantially to the flavors typical of meat and dairy products. Hydrogenation is a chemical process used to change carbon–carbon double bonds to single bonds. The reaction is used on an industrial scale to convert oils into fats with desirable end use properties and greater stability. However, hydrogenation is also responsible for trans-fat, a form of fatty acid in which some carbon–carbon double bonds are isomerized from cis to trans.

Hydrolysis involves breaking the ester bond between fatty acids and glycerol, releasing them into the matrix (Fig. 4.7). The rate of ester bond hydrolysis is faster at the extremes of pH and at high temperature. Chemically, hydrolysis is a reversible reaction in which fatty acids are removed from (hydrolyzed) or linked to (esterified) glycerol's hydroxyl (OH) groups. Hydrolysis reactions have many practical applications. Hydrolysis chemistry is the basis of making soap from fats in a process called saponification. Soap can be made in the kitchen with a few simple ingredients including, fat (tallow), lye (sodium hydroxide), table salt (sodium chloride), and fragrance and/or colorant. Heating the mixture of lye and tallow liberates fatty acids from acylglycerols in the fat. Soap is formed upon cooling, by a combination of positively charged sodium ions with negatively charged carboxyl groups of fatty acids. The combination of an acid and base is technically termed a salt. Hydrolysis of ester bonds and liberation of fatty acids can also be accomplished with lipase enzymes. Lipases are biologically important in the human gut. Liberated fatty acids are absorbed and used as a source of

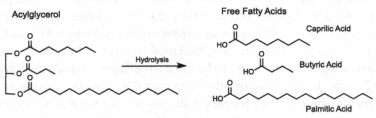

Fig. 4.7 Hydrolysis of fatty acid ester bonds.

energy. Lipase enzymes act on acylglycerol substrates with varying degrees of specificity. Some bacterial lipases are responsible for converting bland tasting milk into the wide variety of flavors found in cheese, yogurt, and other fermented dairy products. Lipase from spoilage bacteria is very effective in releasing short chain fatty acids such as butyric from milk lipids, giving it the sour, offensive taste of spoiled milk.

What is hydrolytic rancidity?

Hydrolytic rancidity is a term that describes off flavors and aromas caused by release of short chain fatty acids from acylglycerols. Hydrolytic rancidity is commonly caused by lipase enzymes of bacterial origin. Short chain fatty acids, such as butyric, are responsible for sharp and unpleasant flavors of sour milk. Hydrolytic rancidity can also result from a chemical mechanism that occurs when foods are cooked at high temperature (frying). Water from the food itself can catalyze hydrolysis of ester bonds, releasing a variety of fatty acids and their off flavor breakdown products. Hydrolytic rancidity is a common problem in fast food restaurants. When frying oils are used too long, the undesirable flavors end up in the food.

Hydrogenation is a chemical reaction that modifies carbon–carbon double bonds in lipids through addition of hydrogen atoms and an inert meat catalyst. The predominant use of this chemistry is to change oils to solid fats by reducing double bonds in polyunsaturated fatty acids. Hydrogenation had its start in the early 20th century when Wilhelm Normal developed and patented the chemical process in 1902. In 1911, Proctor and Gamble acquired the patent rights and used the process to develop the product known as Crisco ("vegetable shortening"). Crisco was so successful that fats from hydrogenated plant oils soon displaced the more expensive and prone to off flavor, animal fats (lard). Fats derived from the hydrogenation process performed well in processed food. They are more stable, lower in cost, and give products with longer shelf life. It is still debated whether a better pie crust is made with Crisco or butter. By the mid 20th century hydrogenation was used to convert a surplus of oils from soybeans and corn to margarine products that were less costly than butter and spreadable at cool temperatures. The texture of hydrogenated fats can be varied from soft and spreadable to solid by controlling the degree of hydrogenation. Margarine is created by incomplete hydrogenation of carbon-carbon double bonds in acylglycerol oils in the process known as partial hydrogenation. Partial hydrogenation is also responsible for causing a conformational change in remaining carbon–carbon double bonds. Specifically, hydrogenation creates a mixture of cis or trans double bonds. Unfortunately, the trans–fat content of these products (labeled as "partially hydrogenated") has been linked to increased risk of coronary heart disease (Katan et al., 1995). The disease risk factor linked to trans–fat is associated with changes in HDL blood levels. Studies have shown that trans–fat results in a decrease of serum HDL level, compared to saturated fats. A large body of evidence summarized by Mozaffarian et al. (2006) demonstrates the link between trans–fat consumption and increased risk of coronary heart disease. In consideration of this evidence, the FDA ruled in January 2006 that nutrition labels for foods and supplements must indicate their content of trans fat.

What is shortening and is it bad for me?

The term shortening has been used since the beginning of the last century. It is predominantly composed of saturated fats. Shortening is often the preferred fat for baking applications. The term shortening can be applied to a variety of fats derived from either animal (lard) or vegetable fat (hydrogenated oils) sources. The chemical process of hydrogenation is used to make shortening. Because hydrogenation unavoidably produces some level of trans–fat, it is recommended that products containing hydrogenated or partially hydrogenated fats, be avoided. Other technologies such as interesterification have been developed to make shortening containing little or no trans–fat. It should be mentioned that US food labeling regulations allow food products to be labeled as containing "zero trans–fat" if they contain less than 0.6g of trans fat per serving. This labeling regulation begs the question, when is zero, zero?

Lipid oxidation: Unsaturated lipid molecules are particularly prone to oxidation reactions resulting in undesirable effects in food. Warmed over flavor in meats, sunlight flavor in milk, and reversion flavor in soybean oil are examples of undesirable flavors resulting from lipid oxidation. Lipid oxidation also affects the nutritional quality of food. Fat soluble vitamins and some nutritionally important polyunsaturated fatty acids are destroyed by this chemistry. Lipid oxidation is a free radical type of chemical reaction whose target is carbon–carbon double bonds in unsaturated lipid molecules. Free radicals are defined as an atom, molecule, or ion with an unpaired electron. Electrons typically exist in pairs within the orbitals of atoms. In a free radical, an electron missing from one of the orbital pairs results in an unstable (high energy) species that will either gain or lose an electron to become stable (low energy) again. Electrons in double bonds are more easily reacted than those in single bonds and therefore are the preferred site of free radical reaction. The pathway of lipid oxidation involves the formation of a carbon based free radical. The reaction is initiated by abstracting a hydrogen atom a carbon adjacent to a double bond. This creates the first lipid radical (L$^\bullet$), called an alkyl free radical (shown below). The initiation phase is the rate limiting step of lipid oxidation. Once formed, alkyl radicals (L$^\bullet$) react with oxygen molecules to create peroxy free radicals (LOO$^\bullet$). Peroxy radicals, in turn, react with carbon–carbon double bonds (LH) to create additional alkyl radicals (L$^\bullet$) and hydroperoxides (LOOH). The cycle of reactions involving generation of peroxyl radicals and hydroperoxides represents the propagation phase of the reaction. The rate of reaction in the propagation phase in exponential because the products of one reaction cycle become re-actants in the next.

$$\text{Initiation} \quad L^\bullet + LH \longrightarrow L^\bullet + H^\bullet$$

$$\text{Propagation} \begin{bmatrix} L^\bullet + O_2 \longrightarrow LOO^\bullet \\ LOO^\bullet + LH \longrightarrow LOOH + L^\bullet \end{bmatrix}$$

$$\text{Termination} \quad L_\bullet^\bullet,\ LO_\bullet^\bullet,\ LOO_\bullet^\bullet \longrightarrow L\text{-}L + LOOL$$

The final phase of a free radical reaction is the termination phase caused by combining various free radical forms into stable species. For example, two alkyl radicals (L$^\bullet$) combine their extra electrons to form (L-L) a dimer of lipid molecules which are linked by a covalent bond. Lipid polymers formed through the termination reaction are noticeable as that sticky, gummy stuff that forms on the rim of a cooking oil bottle. Lipid polymers are a tell-tale sign that the oil should be discarded.

Decomposition Products of Lipid Hydroperoxides (LOOH)

Off Flavors and Aromas derived from;
Short Chain Aldehydes
Ketones
Alcohols

Lipid hydroperoxides (LOOH): Lipid hydroperoxides are unstable prod-ucts of lipid oxidation. They decompose into numerous short chain, volatile products that are the source of off flavors and aromas. Compounds causing off flavor are chemically characterized as aldehydes, ketones, acids, and alco-hols. Aldehyde products are particularly reactive molecules that can diffuse

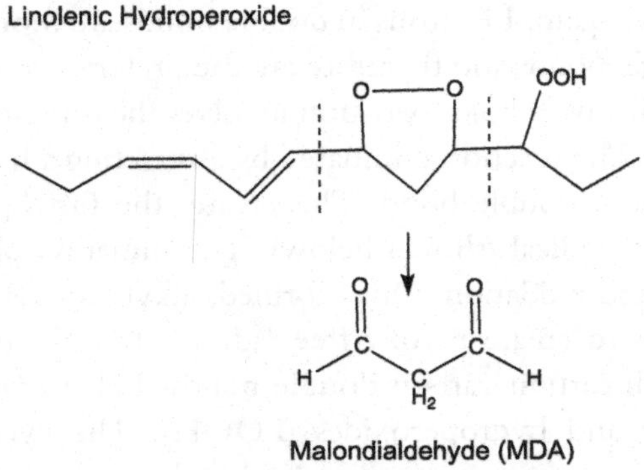

Linolenic Hydroperoxide

Malondialdehyde (MDA)

Fig. 4.8 Linolenic hydroperoxide decomposition.

throughout the food matrix, causing chemical modification of proteins, vitamins, and nucleic acids. Malondialdehyde, resulting from hydroperoxide decomposition, is just one example of a potent chemical modifier. Malondialdehyde (MDA) is a small, 3 carbon molecule containing two aldehyde groups (Fig. 4.8). It is derived from hydroperoxide decomposition of linolenic acid, a fatty acid containing three double bonds. The level of MDA is widely used as a measure of quality, or the lack thereof, in meats. MDA and other aldehyde products of hydroperoxide decomposition represent a potential cause of DNA mutation and essential amino acid destruction (Johnson and Decker, 2015). Consuming food with high levels of hydroperoxides may have negative health consequences.

Factors affecting the rate of lipid oxidation: Several factors affect the rate at which an unsaturated lipid undergo oxidation. The most significant factor is the degree of unsaturation. The more double bonds in an unsaturated lipid, the greater its rate of oxidation. For example, a dienoic fatty acids (2 double bonds) is oxidized 10 times faster than a monoenoic fatty acid (1 double bond). The level of oxygen in the food material or packaging is critical to the rate of lipid oxidation. Methods that lower oxygen level, such as replacing air with nitrogen (modified atmosphere packaging), effectively inhibit lipid oxidation. Exposure of unsaturated lipids to UV light substantially increases rate of oxidation by accelerating the decomposition of lipid hydroperoxides. Therefore, packaging materials, such as dark glass or foil, are an effective means to extend their shelf life. Addition of antioxidants (ascorbic acid), together with chelators (EDTA), can also be effective in limiting lipid oxidation and extending shelf life.

Lipoxygenase: Lipid oxidation in food materials is also caused by a family of enzymes called lipoxygenases. These enzymes catalyze the oxidation of unsaturated fatty acids and have the same negative outcomes as the strictly chemical process. Lipoxygenase specifically oxidizes linoleic (C18:2 $\Delta^{9,12}$), linolenic (C18:3 $\Delta^{9,12,15}$) arachidonic acid (C20:4 $^{\Delta 5,8,11,14}$ an essential fatty acid). The mechanism of lipoxygenase-catalyzed oxidation involves a bound iron atom whose chemistry is essential to the reaction. Lipoxygenase activity is a significant problem in plant foods, such as peas and soybeans. These legumes contain high levels of lipoxygenase and its activity represent a large economic problem. Left unchecked during processing, off flavors produced by lipoxygenase activity can bind to proteins and make them unsuitable for food use. Lipoxygenase activity can be controlled by using a heat treatment prior to processing the beans.

Reactive Oxygen Species (ROS) are molecules that promote free radical formation, or are themselves, a free radical. The following free radical molecules are listed in the order of their ability to cause lipid oxidation, hydroxide radical ($^{\bullet}OH$) > alkoxy radical ($LO^{\bullet}$) > hydroperoxyl radical ($LOO^{\bullet}$) > superoxide anion (O_2^-). Singlet oxygen ($^1O_2$) and hydrogen peroxide (H_2O_2) are molecules that promote free radical formation. The consequence of ROS chemistry in food is most noticeable by the presence of off flavors and loss of color. ROS activity also lowers the nutritional quality of food through the destruction of vitamins and biologically important lipids. To better understand the chemistry of ROS, it is important to know something of oxygen's molecular nature. Oxygen (O_2) is a diatomic molecule whose electrons normally exist in a stable orbital configuration, termed triplet oxygen. The symbol for triplet oxygen is written as ($^3O_2$). The superscript "3" is used to indicate the triplet configuration of its electrons. Oxygen can also exist as singlet oxygen whose symbol is ($^1O_2$). A subtle, but important, difference in electron configuration of these oxygen molecules is the slightly lower negative charge of the singlet oxygen compared to triplet oxygen. A result of this difference Singlet oxygen has a strong tendency to accept electrons from other atoms and form new bonds. Singlet oxygen can directly react with carbon–carbon double bonds in unsaturated lipids but, triplet oxygen and can't. Singlet oxygen is believed to be the primary cause of lipid oxidation in food (Choe and Min, 2006).

How is singlet oxygen produced? Singlet oxygen is formed from triplet oxygen (its normal state), after receiving an energy boost from an external source. Typically, the energy is supplied through an interaction with a sensitizer molecule (S). A sensitizer is a molecule whose electrons are exposed to UV light creating the excited state ($S^{\bullet}$). The energy derived from excitation is passed on to triplet oxygen, converting it to the singlet form.

Singlet oxygen formation

$$\text{UV Light} + \text{Sensitizer} \rightarrow \text{Excited Sensitizer} \,(S^{\bullet})$$
$$O_2^3 \,(\text{Triplet Oxygen}) + S^{\bullet} \rightarrow O_2^1 \,(\text{Singlet Oxygen})$$

Chlorophyll, riboflavin, and heme-containing proteins are notable examples of sensitizers in foods. Chlorophyll, for example, is a sensitizer when excited by UV light. The energy it absorbs is transferred to oxygen, transforming it to the singlet state. Sensitizer molecules are termed pro-oxidants because of their role in generating ROS. Transition metals (copper and iron) are pro-oxidants because they also promote the production of ROS. Copper and iron, while present in at low levels in food, are major contributors to lipid oxidation. Copper is a stronger pro-oxidant compared to iron, but iron is typically present at higher levels in food materials and therefore, a greater problem. Transition metals are also accelerators of hydroperoxide decomposition reactions. The reduced states of copper (cuprous, Cu^{+1}) and iron (ferrous, Fe^{+2}) are more effective than their oxidized states (cupric, Cu^{+2} and ferric, Fe^{+2}). The following equation illustrates the reaction of ferrous iron with lipid hydroperoxide (LOOH) to produce an alkoxy ($LO^{\bullet}$) free radical.

$$Fe^{+2} + LOOH \rightarrow LO\cdot + Fe^{+3} + OH^{-}$$

Singlet oxygen and ROS in food: Biologically, chlorophyll functions in the well-known process of plant photosynthesis. Absorbed light energy is transformed into a form of chemical energy that plants can store. Chlorophyll can cause oxidation of unsaturated lipids through its action as a sensitizer. For example, a defect in soybean oil, known as "reversion flavor" (a grassy or paint-like aroma), is attributed to low levels of chlorophyll that remain in purified oil. Reversion flavor occurs when chlorophyll-containing oil is exposed to light, causing production of singlet oxygen. The subsequent formation and decomposition of lipid hydroperoxides results in the off flavor. Similarly, cow's milk is prone to a defect called "sunlight" flavor that results from UV light excitation of riboflavin (vitamin B2). Cow's milk contains a substantial amount of this vitamin and can be seen when milk is held up to the light. The florescent greenish tint observed is due to emitted light produced by excited riboflavin molecules returning to their ground state. Excited riboflavin transfers the energy from irradiation to dissolved oxygen in milk causing singlet oxygen formation. The subsequent decomposition of milk lipid hydroperoxides results in the sulfurous "sunlight" off flavor. Meat contains heme proteins (myoglobin and cytochromes) that bind oxygen and function in the process of respiration. The oxygen binding property of heme, while important to muscle, can be a problematic in meat because of its ability to act as a sensitizer. Energy absorbed from light is transferred to oxygen which excites to the singlet state. In addition to singlet oxygen, free radicals (i.e., hydroxide radical ($^\bullet$OH) and superoxide anion ($O_2^{\bullet-}$) generated by heme iron atom, contribute to the oxidation of meat lipids. Off flavors from oxidized lipids and are strongest in meat that has been cooked, refrigerated, and reheated. The defect called "warmed over flavor" (WOF) develops quickly from decomposition of lipid hydroperoxides during the reheating step. Reactions involving ROS do more than cause off flavors. They can cause a loss of biological activity in proteins and destroy essential amino acids. ROS destroy vitamins (e.g., A, D, E, B_2, and C) and chemically modify nucleic acids. Mutations to DNA caused by ROS are thought to be linked to cancer (Lucymara et al., 2012).

Antioxidants control free radicals by several chemical mechanisms and prevent the adverse effects of lipid oxidation in food. Antioxidants that neutralize free radicals are termed free radical scavengers (FRS). Free radicals are atoms or molecules with unpaired electrons in their valence orbitals. FRS eliminate free radical activity by donating a hydrogen or taking away an electron from the radical species. Two additional types of antioxidants are free radical quenchers and chelators. A free radical quencher is a molecule that prevents the production of singlet oxygen by absorbing light before it activates a sensitizer molecule. Chelators are molecules that bind transition metals and prevent their participation in the chemistry that produces free radicals. Some antioxidants, such as ascorbic acid, provide both quenching and chelating properties. Many foods, such as cereals, fruit juices, and vegetables, naturally contain antioxidants, but a substantial amount of activity is lost through processing. Separation of the hull and endosperm in grains, for example, greatly reduces the vitamin and antioxidant content. Therefore, it has become common practice to supplement products, such as breakfast cereals with vitamins and add antioxidants to prevent off flavor. Synthetic antioxidants, such as butylated ydroxyanisole (BHA), butylated hydroxytoluene (BHT), tertiary butylhydroxyquinone (TBHQ), propyl gallate (PG), are effective and widely used for this purpose in processed foods. While synthetic antioxidants have economic advantages, there is growing preference to avoid the use of synthetic additives in food.

How do Free Radical Scavengers work? A free radical scavenger is a compound that converts free radicals to the normal or ground state by donating or accepting an electron. The free radical scavenger AH is typically an unsaturated phenolic compound (a six-membered ring containing three double bonds and a hydroxyl group). The (AH) molecule donates a hydrogen atom to a lipid peroxy radical (LOO$^\bullet$), converting it to a hydroperoxide (LOOH). The energy of the unpaired electron in (A$^\bullet$) is dissipated in a process called resonance delocalization. The free radical is distributed among multiple carbons within the ring. The net effect is that the free radical's energy is returned to the ground state.

$$AH + LOO^\bullet \rightarrow LOOH + A^\bullet$$

Vitamins as antioxidants

Ascorbic acid exists in both vitamin and non-vitamin forms and is found in a variety of fruits and vegetables. All forms, including those without vitamin C activity, can act as antioxidants and scavengers of oxygen. Non-vitamin forms of ascorbic acid are less expensive, often used to control lipid oxidation. Ascorbic acid is an effective free radical scavenger and terminates free radical reactions. In this reaction, ascorbic acid donates a hydrogen (circled in Fig. 4.9) to the lipid radical (L$^\bullet$) forming a stable lipid alkyl (LH). Ascorbate radical undergoes resonance delocalization and changes to dehydroascorbate. This form of ascorbate can be returned to ascorbic acid by a reduction reaction.

2 Ascorbic Acid + L$^\bullet$ → Ascorbate Radical ↔ Dehydroascorbate + LH

Ascorbic acid's antioxidant activity also includes binding transition metal ions. For example, chelation of copper (Cu^{+2}) ions by ascorbic acid limits copper's ability to act as pro-oxidants.

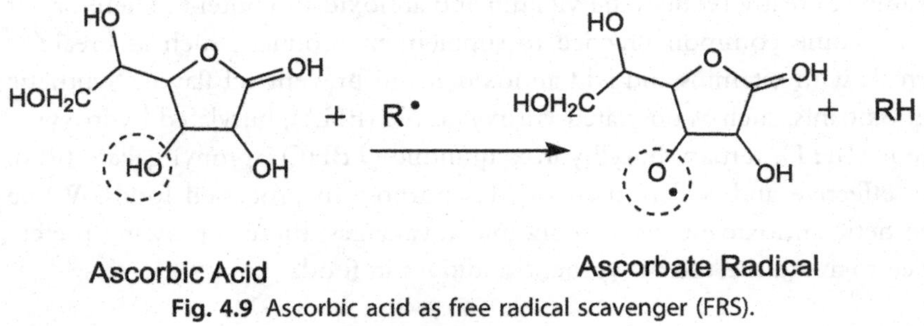

Ascorbic Acid   Ascorbate Radical

Fig. 4.9 Ascorbic acid as free radical scavenger (FRS).

Alpha-tocopherol (vitamin E) is a lipid soluble (hydrophobic) molecule found primarily in plant foods (e.g., oils and nuts). Tocopherol exits in several forms (i.e., alpha, beta, and gamma), but, only alpha tocopherol has vitamin E activity. The structure of alpha tocopherol consists of a hydroxylated ring structure coupled to a twelve-carbon aliphatic chain. All forms of tocopherol are structurally similar with differences limited to the number of methyl groups attached to the aliphatic chain. Alpha has three methyl groups and is the most non-polar. Gamma is slightly less non-polar with only one methyl group. The hydrophobic nature of tocopherols limits their nonpolar regions (e.g., cell membranes and fat depots). The antioxidant activity of tocopherols comes from their ability to act as free radical scavengers. Specifically, tocopherols readily donate a hydrogen atom from the hydroxyl group (circled in Fig. 4.10) to a free radical. In the equation below, the hydrogen from alpha tocopherol is donated to a lipid peroxy radical (LOO$^\bullet$) transforming it to a lipid hydroperoxide LOOH). The other product of this reaction (tocopherol radical) terminates the radical's activity by delocalization of the unpaired electron throughout the ring structure.

$$LOO^\bullet + \alpha \text{ tocopherol} - OH \rightarrow LOOH + \alpha \text{ tocopherol} - O^\bullet$$

Hot dogs, for example, contain as much as 25% fat that is uniformly emulsified. Limiting lipid oxidation in a high fat product such as hot dogs with polar antioxidants such as ascorbic acid is not very effective. However, this difficulty is overcome by using a combination of ascorbic acid (water soluble) and alpha tocopherol (lipid soluble) antioxidants to provide uniform antioxidant activity in emulsified food systems. This combination of antioxidants is especially effective in controlling free radical lipid oxidation because their action is synergistic.

Synergistic antioxidants: When the FSR activity of a combination of compounds is greater than the sum of their individual activity, the system is said to be synergistic. Ascorbic acid and tocopherol represent a synergistic pair of FRS whose overall antioxidant activity is increased through their

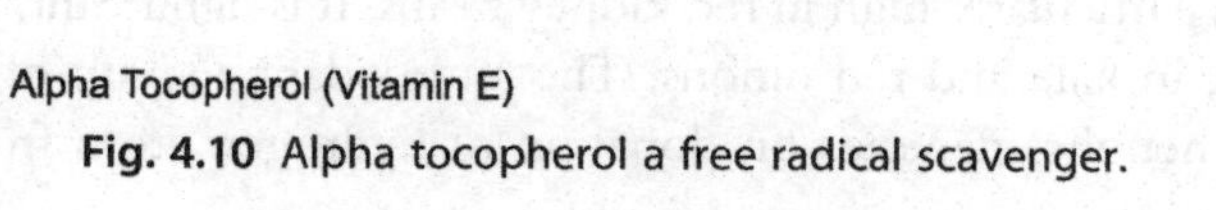

Alpha Tocopherol (Vitamin E)

Fig. 4.10 Alpha tocopherol a free radical scavenger.

chemical interaction. Alpha tocopherol readily terminates lipid radicals by donating a proton. Ascorbic acid then regenerates tocopherol to its native form by accepting the unpaired electron from dehydroascorbate. The combination of tocopherol and ascorbic acid works well in emulsified food systems.

Natural antioxidants-carotenoids

Carotenoids are a diverse family of lipid type compounds that are noted for their properties as colorants (described in Chapter 8) and as source of vitamin A (described in Chapter 5). Carotenoids are known to have health promoting properties, such a preventing eye damage, coronary heart disease, and are proposed to inhibit the progression of cancer (Gallicchio et al., 2008). Much of carotenoid's health promoting effects are derived from their ability to control the production of ROS. Carotenoid molecules are effective quenchers of singlet oxygen because of their large content of double bonds (i.e., 11). A general scheme for the reaction of beta carotene with singlet oxygen is shown in the equation below. Singlet oxygen directly reacts with double bonds in beta carotene and converts oxygen to the triplet form. The target of the reaction (beta carotene) is chemically modified through its interaction with singlet oxygen. When carotenoids are exposed to UV light and singlet oxygen, the loss of color and further chemical decomposition occurs, especially when heated.

$$\beta Carotene + Singlet\ Oxygen \rightarrow Triplet\ Oxygen + \beta Carotene^{\bullet}$$

Natural Antioxidants—Phenolic compounds

Phenolics are a large class of naturally occurring plant compounds that are important to food and health. These water soluble substances are responsible for the yellow and blue-red colors of many fruits, vegetables, and flowers. Phenolics and their polymers (polyphenolics) are found in a variety of foods, including berries, grapes, wine, coffee, tea, chocolate, and legumes. In addition to their properties as colorants (described in Chapter 7), phenolics are potent FRS antioxidants. The structures of plant phenolic compounds are varied, but they commonly contain at least one six-membered, unsaturated ring with one or more hydroxyl group. Quercetin, for example, is a phenolic antioxidant whose structure is shown in Fig. 4.11. The level of quercetin in particularly high in red kidney beans. It is significant, but somewhat lower, in kale and red onions. The antioxidant activity of quercetin and other phenolics depends on donation of hydrogen atom from one of

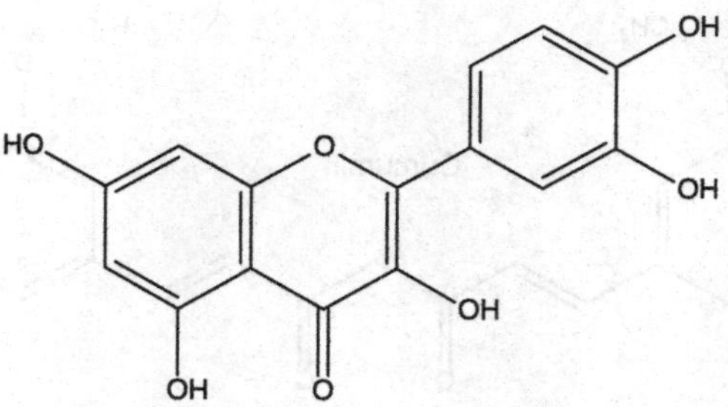

Quercetin, a phenolic antioxidant

Fig. 4.11 Quercetin.

its hydroxyl groups. A generalized scheme for the free radical scavenging activity of phenolics is shown below.

$$POH + LO^{\bullet} \rightarrow PO^{\bullet} + LOH$$

A hydrogen atom from the phenolic hydroxyl group is donated to a lipid alkoxy radical (LO$^{\bullet}$) converting it to the non-radical lipoxy compound (LOH). The phenolic radical (PO$^{\bullet}$) readily reverts to a non-radical form by dissipating the radical's energy through resonance delocalization.

Natural Antioxidants-Curcumin

The phenolic antioxidant curcumin is found at a low level (approximately 2%) in the spice called turmeric. Turmeric is obtained from the dried and ground roots of the *Curcuma longa* plant. It is responsible for the spice's orange-yellow color. The antioxidant activity of curcumin (Fig. 4.12) has been the subject considerable interest and research. While investigations are in early stages, the results regarding its health promoting effects are encouraging. Curcumin's antioxidant activity against ROS may explain its proposed role in health. Specifically, curcumin is suggested to have neuro-protective, anti-inflammatory, and anticancer activities. At present, the biological efficacy of curcumin is somewhat limited by its poor absorption from the gut. Curcumin is a polyphenolic compound that contains two un-saturated phenolic rings. These are responsible for the effective free radical scavenging activity. Curcumin inactivates free radical ROS by donating a proton from one of its phenolic hydroxyl groups (Fig. 4.12). The compound exists as two major isomeric forms that are based on pH of the medium. The keto form predominates below pH 7 and reversible formation of the enol occurs above pH 8. Promising results of curcumin's benefits in animal studies

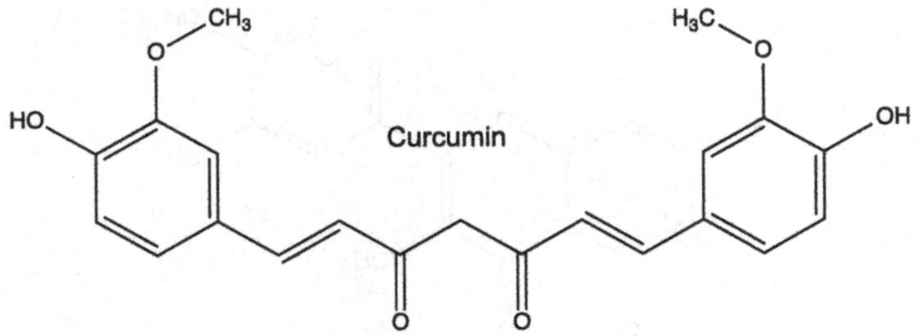

Fig. 4.12 Curcumin.

has spurred a number of over the counter supplements with questionable claims.

Lipids and health

Conjugated Linoleic Acid (CLA) is the common name for the unsaturated fatty acid containing 18 carbons and 2 double bonds. The numerical abbreviation describing it is $C18:\Delta^{cis\ 9,\ trans\ 11}$. The configuration of CLA double bonds are cis at positon 9 and trans at position 11. An additional isomeric form of CLA ($C18:\Delta^{trans\ 10,\ cis\ 12}$) also occurs naturally. The major difference in the later isomer is the location of the double bonds closer to the methyl end of the molecule. When two double bonds occur in a fatty acid, they are typically separated by a methyl group. In contrast, the double bonds of CLA are located on adjacent carbon atoms, thus CLA is called a conjugated fatty acid. CLA is produced in the foregut of ruminant animals by bacterial fermentation of carbohydrates (cellulose). CLA can be found in cow's milk and the fat of grass-fed meat animals. Chickens with CLA in their diet also store a significant amount in their eggs. CLA is most notable for its heath promoting benefit as anti-cancer agent. In mouse studies of breast cancer and human studies of colorectal cancer, administration of CLA has been shown to be effective in activating the process that kills cancerous cells and reduces the size of malignant tumors. It has been proposed that CLA be taken orally as a dietary supplement, but the efficacy of this approach has not been proven in humans.

What is an essential fatty acid?

Essential fatty acids are polyunsaturated fatty acids that provide nutritional value above and beyond their contribution as a source of energy. By definition, they must be included in the diet. Humans can synthesize saturated

Table 4.3 Omega 6 and 3, essential fatty acids.

Omega 6 fatty acids		Omega 3 fatty acids	
Linoleic acid (LA)	C18:2 Δ9,12	α -Linolenic acid (ALA)	C18:3 Δ9,12,15
Gamma (γ) Linolenic acid (GLA)	C18:3 Δ6,9,12	Eicosapentaenoic acid (EPA)	C20:5 Δ5,8,11,14,17
Arachidonic acid (AA)	C20:4 Δ5,8,11,14	Docosahexaenoic acid (DHA)	C22:6 Δ4,7,10,13,16,19

fatty acids and some monounsaturated fatty acids up to about 20 carbons in length. However, they are limited in the ability to synthesize longer chain PUFAs, especially those with double bonds at ω3 and ω6 positions. Humans must rely on dietary sources from plants and animals (mostly fish) to acquire sufficient amounts of ω3 and ω6 (long chain) fatty acids. Biologically, CLA fatty acids are esterified to phospholipids and located incorporated into cell membranes. Long chain polyunsaturated fatty acids are important to our health and have critical roles in metabolism, immune response, prevention of coronary heart disease (CHD), and brain development in the young.

Omega 6 (ω6) and omega 3 (ω3) are two classes of essential fatty acids (EFA) that are made from linoleic acid (LA) and α linolenic acid (ALA), respectively. The designation of a fatty acid as ω6 or ω3 means that a double bond is located 6 or 3 carbons, respectively, from the methyl end. Linoleic acid (LA) is an 18 carbon ω6 fatty acid with two double bonds. Abundant amounts of linoleic are found in olive and corn oils where it makes up over half of the fatty acid content. Other dietary sources of LA include chia, grape, and sunflower seeds. LA is the building block for synthesis of other ω6 fatty acids, such as gamma linolenic and arachidonic acids. Gamma-linolenic acid is formed by addition of a third double bond to linoleic acid. Humans can synthesize gamma linolenic acid, but the amount generally declines with age. Gamma linolenic is recommended in the diet as a supplement or from plant sources. Plant oils (e.g., evening primrose, borage oil, black grape seed oil, and some types of algae) are good sources of gamma linolenic acid. The ω6 arachidonic acid (AA) is synthesized from gamma linolenic by the addition of two carbons, plus one more double bond, to make a total of 20 carbons and 4 double bonds. A major benefit of ω6 fatty acids is their anti-inflammatory activity. For example, ω6 fatty acids are noted for treatment of skin conditions (i.e., dermatitis, eczema) and in

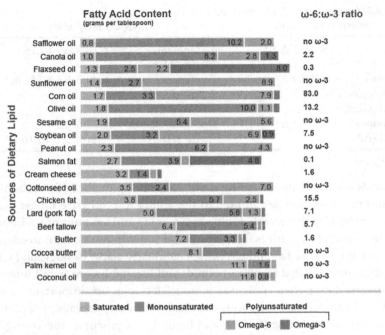

Fig. 4.13 Fatty acid composition in common food lipids. *Image Source: https://www. gbhealthwatch.com/Science-Omega3-Omega6.php.*

controlling arthritis. Arachidonic acid is found in meat, eggs, and dairy sources. Arachidonic acid important to health and considered as a conditionally essential fatty acid. Arachidonic acid paradoxically, can also promote inflammation. Inflammation is linked to arthritis and heart disease. This negative consequence is suggested to result from a diet in which the content of ω6 fatty acids is much greater than that of ω3 fatty acids (Table 4.3).

Alpha linolenic acid (ALA), an omega 3 fatty acid, is also synthesized from linoleic by the addition of a third double bond. Substantial levels of α-linolenic are found in canola and soy oils and green leafy vegetables (e.g., spinach). Alpha linolenic acid serves as a synthetic precursor of eicosapentaenoic acid (EPA) and subsequently docosahexaenoic acid (DHA). EPA and DHA omega 3 fatty acids containing 20 and 22 carbons, respectively. The number of double bonds in EPA and DHA is 5 and 6, respectively. Omega 3 fatty acids can be found in foods such as green leafy vegetables, flax seeds, canola oil. Omega 3 fatty acids serve as a source of eicosanoids which are found at significant levels in oily fish (krill) and algae. Eicosanoids are signaling molecules that physiologically function in inflammation, allergy, and other immune responses. Additionally, DHA occurs at high levels in retinal membranes and is important to maintaining normal vision. DHA is also found in neuronal membranes and its dietary deficit is linked to deficiencies in memory.

Fatty Acid Composition of Food Lipids: The fatty acid composition of acylglycerols is important to nutritional quality. It is generally agreed that diets containing a high level of polyunsaturated fatty acids are healthier than one high in saturated fatty acids. The ratio of omega 6 to omega 3 fatty acids is also important, it is suggested that an an optimum ratio is 4:1. The graphical representation of fatty acid composition in these sources better illustrates these differences as shown in Fig. 4.13. In this figure, the relative amount of saturated, monounsaturated, and polyunsaturated fatty acid (as a percent of the total) is shown by a colored bar for each source listed. This graphical presentation shows at a glance why canola oil is superior in nutritional quality to corn or soybean oil.

Summary

The review of food lipids presented here describes their chemical and nutritional properties. In foods, lipids are a source of vitamins, energy, flavor, and color. Above all other food components, lipids are important to the quality of chocolate and ice cream. In food, lipid chemistry is responsible for three major types of reactions hydrolysis, hydrogenation and oxidation. Hydrolysis of fatty acids in acylglycerols, is a source of desirable and undesirable flavors. Enzymes derived from fermentation release fatty acids and transform milk's bland taste into the desirable flavors of cheese and other dairy products. Conversely, the taste sour milk is also due to a fatty acid (butyric). The chemical process of hydrogenation has been widely used in converting oils into fats that have been a staple in processed foods. Unfortunately, hydrogenation of oils also creates trans fatty acids that are linked to adverse health effects, (i.e., cardiovascular disease). In recognition of their effects on health, FDA has issued a ban on trans fats in food, effective in 2019. Oxidation is a natural chemical process occurring to unsaturated lipids (fatty acids). The presence of pro-oxidant minerals, exposure to oxygen, and UV light, initiate reactions that cause off flavors, degrade natural colors, and destroy vitamins. Additionally, lipid oxidation produces potentially toxic compounds, (e.g., malondialdehyde). As a result, the benefits of antioxidants and their chemistry have been widely recognized. Nutritionally, lipids are high in nutrient density proving more calories per gram than any other food component. We are unable to synthesize alpha linoleic and linoleic fatty acids and thus they are considered to be essential. Others, such as arachidonic, eicosapentaenoic, and docosahexaenoic are also important to health. Several vitamins (A, D, E, and K) are lipid-based molecules needed to prevent conditions caused by deficiency.

Glossary

Acyglycerol A lipid molecule formed by the linkage of glycerol with 1 to 3 fatty acids

Aliphatic Compounds consisting of only carbon and hydrogen atoms.

Amphiphilic Molecules containing both hydrophilic and aliphatic (or lipophilic) properties.

Antioxidant Substance that inhibits or stops free radical reactions.

Aromatic A substance consisting mostly of carbon and hydrogen atoms. Typically small compounds with atoms arranged in ring structure

Colloid Microscopic dispersion of one substance uniformly suspended in another. Milk, for example, is a colloidal suspension of lipid in water.

Eicosanoids Signaling molecules derived from polyunsaturated fatty acids.

Electrophilic A molecule or group with the tendency to attract electrons

Endothermic A process that absorbs heat from the environment as a substance changes in state. For example, water changing from liquid to gas.

Free Radical Scavenger A substance that stabilizes a free radical by donating a hydrogen to an atom or molecule with an unpaired electron.

Free radical An atom, ion, or molecule that has unpaired valence electrons. Free radicals must gain or lose an electron and achieve a stable configuration

Hydrogenation Chemical reaction using hydrogen and a catalyst to convert carbon-carbon double bonds to single bonds. The reaction is used commercially in making margarine from oils.

Hydrolytic rancidity Chemical or enzymatic process resulting in the release of fatty acids.

Hydroperoxide A reactive substance having the general structure LOOH, formed by oxidation of unsaturated lipids.

Interesterification Chemical process by which the composition of fatty acids in acylglycerols, are reordered

Ketosis Metabolic condition characterized by elevated levels of ketones in the blood. It can occur in diets containing very low levels of carbohydrate.

Lipase Enzyme that catalyzes the hydrolysis of ester bonds in acylglycerol molecules resulting in the liberation of free fatty acids.

Melting point The temperature at which a solid becomes liquid. The change from solid to liquid is an endothermic process meaning that heat is absorbed from the environment

Micelle A spherical assembly of surfactant molecules with polar and non-polar groups facing their respective environments.

Oxidation Chemical process involving the loss of an electron from an atom, molecule, or compound.

Phospholipid An acylglycerol containing an esterified phosphate group giving the molecule both polar and non-polar regions.
Prooxidant Substance that promotes or accelerates oxidation of another.
Shortening A type of fat predominantly containing saturated fatty acids. Its major use is in baking applications pie crusts and pastry
Surfactant A substance containing both polar and nonpolar groups.
Tempering A process used to create the desired crystalline form of acylglyceride (fat).

References

ADM Cocoa Products Manual, 1999. The DeZaan Cocoa Products Manual. 1999. ADM Cocoa B.V.
Beckett, S.T., 2000. The Science of Chocolate. The Royal Society of Chemistry, Cambridge CB4OWF. UK.
Choe, E., Min, D.B., 2006. Chemistry and reactions of reactive oxygen species in foods. Crit. Rev. Food Sci. Nutr. 46, 1—22.
Gallicchio, L., Boyd, K., Matanoski, G., et al., 2008. Carotenoids and the risk of developing lung cancer: a systematic review. Am. J. Clin. Nutr., 88 (2), 372—383.
Genser, B., Silbernagel, G., De Backer, G., Bruckert, E., Carmena, R., Chapman, M.J., Deanfield, J., Descamp, O.S., Reitzschel, E.R., Dais, K.C., Marz, W., 2012. Plant sterols and cardiovascular disease: a systematic review and meta-analysis. Eur. Heart J. 33, 444—451.
IUPAC, 1997. Compendium of Chemical Terminology, second ed. International Union of Pure and Applied Chemistry.
Johnson, D.R., Decker, E.A., 2015. The role of oxygen in lipid oxidation reactions: a review. Ann, Review Food Sci.Tech. 6, 171—190.
Katan, M.B., Mensink, R.P., Zock, P.L., 1995. *Trans* fatty acids and their effect on lipoproteins in humans. Annu. Rev. Nutr. 15, 473—493.
Lucymara, F., Julliane, A.L., Melo, T.A., et al., 2012. DNA damage by singlet oxygen and cellular protective mechanisms. Mutation Res/Reviews in Mutation Res. 751, 15—28.
Marventano, S., Kolacz, P., Castellano, S., Galvano, F., Buscemi, S., Mistretta, A., Grosso, G., 2015. A review of recent evidence in human studies of n-3 and n-6 PUFA intake on cardiovascular disease, cancer, and depressive disorders: does the ratio really matter? Int. J. Food Nutr. Sci. 66, 611—622.
McCance, Widdowso, 2002. The Composition of Foods, sixth ed. Foods Standards Agency and Royal Society of Chemistry.
Mozaffarian, D., Katan, M.B., Ascherio, A., Stampfer, M.J., Willett, W.C., 2006. Trans fatty acids and cardiovascular disease. N Engl J. Med. Apr 13 (354), 1601—1613.
Załęski, A., Banaszkiewicz, A., Walkowiak, J., 2013. Butyric acid and irritable bowel syndrome. PZ Gastroenterol. 8:350-353 (look up authors) the Nomenclature of Lipids. Recommendations, 1976. Eur. J. Biochem. 79 (1), 11—21, 1977.

Further reading
Wikipedia

Fat: the facts National Health Service, UK https://www.nhs.uk/live-well/eat-well/different-fats-nutrition/[Accessed April 2019].
Fatty acid nomenclature (summary) https://en.wikipedia.org/wiki/Fatty_acid#Nomenclature.

Forouhi, N.G., Krauss, R.M., Taubes, G., Willett, W., 2018. Dietary fat and cardiometabolic health: evidence, controversies, and consensus for guidance. Br. Med. J. Int. Ed. 361, k2139.
List of unsaturated fatty acids https://en.wikipedia.org/wiki/List_of_unsaturated_fatty_acids.
Liu, A.G., Ford, N.A., Hu, F.B., Zelman, K.M., Mozaffarian, D., Kris-Etherton, P.M., 2017. A healthy approach to dietary fats: understanding the science and taking action to reduce consumer confusion. Nutr. J. 16 (1), 53.

Review questions

1. Describe the structure of fatty acid and acylglycerol molecules. Give an example of each.
2. What are Short Chain Fatty (SCF) acids? How are they important to food and to health?
3. What is an essential fatty acid? Give examples of omega 3 and omega 6 fatty acids. What foods are good sources?
4. Describe how fat crystalline structure affects melting point.
5. What is the fat in real chocolate? How does it affect eating quality?
6. Describe the process of tempering used in making chocolate.
7. What is "Dutched" cocoa? What are some advantage of the process?
8. What is margarine and how is it made?
9. What are LDL and HDL? Why are they important?
10. Describe the structure of a phospholipid. Why is egg yolk good for making salad dressings?
11. What is the difference between shortening and butter? Which is better for biscuits?
12. What does the term "partially hydrogenated" mean on food labels?
13. Why are phospholipids good emulsifiers?
14. What is the difference between oleic and elaidic fatty acids?
15. What is hydrolytic rancidity? What is the chemistry responsible for it?
16. What type of chemical reaction causes lipid oxidation?
17. Which lipids are most susceptible to oxidation?
18. List the major factors influencing the rate of lipid oxidation.
19. What is malondialdehyde and how is it important to food?
20. Name three nutritionally important substances that are destroyed by lipid oxidation.
21. What is an antioxidant? How do they control lipid oxidation?
22. Describe ascorbic acid's function in controlling free radical reactions.
23. Why is the combination of ascorbic acid and tocopherol effective in controlling oxidation in processed meats?

24. What is EDTA and how does it inhibit lipid oxidation?
25. What are ROS? Give an example of their nutritional importance and impact on food quality.
26. Give examples of ω6 and ω3 fatty acids and foods that are comparatively high in each form.
27. How do carotenoids contribute to the nutritional quality of food?
28. Which of the cooking oils listed in Fig. 4.13, would you rate as best an worst on the basis of nutritional quality. Defend your answer.

CHAPTER FIVE

Vitamins and minerals

Learning objectives

This chapter will help you describe or explain:

- Properties of vitamins and minerals
- Food sources of vitamins and minerals
- Biological function of vitamins and minerals
- Nutritional labeling of foods

Introduction

Vitamins and minerals have important, unique, and specific functions in the body. Vitamins are small organic molecules required at low amounts in the diet to maintain optimal health and growth. The discovery of vitamins was driven by investigations of diseases that were ultimately caused by diets deficient in these compounds. The word vitamin is derived from the term, vital-amine, coined early in the twentieth century. Naming these compounds as vital-amines was based on the mistaken belief that they all contained an amine (NH_2) group. Vitamins are not used by the body as a source of energy, but serve supporting roles in biochemical and physiological processes. While vitamins are traditionally thought of as prevention for deficiency diseases, they also have a variety of chemical properties that support the health of cells, tissues, and organs. For example, vitamin C (ascorbic acid) is noted for its antioxidant chemical activity. Vitamins A and E are potent inhibitors of free radical reactions. Vitamins are traditionally divided into water-soluble and fat-soluble groups. Water-soluble vitamins include all the B vitamins, plus vitamin C. Once absorbed from the gut, they enter the blood stream and any excess is eliminated in the urine. In contrast, fat-soluble vitamins (A, D, E, and K) known by the acronym ADEK, can be stored in the liver or fat depots of the body.

Minerals, such as calcium, magnesium, and iron, are ultimately derived from the soil and supplied to humans via the food chain and water. Microorganisms and plants are primary absorbers of minerals from the environment,

Introduction to the Chemistry of Food
ISBN: 978-0-12-809434-1
https://doi.org/10.1016/B978-0-12-809434-1.00005-0

they are subsequently transfered to us through food. In animals, minerals function as cofactors for enzymes and contribute to bone strength. They are essential to metabolism, muscle contraction, and blood pressure regulation. Minerals are grouped into two classes: major- and trace-elements. Major minerals are represented by calcium, phosphorous, potassium, sodium, and magnesium. The trace element group is represented by cobalt, copper, chromium, iodine, iron, manganese, molybdenum, selenium, and zinc. The human requirement for trace elements is small, typically 1 mg per day or less. A reference guide to vitamins and minerals in food can be found in the work of Berdanier et al. (2013).

This chapter includes questions that will help you explore and better understand the importance of vitamins and minerals, especially related to the chemistry of food.

- What are micronutrients?
- What does the Nutrition Facts label tell me?
- Salty Questions: Where does most sodium come from? What foods contain the most salt? How is sea salt different from ordinary table salt?
- How is vitamin activity lost in processed food?
- Can vitamin supplements be harmful?

What are micronutrients?

Micronutrients are dietary components required in small amounts to promote growth and development, maintain health, and support cognitive function. The term micronutrient obviously includes vitamins and minerals, but also extends to non-vitamin substances. For example, essential amino acids, essential fatty acids, and antioxidants belonging to the phenolic and carotenoid families of compounds are also included in the definition of micronutrient (Vitamins and minerals, OSU Micronutrient Information Center, 2018).

Nutritional labeling and food regulation

The importance of vitamins and minerals to health is undisputed, but it is difficult to know the level of these nutrients in processed food. The steps required to transform raw commodities into packaged foods invariably causes loss of vitamins and minerals. In an effort to provide consumers with accurate and easy to understand nutrient information, the FDA mandates the inclusion of a nutrition facts label on most foods in the United States. The European Union, China, India, other countries have similar food labeling requirements.

What does the nutrition facts label tell me?

FDA requires most packaged foods to prominently display nutrient information on Nutrition Facts labels (Fig. 5.1) (Food Labeling Guide FDA, 2018). The requirement for a nutrition facts label was established by Nutrition Labeling and Education Act of 1990. In addition to the nutrition facts label, the law gave FDA authority to require that health claims be substantiated. Nutrition Facts label lists in bold print the calories per serving and per container at the top. Major and minor nutrients are listed as percent of daily value (%DV) per serving and in the whole container. Nutrients, such as protein, carbohydrate, fat, vitamins, minerals, and cholesterol, are listed from top to bottom in order of decreasing content. Protein, carbohydrate, sodium and total fat are always listed, but other components are omitted if they are absent or below a threshold of 0.5 g per serving. Fat content is further specified on the label as total fat, saturated fat, and trans-fat. A provision in FDA's nutrition labeling regulation allows foods with less than 0.5% trans-fat per serving to be labeled as zero. Carbohydrates are listed on the label as total carbohydrate, dietary fiber, and total sugars. The Nutrition Facts label shown in Fig. 5.1 represents the recently adopted version (June 2019).

Nutrition Facts

2 servings per container
Serving size   **1 cup (255g)**

	Per serving		Per container	
Calories	**220**		**440**	
	% DV*		% DV*	
Total Fat	5g	6%	10g	13%
Saturated Fat	2g	10%	4g	20%
Trans Fat	0g		0g	
Cholesterol	15mg	5%	30mg	10%
Sodium	240mg	10%	480mg	21%
Total Carb.	35g	13%	70g	25%
Dietary Fiber	6g	21%	12g	43%
Total Sugars	7g		14g	
Incl. Added Sugars	4g	8%	8g	16%
Protein	9g		18g	
Vitamin D	5mcg	25%	10mcg	50%
Calcium	200mg	15%	400mg	30%
Iron	1mg	6%	2mg	10%
Potassium	470mg	10%	940mg	20%

* The % Daily Value (DV) tells you how much a nutrient in a serving of food contributes to a daily diet. 2,000 calories a day is used for general nutrition advice.

Fig. 5.1 Nutrition facts label.

Dietary intake recommendations

Dietary reference intake (DRI) is a system of nutrition recommendations intended for use as guidelines to assess the adequacy of nutrient intake and lower the risk of excessive intake (Nutrient recommendations NIH, 2018). DRI is used in food labeling, governmental food assistance programs, and the military. DRI values are also used to broaden and update other measures of dietary intake, such as Recommended Dietary Allowance (RDA), Adequate Intake (AI), Estimated Average Requirement (EAR), and Tolerable Upper Intake Level (UL). A history of dietary reference intake (DRI) in the United States and Canada is provided by Murphy et al. (2016). Recommended Dietary Allowance (RDA) is defined as the nutrient intake level that is sufficient to meet the nutrient requirements of nearly all (97%) healthy individuals by gender and life stage. RDA values are established by the Food and Nutrition Board of the Institute of Medicine of the National Academies. These were initially set for protein and eight vitamins in 1941 in response to World War II defense needs and international food relief efforts. Expressing nutrient needs as RDA on food labels was replaced in 1997 by percent daily value (%DV) that is based on a 2000 calorie per day diet for healthy individuals. The change to %DV also broadened the scope of nutrients included on nutrition labels. %DV now provides a threshold for use of terms such as high-fiber or low fat on food labels. For example, a claim for high fiber means that the food must contain at least 20% of the DV. Similarly, a low-fat claim means that the food contains no more than 5% of the DV. Adequate Intake (AI) is the recommended average daily intake based on experientially determined estimates of nutrient intake by a group of healthy people. AI is used when there is insufficient evidence to determine an estimated average requirement. Estimated Average Requirement (EAR) is defined as the average daily nutrient intake that is estimated to meet the requirements of 50% of the healthy individuals in a life stage and gender group. Tolerable Upper Intake Level (UL) is defined as the highest average daily nutrient intake that is likely to pose no risk of adverse effects to almost all individuals in the general population.

Taking a closer look at vitamins important to diet and health

Table 5.1 (Vitamins) summarizes the biological function, common sources, and daily value (DV) for the vitamins discussed in the text. It can

Table 5.1 Vitamins.

Vitamin	Function	Sources	Daily value
B1 Thiamin	Metabolism & Nervous system	Beans, peas, whole grains, nuts, pork	1.5 mg
B2 Riboflavin	Metabolism, growth & development, red blood cell formation	Egg, meat, poultry, milk, mushrooms, sea food, spinach	1.7 mg
B3 Niacin	Cholesterol production, metabolism, digestion	Beans, beef, pork, poultry, sea foods, whole grains,	10 mg
B5 Pantothenic Acid	Fat metabolism, hormone production, red blood cell formation, Nervous system function	Avocados, beans, peas, eggs, milk, mushrooms, poultry, sea food sweet potatoes, whole grains	10 mg
B6 Pyridoxal Phosphate	Immune & nervous system function, metabolism, red blood cell formation	Chickpeas, fruits, potatoes, salmon	2 mg
B7 Biotin	Metabolism & Nervous system function	Avocados, cauliflower, fruits, eggs, milk, pork, salmon, whole grains	300 μg
B9 Folic Acid	Prevention of birth defects, protein metabolism, red blood cell formation	Avocados, asparagus, beans & peas, green leafy vegetables, orange juice	400 μg
B12 Cobalamin	Metabolism, Nervous system function, red blood cell formation	Dairy products, Eggs, meat, poultry, seafood	6 μg
Vitamin C Ascorbic acid	Antioxidant, collagen & connective function, immune function	Broccoli, Brussel sprouts, cantaloupe, fruits & fruit juices, peppers	60 mg
Vitamin A, Retinol	Growth & development, immune function, reproduction, red blood cell & skin formation	Cantaloupe, carrots, dairy products, eggs, green leafy vegetables, pumpkin, sweet potatoes	5000 IU
Vitamin D_3 Calcitriol	Blood pressure regulation, bone growth, hormone production, calcium balance, immune & nervous system function	Eggs, fish, fortified dairy products	400 IU

(Continued)

Table 5.1 Vitamins.—cont'd

Vitamin	Function	Sources	Daily value
Vitamin E, Alpha Tocopherol	Antioxidant, formation of blood vessels, immune system function	Green leafy vegetables, nuts & seeds, peanuts, vegetable oils	30 IU
Vitamin K1, Phylloquinone	Blood clotting, bone strength	Green leafy vegetables,	80 μg

Adapted from Vitamins (FDA, 2018).

be referred to as each vitamin is presented. The biological importance and supplementation section for each vitamin provides a brief overview of the physiological function that they support. The means of loss describes prominent ways in which vitamin loss occurs during the transition of farm to fork.

B vitamins

Most B vitamins are essential to health because of their role in metabolism. Generally, B vitamins function as co-enzymes in metabolic processes, such as glycolysis (utilization of glucose to make energy) and degradation (recycling of amino acids for energy) The eight members of this group (B_1, B_2, B_3, B_5, B_6, B_7, B_9, and B_{12}) are often referred to as the B complex and are commonly available as dietary supplements.

Vitamin B_1, thiamin

Thiamin (Table 5.1) can be obtained from a variety of plant and animal sources, including legumes (lentils), nuts (pecans), and roasted pork. Rice, especially the unpolished brown variety, is a good source of thiamin. Wheat germ breakfast cereal is a rich source of this vitamin. Thiamin daily value varies with age and gender and the amounts for adult men and women are about 1 and 2 mg per day, respectively. Infants and children have a lower RDA for thiamin (Vitamins and Minerals OSU MIC, 2018).

Biological importance and supplementation

Thiamin functions in carbohydrate and branched chain amino acid metabolism (Fig. 5.2). It is a coenzyme (an essential factor for enzyme activity) called thiamin pyrophosphate (TPP). TPP is essential to the activity of several

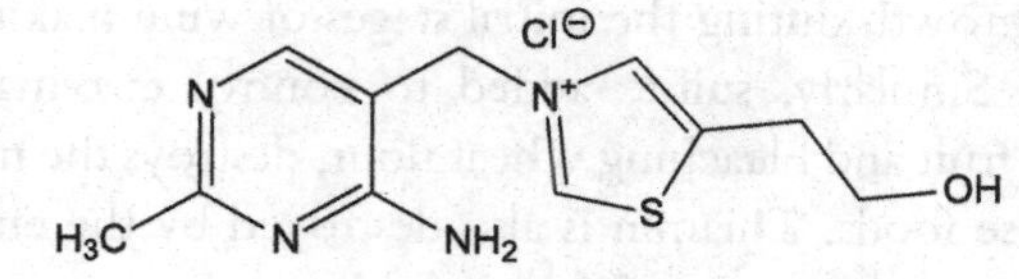

Vitamin B1, Thaimin

Fig. 5.2 Vitamin B$_1$, thiamin.

enzymes in the glycolysis pathway responsible for metabolizing glucose and making energy. TPP is also a cofactor in the enzymatic breakdown of branched chain amino acids and fatty acids. Both of these are ultimately used as sources of energy production. Thiamin deficiency has multiple adverse effects in a condition known as Beriberi. Several recognized forms of Beriberi are recognized that affect gastrointestinal digestion, brain, and muscle function. Gastrointestinal symptoms of deficiency include severe abdominal pain and vomiting that result from the altered metabolism of glucose. When TPP is insufficient, the breakdown of glucose is diverted to a secondary product (lactate) that causes the unpleasant symptoms. Gastrointestinal Beriberi is often found in diabetics. Chronically elevated glucose level occurring in diabetics reduces thiamin uptake from the gut and makes them prone to this disease. Similarly, thiamin deficiency can cause abnormal motor function and a type of dementia known as cognitive Beriberi. The later deficiency condition is associated with severe alcoholism. It occurs because ethanol metabolism creates toxic products (acetaldehyde) and affected individuals may not consume enough thiamin-containing foods. An advanced form of Beriberi affects the heart and circulatory system, particularly in diabetic individuals with elevated blood glucose levels. Heart failure can result for those with chromic thiamin deficiency.

Means of loss

Thiamin destruction occurs in several ways. Foods, such as coffee and tea, contain anti-thiamin factors that inactivate it. Tea, for example, is especially rich in polyphenolic compounds called tannins, that irreversibly bind thiamin and make it unavailable. Chewing tea leaves, as practiced in some cultures, releases tannins into the digestive system causing a loss of vitamin activity. Thiamin is moderately stable under conditions of acidic or neutral pH. However, it can be chemically destroyed by interactions with other food components and by processing operations. For example, sulfite added to control bacterial growth during the initial stages of wine making destroys its thiamin content. Similarly, sulfite added to control enzymatic browning reactions in dried fruit and bleaching wheat flour, destroys the natural content of thiamin in these foods. Thiamin is also destroyed by the enzyme thiaminase, a natural component of fish, shellfish, and some insect species. Thiaminase is highest in raw fish (sushi) and some insects (silkworms). Eating large amounts of these foods in uncooked form can cause deficiency symptoms.

Vitamin B₂, riboflavin

Riboflavin (Table 5.1) occurs in a variety of foods including milk, cheese, almonds, meat, and yeast. Spinach, asparagus, and broccoli also contain significant amounts of riboflavin. Cow's milk represents one of the richest sources of riboflavin and contains approximately 0.5 mg/cup. Wheat flour in the United States is fortified with riboflavin, thiamin, niacin, and iron. This practice provides some, but not all, of the %DV for these vitamins. Riboflavin is one component of the B vitamin complex commonly found in dietary supplements. The daily value for riboflavin is 1.7 mg per day for both men and women (Vitamins and Minerals OSU MIC, 2018).

Biological importance and supplementation

Riboflavin's (Fig. 5.3) most important biological role is as precursor of the coenzyme molecules, Flavin Adenine Dinucleotide (FAD) and Flavin

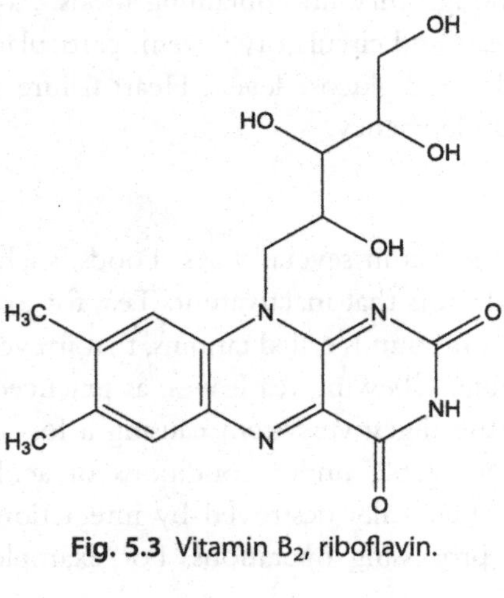

Fig. 5.3 Vitamin B₂, riboflavin.

Adenine Mononucleotide (FMN). Coenzymes are molecules required for enzymes to perform their catalytic functions. The complex of FAD or FMN with enzymes is called a flavoenzyme. They are essential to catabolism (breakdown) of basic nutrients such as carbohydrates, proteins, and lipids in the production of cellular energy. FAD functions to accept or lose electrons in reduction–oxidation (REDOX) reactions occurring in energy producing pathways. Riboflavin participates in the pathway that converts the amino acid tryptophan to niacin (vitamin B_3). Flavoenzymes affect the production of other B vitamins (folate, and vitamin B_6) required for synthesis of nucleic acids and the oxygen carrying hemoglobin protein. Riboflavin deficiency is rarely found without a parallel deficiency in vitamin B_6 and niacin. Anemia, sore throat, and skin inflammations are common symptoms of riboflavin deficiency. Chronic riboflavin deficiency is more serious because it is linked to increased risk of cardiovascular disease and cancer. Riboflavin deficiency is also linked to preeclampsia in pregnant women. This condition can escalate to eclampsia and cause severe bleeding and death. Riboflavin supplementation at very high levels, 400 mg per day, has been suggested to reduce the frequency and severity of migraine headache (Boehnke et al., 2004). There appears to be no adverse effect of consuming riboflavin at this level, other than a pronounced yellow color in the urine.

Means of loss

The greenish fluorescent color in a glass of milk held up to the light is due to its riboflavin content. However, exposure to light can cause substantial loss (e.g., 70%) of riboflavin activity. Its light absorbing properties can also generate free radicals that react with milk lipids and cause off flavors. Riboflavin is also destroyed in this process. This chemistry causes the defect known as sunlight flavor. Heating i.e., boiling also causes a substantial (up to 75%) loss of riboflavin.

Vitamin B_3 niacin

Niacin (Table 5.1) and nicotinamide are two forms of nicotinic acid (vitamin B_3). Nicotinic acid is the predominant form in plants and nicotinamide is the predominant form in animals. The acid is a small water-soluble molecule containing a nitrogen atom in its ring structure (Fig. 5.4). While similar in pronunciation, neither form of the vitamin is related to nicotine, the compound found in tobacco. Meats and fish, such as chicken, turkey, beef, salmon, and tuna, represent the greatest nutritional sources of

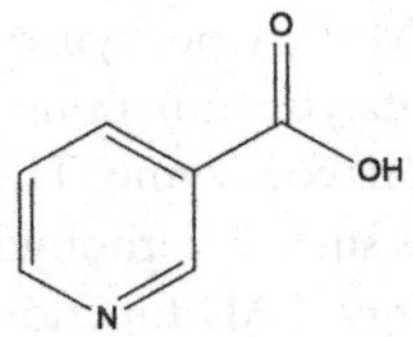

Fig. 5.4 Vitamin B_3, niacin.

niacin. Cereals, peanuts, mushrooms, green leafy vegetables, and sun flower seeds are also good plant food sources of niacin. The daily value for niacin in adult males and females is 20 mg/day. The term NE (niacin equivalent) is defined as the amount of preformed niacin plus the amount derived from conversion of its tryptophan precursor. Most of the requirement for niacin must be supplied from the diet because of the inability ability to synthesize it.

Biological importance and supplementation

Biologically, the nicotinamide form of niacin is converted into two cofactors: nicotinamide adenine dinucleotide (NAD) and its phosphorylated form, nicotinamide adenine dinucleotide phosphate (NADP). NAD is utilized in the catabolism of proteins, carbohydrates, and lipids into energy; NADP is utilized in the synthesis of fatty acids. NAD also participates as a coenzyme in reactions in repairing DNA strand breaks and for its transcription (making a copy of DNA). This activity suggests that niacin has a role in the prevention of cancer. A severe deficiency of niacin in the diet causes pellagra. This disease can be manifested by skin irritations, gastrointestinal, and neurological symptoms. Left untreated, severe niacin deficiency can result in death. Pellagra results from chronic malnutrition that can occur in ways beyond a simple lack of calories. For example, diets that are low in meat, but supplemented with starchy, low protein foods like corn, can cause deficiency disease. Corn contains niacin, but is nutritionally unavailable unless first treated with alkali. Treating ground corn with lye is a common practice used in making tortilla flour. This makes some bound niacin available for uptake.

Means of loss

Niacin is stable to most food processing conditions with exception of prolonged cooking. For example, boiling or stewing can reduce its level by 50%. Much of that loss is due to leaching from the food into the cooking water because of niacin's water-soluble properties.

Vitamin B₅, pantothenic acid

Pantothenic acid is a small water–soluble molecule essential to all forms of life (Fig. 5.5). It is found in most foods. Its highest level occurs in beef liver, but is also plentiful in fish, shellfish, pork, milk, egg yolk, legumes, and yeast (Table 5.1). Bacteria in the large intestine are able to synthesize pantothenic acid, but additional amounts are needed to meet minimum requirements. Individual variation in the level of pantothenic acid synthesized by gut bacteria complicates determination of an RDA value. Therefore, the adequate intake (AI) value is used for its recommended dietary level. The daily value is 10 mg/day for both men and women. Pantothenic acid is a chiral molecule whose biological activity depends on the D enantiomer. The L form has no activity and may infer with the activity of the D form.

Biological importance and supplementation

Pantothenic acid is a precursor of coenzyme A (Co-A) that functions in the synthesis and oxidation of fatty acids. Like other coenzymes, Co-A is also essential for the metabolism of glucose. The glycolysis pathway in combination with the citric acid cycle results in the production of substantial amounts of energy. Co-A is indirectly involved in other pathways responsible for synthesis of nucleic acids. All genomes sequenced thus far encode for enzymes requiring Co-A. Pantothenic acid is therefore essential to all life forms. It occurs in both free and bound forms. The bound form of pantothenic is made available following enzymatic hydrolysis in the gut. Deficiency of pantothenic acid is extremely rare because of its availability in a variety of foods. The primary evidence of deficiency in humans comes from the well know case of WWII prisoners in the Philippines and Japan. Prisoners were severely malnourished and experienced symptoms (e.g., burning sensations in the feet) that were reversed by administration of pantothenic acid. Animals models of pantothenic acid deficiency reported a

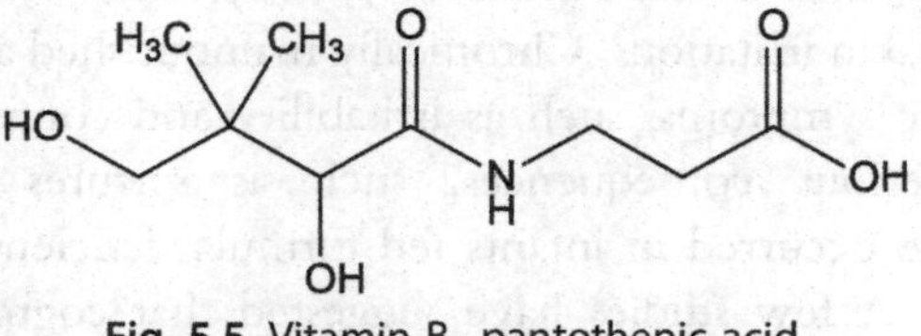

Fig. 5.5 Vitamin B₅ pantothenic acid.

range of symptoms, including anemia and adrenal gland disorders affecting metabolic function. The diversity of symptoms associated with pantothenic acid deficiency reflects its wide biological importance. In contrast, taking pantothenic acid supplements (e.g., pantothenol, pantetheine) are suggested to lower serum cholesterol levels (Vitamins and Minerals OSU MIC, 2018).

Means of loss

Pantothenic acid is stable in the pH range of 5—7. It is stable to boiling and other forms of wet heat. It has been found that roast beef retained approximately 90% of its pantothenic acid after cooking. Acidic or basic pH conditions, however, increase its rate of destruction. The most significant loss of pantothenic acid in food results from leaching into cooking water.

Vitamin B$_6$, pyridoxine, pyridoxamine, pyridoxal

Vitamin B$_6$ (Table 5.1) exists in three chemical forms: pyridoxine, pyridoxamine, and pyridoxal. All forms have vitamin B$_6$ activity. The basic structure common to all forms is a six-membered ring with a nitrogen atom (Fig. 5.6A). Three forms of vitamin B$_6$ are created when other chemical groups are substituted on the molecule as shown by the circled areas in Fig. 5.6A. Specifically, substitution of hydroxyl, aldehyde, amine, or amine groups in the figure creates pyridoxine, pyridoxamine, and pyridoxal forms of vitamin B$_6$. Addition of phosphate ester to pyridoxal creates pyridoxal phosphate, the most biologically important form (Fig. 5.6B). Vitamin B$_6$ can be obtained from a variety of plant and animal foods. Animal sources typically contain higher levels of vitamin B6 (as pyroxamine) compared to plant sources (as pyridoxal). Turkey and salmon contain the highest level at about 0.7 mg per 3oz serving. Whole grain cereals, spinach, and nuts are also good plant-based sources of vitamin B$_6$ (Table 5.1). The daily value for vitamin B$_6$ is approximately 2 mg/day for men and women. Most of vitamin B$_6$ in animal sources is bound to protein. Deficiency disease is rare for vitamin B$_6$ because of its availability. Early disease symptoms include mouth sores and skin irritations. Chronically malnourished adults can experience neurologic symptoms, such as irritability and confusion. Unfortunately, more serious consequences, such as seizures and cognitive impairment, have occurred in infants fed formula deficient in vitamin B$_6$ (Clayton, 2006). A few studies have suggested that cognitive decline in the elderly may be linked to inadequate levels of vitamin B$_6$ and other B vitamins (Selhub et al., 2000).

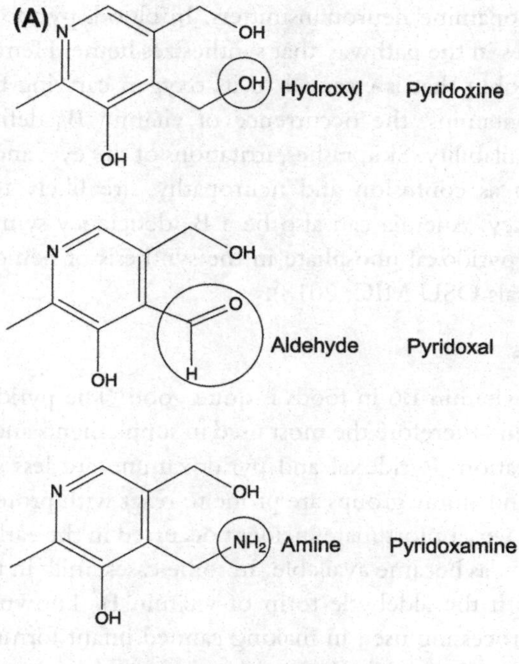
(A)
N
OH
OH
Hydroxyl
Pyridoxine
OH
N
OH
O
Aldehyde
Pyridoxal
H
OH
N
OH
NH2
Amine
Pyridoxamine
OH

Vitamin B6 Forms

Fig. 5.6A Vitamin B_6 forms.

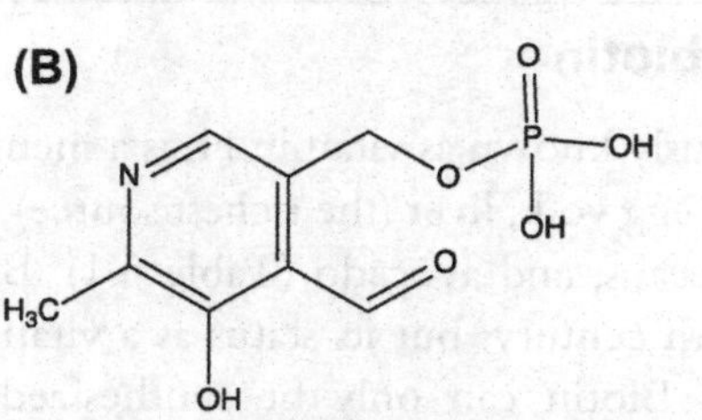
(B)
O
P
OH
O
N
OH
O
H3C
OH

Fig. 5.6B Vitamin B_6 pyridoxal phosphate.

Biological importance and supplementation

Pyridoxal phosphate (Fig. 5.6B) is a coenzyme essential to the activity of many enzymes. It is estimated that pyridoxal phosphate serves as an essential cofactor in 4%–5% of all enzymes in the body. Pyridoxal phosphate functions in amino acid synthesis and fatty acid metabolism. Pyridoxal phosphate is required for breakdown of stored glycogen in muscle. Released glucose provides muscle with fuel to make the energy needed for contraction. In the brain, pyridoxal phosphate activates enzymes that synthesize the serotonin and dopamine neurotransmitters. In blood, pyridoxal phosphate activates enzymes in the pathway that synthesizes heme. Heme is a component of hemoglobin that is essential to its oxygen carrying function. Like some other B vitamins, the occurrence of vitamin B_6 deficiency is rare because of its availability. Skin rashes, irritations of the eye, and neurological symptoms, such as confusion and neuropathy, are likely to result from chronic deficiency. Anemia can also be a B_6 deficiency symptom because of the need for pyridoxal phosphate in the synthesis of hemoglobin (Vitamins and Minerals OSU MIC, 2018).

Means of loss

The stability of vitamin B6 in foods is quite good. The pyridoxine form is the most stable and therefore the most used in supplements and food enrichment or fortification. Pyridoxal and pyridoxamine are less stable because their aldehyde and amine groups are prone to react with proteins, especially when heated. A very unfortunate incident occurred in the early 1950s when infant milk formulas became available. In some cases, milk in these formulas was fortified with the aldehyde form of vitamin B_6 known as pyridoxal. High thermal processing used in making canned infant formula resulted in destruction of vitamin B_6. The lack of this vitamin caused permanent brain damage to a number of infants as a result of this mistake.

Vitamin B_7 biotin

Biotin, previously known as vitamin H, is a member of the B vitamin family that occurs in egg yolk, liver (the richest source), salmon, whole grain cereals, soybeans, pecans, and avocado (Table 5.1). Biotin was discovered early in the twentieth century, but its status as a vitamin was not accepted for several decades. Biotin can only be synthesized by microorganisms such as brewer's yeasts and some bacteria. Some biotin is synthesized by intestinal bacteria, but its contribution is small. A daily value for biotin is about 300 µg/day (µg = micrograms) for most adults. Pregnant women are advised to take a supplement containing biotin and other B vitamins to prevent deficiencies that are known to result in birth defects.

Biological importance and supplementation

Biotin is required for cell growth and nutrient metabolism in all organisms. Specifically, biotin is an essential cofactor in fatty acid synthesis. This synthesis pathway is activated when biotin is covalently linked to key enzymes

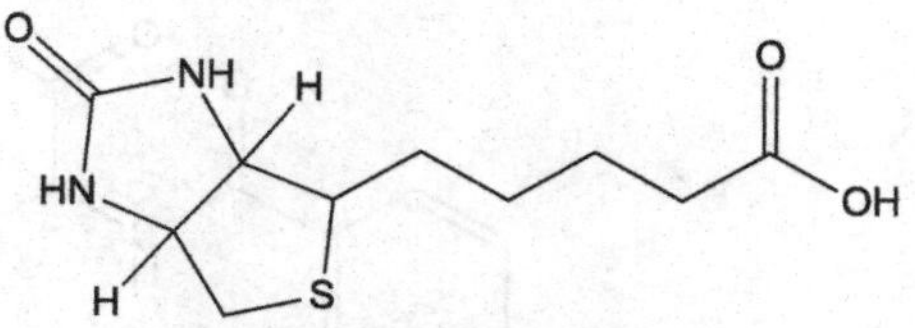

Fig. 5.7 Vitamin B$_7$ Biotin.

through a process called biotinylation. The pathway is turned off when biotin is removed (Fig. 5.7). Biotin is also a cofactor in the process known as gluconeogenesis. This pathway occurs in the liver and is responsible for generating glucose from other metabolites and contributes to the organism's energy pool. Additionally, biotin is required for the breakdown (catabolism) of branched chain essential amino acids, such as leucine, isoleucine, and valine. The metabolism of branched chain amino acids is yet another way biotin contributes to the overall energy pool.

Means of loss

Biotin deficiency is rare, but it can occur under conditions of severe malnutrition or parenteral feeding in which there is insufficient biotin. Deficiency can also occur in individuals who consume raw eggs. Hens eggs contain an egg white protein called avidin that irreversibly binds biotin. Cooking eggs to the point where the whites become solid denatures avidin and destroys its ability to bind biotin. Biotin deficiency symptoms include hair loss, brittle nails, and skin rash. Neurological symptoms can also result and include depression, fatigue, and numbness. Biotin is stable under conditions associated with food processing (e.g., exposure to heat, pressure and light) and the chance of becoming biotin deficient is in very low.

Vitamin B$_9$ folates

Folic acid is a water-soluble molecule that is chemically termed an N-acyl-amino acid (Fig. 5.8). The common name, folate acid, refers to several forms of this vitamin that have biological activity (Table 5.1). Specifically, folates occur as folic acid, tetrahydrofolic acid, and tetrahydrofolate. Folates can be found in a variety of foods, including green leafy vegetable, lentils, nuts, and beans. Avocado and spinach are other good sources of folates. Folic acid is used in food enrichment or fortification. The daily value for folate is 400 μg (0.4 mg) per day. While this level is small, there is some

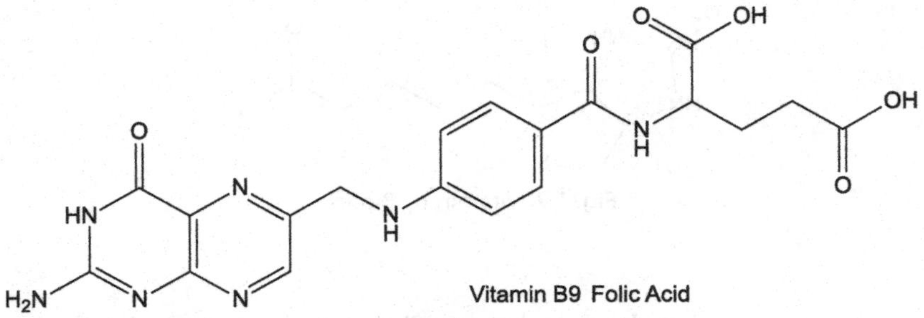

Vitamin B9 Folic Acid

Fig. 5.8 Vitamin B₉ folic acid.

concern by those in the nutrition community that many in Western countries may not be receiving an adequate level of folate in the diet. In these instance a supplement is advised.

Biological importance and supplementation

Folic acid is the major form used to fortify foods and in vitamin supplements. The major function of folate is to serve as a coenzyme in reactions that transfer single carbon units (methyl groups). This function is biologically important in the synthesis of amino acids and nucleic acids. Folate, together with other B vitamins, participates in the synthesis of methionine from homocysteine. A folate deficiency can result in elevated blood levels of homocysteine, an indicator of cardiovascular disease. Similarly, folate functions as a coenzyme in the DNA synthesis pathway. It is critical for reactions that methylate DNA and RNA and ultimately affect gene expression. Folate, therefore, is an important to cancer therapy. A non-functional analog of folic acid, called methotrexate, is commonly used in cancer chemotherapy to slow the replication and spread of cancerous cells. Folic acid is routinely given to pregnant women because it prevents occurrence of neuro tube defects in the developing fetus. Untreated deficiencies can result in birth of a child with spina bifida. Folate deficiency is commonly manifested as anemia because of its negative impact on protein synthesis. Red blood cells are turned over quickly and high levels of synthesis are needed to replace them (Vitamins and Minerals OSU MIC, 2018).

Means of loss

Folate is reasonably stable to most conditions encountered in the processing of food. Major losses in folic acid occur at high temperature, oxidizing conditions, and exposure to light (especially UV light). Chemical damage occurs to folate when sulfite or nitrite is present in the food matrix. Sulfite is added for its ability to control growth of microorganisms and/or to bleach food like wheat flour. The destruction of naturally occurring folate in flour by sulfite treatment is corrected by adding it back at appropriate levels (enrichment). Folate can be protected from oxidative damage by the addition of ascorbic acid.

Vitamin B$_{12}$ cobalamin

Vitamin B$_{12}$ is a large water-soluble molecule that is similar in structure to chlorophyll. Both B$_{12}$ and chlorophyll are composed of a planar ring structure, but B$_{12}$ differs in that it contains an essential cobalt (Co) ion in its center (Fig. 5.9). Vitamin B$_{12}$ also contains ligands, a nucleotide (adenosine) molecule on one side, and a cyanide group on the other. The cyanide

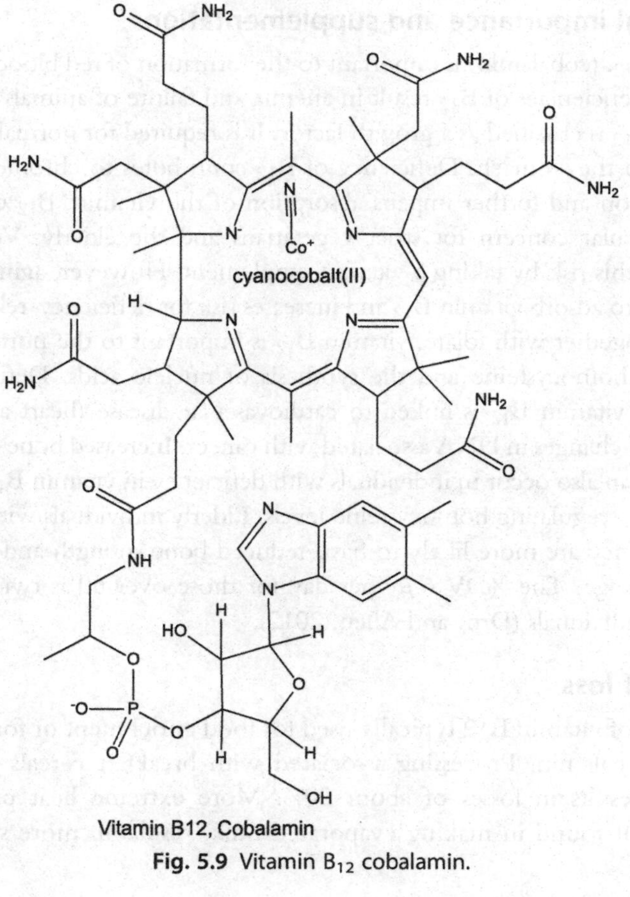

Vitamin B12, Cobalamin

Fig. 5.9 Vitamin B$_{12}$ cobalamin.

containing form (cyanocobalamin) is typically used in dietary supplements and food fortification. It was discovered in the mid-part of the last century as a factor that cured anemia in chickens fed a solely plant-based diet. The chickens recovered when their diets were supplement with animal-based fed. This effect occurred because vitamin B_{12} is only synthesized by gut bacteria (especially in ruminants) and thus solely plant-based diets can be deficient in B_{12}. While bacteria that inhabit the human gut synthesize vitamin B_{12}, the mechanism for absorption is lacking. Humans must obtain vitamin B_{12} from dietary sources such as liver, milk, meat, and eggs. Fish, especially sardines, also contain significant amounts of B_{12} (Table 5.1). Typical levels of vitamin B_{12} in these foods is about $1-2$ μg per 100 g. Given the low daily value of 2 μg per day for vitamin B_{12}, these foods represent good dietary sources.

Biological importance and supplementation

Vitamin B_{12}, (cobalamin) is important to the formation of red blood cells and growth. Deficiencies of B_{12} result in anemia and failure of animals to thrive. Vitamin B_{12} is classified as a growth factor. It is required for normal digestive function in the stomach. Deficiency of B_{12} contributes to chronic stomach inflammation and further impairs absorption of the vitamin. B_{12} deficiency is of particular concern for strict vegetarians and the elderly. Vegetarians can avoid this risk by taking a vitamin supplement. However, aging reduces the ability to adsorb vitamin B_{12} and increases risk for deficiency-related conditions. Together with folate, vitamin B_{12} is important to the normal metabolism of homocysteine and the synthesis of nucleic acids. Deficiency of folate and vitamin B_{12} is linked to cardiovascular disease (heart attack and stroke) and changes in DNA associated with cancer. Increased bone loss (osteoporosis) can also occur in individuals with deficiency in vitamin B_{12} because of its role in regulating homocysteine levels. Elderly individuals with vitamin B_{12} deficiency are more likely to have reduced bone strength and increased bone breakage. The %DV 6 μg per day for those over 60 is twice that of younger individuals (Dror and Allen, 2012).

Means of loss

The form of vitamin B12 typically used for food enrichment or fortification is cyanocobalamin. Processing associated with breakfast cereals and milk typically results in losses of about 20%. More extreme heat processing, such as that found in making evaporated milk, results in more substantial losses. Acidic or alkaline conditions is responsible for increased destruction of vitamin B_{12}. The presence ascorbic acid (vitamin C) causes destruction of vitamin B_{12}, but the mechanism by which it occurs, is unknown.

Vitamin C, ascorbic acid

Vitamin C is a water-soluble vitamin also known by its chemical name, L ascorbic acid. Vitamin C exists in two forms that are readily reversible via chemical reduction or oxidation. The reduced form is L ascorbic acid and its oxidized form is L dehydroascorbic acid (Fig. 5.10). Vitamin C is perhaps the best-known as the cure for scurvy. This deficiency disease causes bone weakness, bleeding gums, tooth loss, and poor wound healing. Death can result in cases of chronic deficiency. These conditions have existed since ancient Egyptian and Roman times, when the cause was unknown. The first important discovery was in 1747 when James Lind of the British Royal Navy found that scurvy symptoms could be reversed by eating citrus fruit, such as oranges, lemons, and limes. The Royal Navy adopted routine practice of giving sailors lime extracts as a preventative measure. As a result, the sailors acquired the nickname limees. It was not until the early twentieth century that the compound (L ascorbic acid) responsible for curing scurvy was characterized. Albert Szent-Gyorgyi and Walter Haworth were credited with the discovery and characterization of this compound that they named a-sorbic or ascorbic acid. They received the Nobel prize for their work in 1937. Most plants and many animals have the ability to synthesize L ascorbic acid, but mammals have lost this ability. The daily value for adult individuals is 60 mg per day. Vitamin C can be obtained in various citrus fruits. Kiwifruit, grapefruit, and oranges contain the highest level (60–80 g of ascorbic acid per serving).

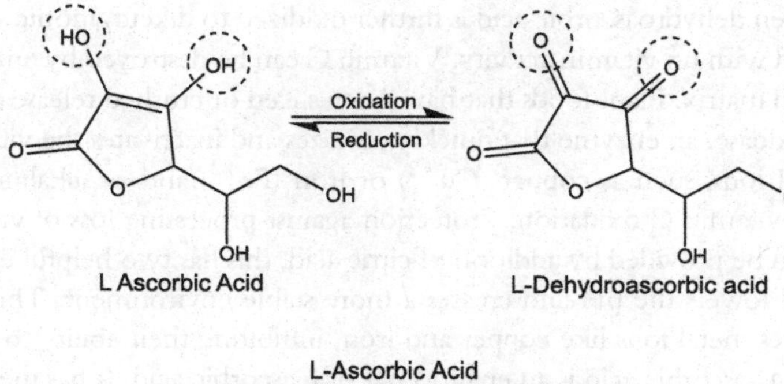

L Ascorbic Acid   L-Dehydroascorbic acid

L-Ascorbic Acid

Fig. 5.10 Vitamin C, L ascorbic acid.

Biological importance and supplementation

L ascorbic acid is a potent antioxidant. It functions in reduction–oxidation (REDOX) reactions that occur in all cells and is a scavenger for free radicals. As such, ascorbic acid provides proteins, lipids and nucleic acids (DNA and RNA) with protection against damage from these reactants. L ascorbic acid has a major role in the synthesis of collagen. It is necessary for enzyme catalyzed reactions that add hydroxyl groups to the amino acids, proline and lysine. The products hydroxyproline and hydroxylysine are essential to formation of strong, stable collagen polymers in connective tissue and in bone. L ascorbic acid functions in synthesis of collagen that is a component of blood vessel walls. It provides an explanation for scurvy symptom of excessive bleeding. L ascorbic acid improves iron absorption that forms in the gut as minerals are kept in the chemically reduced state. It contributes to normal functioning of the immune system by stimulating the production of white blood cells. It has health promoting effects, specifically in the prevention of coronary heart disease (CHD). While there is conflicting information on whether dietary or supplement ascorbic acid is responsible, there is evidence that it lowers the risk of heart attack. It is thought that vitamin C provides protection to the endothelial layer of cells that line the interior of blood vessels. (Erdman et al., 2012).

Means of loss

Ascorbic acid is very sensitive to oxidation, a chemical process that can result in its destruction. Oxidation can be caused by several factors including sunlight, heat, oxygen, and the presence of metal ions. Oxidation of hydroxyl groups in L-ascorbic acid (dashed circles in Fig. 5.10), converts the molecule to dehydroascorbic acid (Fig. 5.10). Fortunately, reversion to L-ascorbic acid occurs readily if a reducing agent is present. Irreversible destruction of vitamin C occurs when dehydroascorbic acid is further oxidized to diketogluonic acid, a compound with no vitamin activity. Vitamin C can be destroyed by enzymes in the food matrix. Plant foods that have been sliced or crushed release ascorbic acid oxidase, an enzyme that quickly oxidizes and inactivates the vitamin. Free metal ions, such as copper (Cu^{+2}) or iron (Fe^{+3}) and/or alkaline pH, accelerate vitamin C oxidation. Protection against processing loss of vitamin activity can be provided by addition of citric acid, this has two helpful effects. Citric acid lowers the pH and creates a more stable environment. This acid also chelates metal ions like copper and iron, inhibiting their ability to cause oxidation. D-ascorbic acid is an enantiomer of L ascorbic acid. It has the same chemical properties as the L form, but it does not have vitamin activity. D ascorbic acid's antioxidant properties and low cost have made it a widely used food additive. For example, a combination of citric acid and D ascorbic acid are used to control enzymatic browning in fresh fruits and vegetables. D ascorbic acid is also used to control lipid oxidation reactions that cause off flavors. It is added to cured meats, such as bacon and ham, to preserve color and prevent off flavors.

Fat soluble vitamins

In addition to the obvious differences in solubility, fat and water-soluble vitamins are physiologically different. Water soluble vitamins circulate through the blood system after absorption. They are readily eliminated by excretion in the urine. Water soluble vitamins, therefore, have lower potential to be toxic in cases of excess consumption. In contrast, fat soluble vitamins can be stored in the liver and/or fat depots of the body after being absorbed. Excess consumption of fat-soluble vitamins can result in a toxic build up. The recommended intake level (daily value) for fat-soluble fats is most often specified in international Units (IU).

Vitamin A, retinol

Vitamin A is a fat-soluble vitamin belonging to the carotenoid family of compounds (Fig. 5.11). Numerous carotenoids (20+) occur in fruits and vegetables, but only beta carotene and the closely related compounds alpha carotene and beta cryptoxantin have vitamin A activity (Weber and Grune, 2012). Vitamin A exists in three chemical forms with equal activity: retinol, retinal, and retinoic acid. Carotenoids are responsible for the red, orange, and yellow colors of fruits and vegetables. Vitamin A is obtained in diet from two major sources. The first and most direct source comes from animal foods like liver,

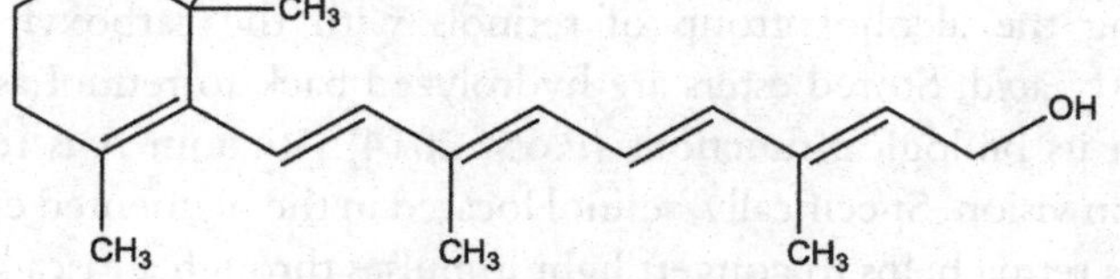

Fig. 5.11 Vitamin A, retinol.

meat, milk, butter, and eggs (Table 5.1). Vitamin A obtained via this route is called preformed vitamin A and is in the retinol form. The second source comes from fruits and vegetables containing beta carotene. Plant foods such as sweet potato, pumpkin, carrots, spinach, and broccoli are good sources of beta carotene (Table 5.1). This is the inactive form called pro-vitamin A. It is converted to active form in the small intestine by enzymes that split beta carotene into two vitamin A molecules of retinol. Retinol is absorbed from the gut and transported to the liver that is its site of storage.

The conversion efficiency of beta carotene (provitamin A) to retinol is variable and affected by the food matrix. For example, conversion of beta carotene supplement in oil to retinol is about 2:1. It takes 2 μg of beta carotene to yield 1 μg of retinol. The conversion rate is much lower in food. Approximately 12 μg of beta carotene in food is required to yield 1 μg of retinol (a ratio of 12:1). These ratios are termed Retinol Activity Equivalents (RAE) and are used as a standard method to report retinol activity for both preformed and pro-vitamin A. The %DV for vitamin A was defined by the Food and Nutrition Board of the United States Institute of Medicine in 2001. Specifically, the RDA for vitamin A is the recommended intake needed by nearly all of the population to ensure adequate hepatic stores of vitamin in the body to support normal reproductive function, immune function, gene expression, and vision. The RDA for adult males and females is 900 and 600 μg per day, respectively. It should be cautioned that excessive intake of vitamin A from supplements can do more harm than good. Specifically, excess vitamin A can lead to damage of the liver and kidneys (Vitamins and Minerals OSU MIC, 2018).

Biological importance and supplementation

Beta carotene and its vitamin active product retinol are aliphatic (no polar groups) molecules and poorly soluble in water (Fig. 5.11). The beta caroteen is highly unsaturated. This means it contains several carbon–carbon double bonds that are in the trans configuration (e.g., all-trans-retinol). Vitamin A is stored in the liver in retinyl esters formed through a biochemical reaction linking the alcohol group of retinols with the carboxyl group of palmitic fatty acid. Stored esters are hydrolyzed back to retinol as required to perform its biological functions (Ross, 2014). Vitamin A is recognized for its role in vision. Specifically, retinol located in the pigmented epithelium of the eye's retina helps to convert light impulses through a cascade of reactions into a nerve impulse that is sent to the brain. Chronic vitamin A deficiency can result in impaired night vision and even cause blindness. Vitamin

A in the form of retinoic acid is important to the proper functioning of the immune system. Retinoic acid acts on cellular components of the immune system to activate its action on foreign substances such bacteria. Children with even moderate vitamin A deficiency have been shown to have a higher incidence of respiratory infections. Vitamin A also supports growth and differentiation and has a role in formation of vital organs. There is increased recognition for the importance of vitamin A in the diet. Deficiency is a major cause of childhood blindness in populations where it is most severe. The introduction of golden rice, a genetically modified strain of rice containing provitamin A, has the potential to alleviate this problem.

Means of loss

Retinol is the most chemically stable form of vitamin A and used to fortify or restore its level in food. Retinol is fairly stable under the most common food processing conditions (e.g., heat). Because of retinol's high degree of unsaturation (double bonds), substantial destruction occurs in the presence of free radicals, such as hydroxyl radicals produced by lipid oxidation. Light can catalyze changes in the retinol structure through a process known as photoisomerization. Exposure of whole milk to fluorescent light for one day causes a two-thirds loss in vitamin A activity. The loss is even greater for milk with lower fat content. The cause of vitamin A destruction in milk is riboflavin. When exposed to UV light, this molecule produces reactive oxygen species (ROS) that promotes free radical reactions. Similarly, light-induced oxidation reactions cause the loss of natural carotenoid color in foods, flowers, and plants.

Vitamin D, calcitriol

Vitamin D (Fig. 5.12) is a fat-soluble molecule that functions in the absorption of calcium, magnesium, and phosphate. It is created when the skin is exposed to sun light. Irradiation of the skin by the UV portion of sunlight causes conversion of a cholesterol derivative (7-dehydrocholesterol) to previtamin D_3. Subsequently, previtamin D_3 is chemically converted in the skin to vitamin D_3 (cholecalciferol) that can be stored in fat tissue. The active forms of vitamin D (25-hydroxyvitamin D, Calcidiol and 1α 25-Dihydroxyvitamin D, Calcitriol) are created by hydroxylation of cholecalciferol occurring in the liver and kidney (Bikle, 2014). Naturally occurring food sources of vitamin D are few. The best food sources are fatty fish, including salmon, mackerel, and sardines (Table 5.1). Additional sources of vitamin D can be obtained

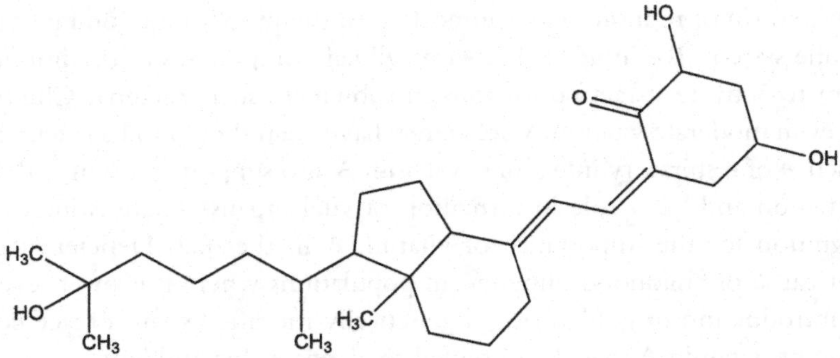

Vitamin D₃, Calcitriol
(1,25-dihydroxycholecalciferol)
Fig. 5.12 Vitamin D₃ Calcitriol.

from beef liver, egg yolk, cheese, mushrooms, and yeast. Vitamin D is most often obtained in the diet from fortified foods. Milk, for example, is fortified to contain about 400 IU (International Units) per 8 ounce serving. The daily value for vitamin D varies by age and gender. The daily value for children and adolescents is 600–1000 IU (15–25 µg) and for adults is 2000 IU (50 µg). Because vitamin D synthesis in vivo, it depends on exposure to available sun light and limited food sources. Thus its deficiency is a concern for many. Individuals living in northern regions and those with darker skin may not be receiving sufficient sunlight to synthesize adequate amounts of vitamin D.

Biological importance and supplementation

A major function of Vitamin D is the regulation of calcium in the body. It increases calcium absorption from the gut and decreases urinary excretion (Holick, 2003). Calcium is needed for bone health and deficiency of this vitamin can cause rickets that results in deformed bones (bowed legs) in children. Long term calcium deficiency can result in osteoporosis (bone fragility) in adults. When vitamin D levels are low, the incorporation of calcium and phosphate is reduced and bone structure is weakened. Vitamin D is important to the growth and differentiation of cells throughout the body. Most, if not all, of vitamin D's action involves activation of gene expression. These processes are mediated by vitamin D binding to receptor molecules (proteins) that subsequently interact with DNA and activate gene expression. Examples of vitamin D's role in gene expression number in the hundreds (Sutton and MacDonald, 2003). Calcium balance, the process of maintaining proper calcium levels between absorption, excretion, and storage (bone), is mediated by vitamin D activation of target genes. Vitamin D also has a role in disease prevention. It is essential to the normal function of the immune system. Supplementation in the first year of life may be protective against development of type 1 diabetes (an autoimmune form of the disease) in later years. There is hope that vitamin D may protect against other autoimmune diseases such as rheumatoid arthritis. The data have not yet been conclusive.

Means of loss

Vitamin D is fairly stable to most forms of thermal processing. Boiled eggs were found to retain 89% of their vitamin D. Additionally, pasteurization and sterilization of milk was found to cause little or no loss of vitamin D. Exposure of milk in plastic containers to UV light resulted in a small, but significant loss of vitamin D.

Vitamin E, alpha tocopherol

Vitamin E is a fat-soluble vitamin and a member of the tocopherol family of compounds. Four forms of tocopherol exist (α, β, γ, and δ) and each form represents a different optical isomer. α-Tocopherol is principally responsible for vitamin E activity in humans (Fig. 5.13). The other tocopherols together with the related tocotrienols contribute only a minor part of total vitamin E activity. α-Tocopherol is the form used in fortifying foods and in vitamin supplements. The tocopherol molecule is composed of a chromane ring (i.e., a heterocyclic structure) and a long aliphatic tail that is indicated by the dotted circle in Fig. 5.13. All tocopherols, including non-vitamin forms, are noted for their antioxidant activity provided by the hydroxyl (OH) group attached to the chromane ring. Free radical reactions are terminated by donation of its hydrogen atom to the radical molecule. The aliphatic (hydrophobic) tail allows the molecule to be membrane anchored. Plant foods are the best sources of α-tocopherol (Table 5.1).

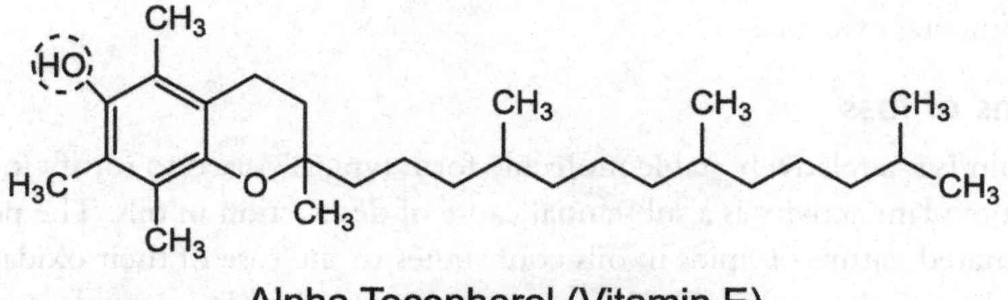

Alpha Tocopherol (Vitamin E)
Fig. 5.13 Vitamin E alpha tocopherol.

The content of α-tocopherol in various foods as given by the USDA Food Composition data base (listed from highest to lowest) is wheat germ oil (141 mg per 100 g), sunflower seeds (36 mg per 100 g), almonds (23 mg per 100 g), olive, canola, soybean, and corn oils (12–17 mg/100 g), peanuts (5 mg per 100 g), and spinach (2 mg per 100 g) (Table 5.1). Given that the daily value for α-tocopherol is 15 mg (22.5 IU) per day, it is easy to understand why an estimated 90% of adults in the United States do not meet that level of intake.

Biological importance and supplementation

Alpha-tocopherol is a fat-soluble molecule with potent antioxidant activity that protects membrane components from free radical damage. It terminates free radicals by donating a hydrogen from the hydroxyl group on its chromane ring (dashed circle in Fig. 5.13). The antioxidant activity of α-tocopherol is also enhanced by the antioxidant, vitamin C. Donation of α-tocopherol's hydrogen atom results in its oxidation and loss of antioxidant activity. However, α-tocopherol's antioxidant activity is restored by the action of vitamin C. These two antioxidants work in concert to reduce the oxidative stress. The level of oxidized low-density lipoproteins (LDL) in the blood is a risk factor for cardiovascular disease. Tocopherols may also play a role in reducing inflammation caused by the action of free radicals. Deficiency of α-tocopherol is linked to several conditions including muscle weakness, unsteadiness and lack of coordination, impaired immune response, and damage of the eye's retina. The antioxidant activity of α-tocopherol, especially in providing protection against DNA damage and subsequent mutations, suggests that it may be a factor in preventing cancer of various types. However, studies of the effect of tocopherol supplementation on breast and lung cancer have not shown beneficial effects (Higdon and Drake, 2012, 2013). Similarly, the hypothesis that α-tocopherol supplementation could provide a protective effect against development of cataracts, a vision impairment caused by oxidation of proteins in the lens, is not supported by experimental evidence.

Means of loss

Vitamin E is a relatively stable molecule form typically used to fortify foods. Its antioxidant activity is a substantial cause of destruction in oils. The polyunsaturated nature of lipids in oils contributes to the ease of their oxidation and subsequent formation of hydroperoxides. Lipid hydroperoxides represent a significant cause of vitamin E destruction. This is an unfortunate circumstance because plant oils like canola, soybean, and canola are among the richest sources of vitamin E however, they may also contain high levels of hydroperoxide.

Vitamin K₁ phylloquinone

Vitamin K is a fat-soluble vitamin essential to blood coagulation and calcification of bone. Naturally occurring forms are members of the phylloquinone and menaquinones family of compounds that correspond to vitamins K_1 and K_2 (Fig. 5.14). Vitamin K was identified in the 1930s as a factor required for normal blood clotting. It was named Koagulationsvitamin, a German word meaning coagulation vitamin. The molecular structure of vitamin K and its mechanism of action were characterized by Henrik Dam and Edward Doisy for which they received the Nobel prize in 1943. The greatest source of vitamin K_1 occurs in green leafy plants such as kale, Swiss chard, broccoli, and spinach (Table 5.1). The level of vitamin K_1 parallels the chlorophyll content of the plant because of its role in photosynthesis. The outer and greener leaves of these plants contain the highest level of vitamin K. Plant oils, such as canola and olive, are also good sources of vitamin K_1. Vitamin K_1 is synthesized by bacteria and occurs at highest level in fermented foods, such as cheese and fermented soybeans. Liver is also a good source of vitamin K_2. The amount of vitamin K recommended in the diet of healthy individuals is given as a daily value that varies by age and sex. Male and female newborns have an AI of 2 μg per day through

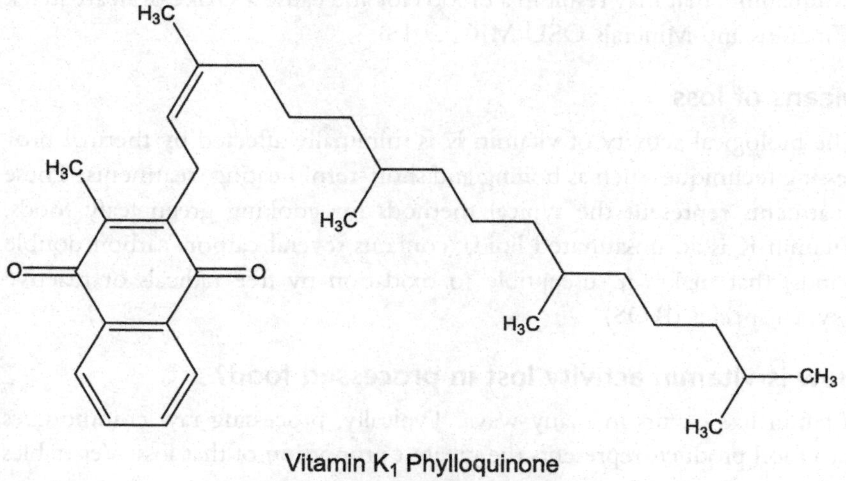

Vitamin K₁ Phylloquinone

Fig. 5.14 Vitamin K1, phylloquinone.

the first year. The requirement for all children age 1—18 years, progressively increases from 30 to 75 μg per day. The daily value for adults 19 and older is 90 and 120 μg per day for female and male adults, respectively.

Biological importance and supplementation

Vitamin K functions by modifying glutamic acid residues in proteins. Specifically, vitamin K_1 is a cofactor of an enzyme catalyzing the carboxylation (addition of a carboxyl group) of glutamic acid and converting it to gamma glutamic acid. This form of the amino acid has two negative charges that enhances its ability to bind positively charged ions (calcium). Calcium is essential to the activation of clotting factors in the blood clotting cascade. Deficiency in vitamin K causes increased blood clotting times that result in excessive external and internal bleeding. Proteins with the gamma glutamic acid modification provided by vitamin K are similarly important for bone strength. Calcium carried by these proteins are also important to bone mineralization and strength. Vitamin K deficiency is a contributing factor to osteoporosis and bone weakness and breaking. Vitamin K deficiency may also result in soft tissue (arterial) calcification that contributes to cardiovascular disease. Fortunately, the incidence of vitamin K deficiency is low because the dietary requirement is low. Natural forms of vitamin K (K_1 and K_2) have low toxicity. However, individuals taking the drug Coumadin™ (warfarin) should not also take vitamin K supplements. Coumadin™ is an anticoagulant that is prescribed to prevent stroke and heart attack. Additional vitamin K could create an antagonistic action against Coumadin™ that may result in a blood clot and cause a stroke or heart attack (Vitamins and Minerals OSU MIC, 2018).

Means of loss

The biological activity of vitamin K is minimally affected by thermal processing techniques such as boiling and short-term heating treatments. These treatments represent the typical methods for cooking green leafy foods. Vitamin K is an unsaturated lipid, (contains several carbon-carbon double bonds) that makes it susceptible to oxidation by free radicals or reactive oxygen species (ROS).

How is vitamin activity lost in processed food?

Vitamin loss occurs in many ways. Typically, processing raw commodities into food products represents the greatest proportion of that loss. Vegetables to be packaged as frozen product are first blanched for 2—3 min in boiling water. This results in substantial loss of nutrients by leaching into water, especially if there are cut surfaces. Dehulling and grinding grains removes the bran and germ portions containing the majority of its vitamin content. Fat-soluble vitamins in wheat flour, for example, are damaged by exposure to oxygen and light. UV light initiates free radical reactions, oxidizing unsaturated lipids, destroying vitamin activity, and generating off flavors. Endogenous enzymes breakdown vitamins in raw commodities. Thiamin (vitamin B_1) in fish, for example, is degraded by the enzyme thiaminase. Finally, cooking steps essential to creating flavor, color, and microbiologically safe food products, causes additional vitamin degradation. In recognition of vitamin content lost in processed foods, many products (e.g., milk, breakfast cereals, wheat flour) contain added vitamins and/or minerals through a process known as enrichment. In enrichment nutrients are added to replace those lost in processing. Milk, wheat flour, and table salt contain added vitamin D, B vitamins, and iodine, respectively, and are examples of enriched foods.

Minerals

The ultimate source of minerals is the soil. They are selectively absorbed from soil by plants and passed up the food chain to humans. In food, minerals provide functional properties and fulfill biological needs. They are added to food to achieve a desired function. For example, sodium chloride (NaCl) is added to give food a salty taste and/or preserve it by inhibiting the growth of spoilage organisms. Biologically, minerals are essential to biochemical reactions and physiological processes. For example, minerals are cofactors required to activate enzymes. Heme iron in myoglobin and hemoglobin is essential to the oxygen binding function of these proteins. Approximately 25 minerals are recognized as needed in the human diet. The 14 listed in Table 5.2 represent those considered to be most important. It should be noted that the Daily Values in this table are frequently revised and anyone seeking the most accurate information should consult the appropriate nutritional reference. Dietary minerals are divided into two groups. Macro and trace minerals are based on the amount required in the diet. Macro minerals include calcium, magnesium, sodium, chloride, potassium, and phosphorus. Trace minerals include copper, iron, fluorine, iodine, manganese, chromium, cobalt, zinc, selenium, and molybdenum (Nutrient recommendations NIH, 2018).

Table 5.2 Minerals.

Mineral	Function	Source	Daily value
Calcium (Ca)	Blood clotting, bone & teeth formation, hormone secretion, muscle contraction, nervous system function	Dairy products, almond & coconut milk, green leafy, vegetables	1000 mg
Magnesium (Mg)	Blood pressure & sugar level regulation, bone formation/strength immune & nervous system function, heart rhythm	Avocados, bananas, beans, peas, Dairy products, nuts, potatoes, raisins, whole grains	400 mg
Sodium (Na)	Acid/base balance, blood pressure regulation, muscle contraction, nervous system f unction	Breads, cheese, processed meats, soups, table salt	2400 mg (lower amounts advised)
Chloride (Cl)	Acid/base balance, metabolism, fluid balance, nervous system function	Celery, lettuce, olives, table salt, tomatoes	3400 mg
Potassium (K)	Blood clotting, bone strength	Green leafy, vegetables	80ug
Phosphorous (P)	Acid-base balance, bone strength, hormone secretion,	Beans, peas, dairy products, meat, poultry, seafood, whole grains	1000 mg
Chromium (Cr)	Insulin function, protein carbohydrate, & fat metabolism	Broccoli, fruits, bananas, meats, garlic, basil, whole grains	120 μ g
Copper (Cu)	Antioxidant, bone formation, collagen 7 connective tissue formation, iron metabolism	Chocolate, cocoa, shellfish, lentils, whole grains	2 mg
Florine (F)	Bone & tooth strength	Fish (sardines), poultry	1−2 mg (as AI level)
Iodine (I)	Thyroid gland function,	Seafood and seaweed, fortified table salt,	90-150ug (as AI level)

Table 5.2 Minerals.—cont'd

Mineral	Function	Source	Daily value
Iron (Fe)	Metabolism, growth & development, immune function, reproduction, red blood cell formation	Beans, peas, dark green vegetables, meats, poultry, prunes, seafood, whole grains	18 mg
Manganese (Mn)	Carbohydrate, protein & cholesterol metabolism, cartilage & bone formation	Beans, nuts, pineapple, spinach, sweet potato, whole grains	2 mg
Selenium (Se)	Antioxidant, immune function, reproduction, thyroid function	Eggs, meats, nuts, poultry, seafood, whole grains	70 μ g
Zinc (Zn)	Growth & development, immune & nervous system function, protein synthesis, reproduction, taste & smell,	Beans, peas, beef, dairy products, nuts, poultry, seafood, whole grains	15 mg

Adapted from Vitamins and Minerals (FDA, 2018).

Calcium (Ca)

Calcium (Table 5.2) is an essential mineral whose principal sources are milk, cheese, and other dairy products. The requirement for calcium in adults is 1000—1200 and mg/day. Cow's milk (whole or skimmed) contains about 240 mg of calcium per 8oz (225 g) serving. Cheese contains 400—800 mg (depending on type) of calcium per 100 g per (4oz) serving and yogurt about 13 mg per 100 g serving. Other animal foods, such as egg, meat, and fish, are calcium sources. Egg contains 50 mg of calcium per 100 g serving (2 eggs). Canned sardines are particularly high in calcium and contain about 400 mg per 100 g serving. Plant foods, such as beans, water cress, kale, broccoli, nuts, and seeds are also sources of calcium. Almonds, for example, contain 40 mg of calcium per 100 g of nuts. Almond milk is a relatively new product that has a calcium content of 90 mg per 8oz serving. This is about one-half the amount in cow's milk. A number of plant foods, such as soybeans, spinach, and nuts, contain anti-nutritional factors (phytic and oxalic acid) that bind minerals and potentially limit their availability. Phytic acid complexes with positively charged calcium, magnesium, and zinc ions in the stomach. The complex between mineral and phytic acid

is difficult to breakdown in digestion. Oxalic acid also complexes with calcium to form calcium oxalate that is usually eliminated via the kidney.
Excess amounts of the complex can form stones at the site of excretion.
The likelihood of stone formation from eating foods with a high oxalate
level is reduced when these are consumed with high calcium foods. Calcium
oxalate is then formed in the stomach and not in the kidney. Typical soybean
processing does not remove or destroy its phytic or oxalic acid content. Excessive consumption of products, such as soy milk and tofu, could result in
decreased calcium levels. The process of fermenting soy to make products
such as soy sauce, tempeh, and miso, effectively destroys phytic and oxalic acids.

Biological importance and supplementation

Calcium is the most plentiful body mineral. Almost all (99%) calcium is
stored in bones and teeth and the remainder is found in blood. The balance
of calcium level between blood and bone is under a hormonal regulation in
a process called homeostasis. Vitamin D participates in maintaining calcium
homeostasis. Calcium in bone exists in a form known as hydroxyapatite that
is a complex of the mineral and phosphate. When the level of calcium is low
in blood, it is reabsorbed from bone through the action of parathyroid hormone (PTH). PTH causes bone cells (osteoclasts) to secrete enzymes that
dissolve the tissue. Circulating PTH also acts in the kidney to convert
vitamin D to its active form, 1,25–dihydroxvitamin D. Vitamin D, in its activated form, increases absorption of calcium from the small intestine and thus
increases its level in the blood. When a normal level of calcium is reached in
the blood, PTH synthesis is turned off (down regulated) and resorption of
calcium from bone stops. Under conditions of chronic calcium deficiency
resulting from diet and/or age, losses from bone can be significant and result
in a condition known as osteoporosis. Calcium loss weakens bone and increases the risk of fracture. Similarly, calcium deficiency can exacerbate muscle atrophy and loss of strength in a condition known as sarcopenia.
Additionally, calcium is essential in signaling processes that control many
functionalities, including contraction of smooth and skeletal muscle. Smooth
muscle is a component of arteries and its contraction or relaxation results in
increased or decrease blood pressure, respectively. Calcium concentration is
essential to enzymes that have a role in metabolism and catabolism (breakdown)
of proteins, carbohydrates, and lipids. Calcium is essential to the transmission of
neural impulses. It is an activator of several enzymes in the blood clotting process. EDTA, a chemical chelator of calcium, is added to donated blood to

prevent it from clotting. Calcium supplements containing vitamin D represent a good choice for optimal bone health because they act in concert to maximize bone mineralization. Supplements containing 1000—1200 mg of calcium and 400 units of vitamin D are recommended for older adults, (women > 50 years and men > 70 years). Supplementation at this level has been shown to reduce the risk of bone fracture, especially in older women (age>70 years). There are conflicting reports from large studies that question if calcium supplementation increases the risk of cardiovascular disease and heart attack. Based on best available evidence, the consensus is that total calcium intake (diet plus supplementation) should be equal to the RDA set for the appropriate gender and age (Higdon and Drake, 2012).

Magnesium (Mg)

The total amount of magnesium in the body is about 25 g. The majority (60%) of that is found in bone (Table 5.2). About one-fourth of the total magnesium is in muscle and the remainder in other tissues and cells. Magnesium is found in a variety of plant foods, milk, meat, and fish. Whole grain cereals, nuts, and sardines are relatively high in magnesium. Green leafy plants contain magnesium as an integral part of their chlorophyll which is essential to photosynthesis. Therefore, foods such as spinach are good sources of magnesium. The requirement for magnesium varies with age and gender. Infants (both male and female) under one year have an Adequate Intake (AI) level of 30—75 mg per day. Children under 13 years have an RDA that increases with age from 80 to 240 mg per day. The RDA for adults is 420 and 320 mg per day for men and women, respectively. However, survey data suggest that magnesium intake level is significantly below the RDA for many in the United States (Moshfegh et al., 2009).

Biological importance and supplementation

Magnesium is an essential mineral functioning in numerous physiological roles, including metabolism of carbohydrates and fat to make energy. Energy produced from metabolism is stored in a compound called adenosine triphosphate (ATP). Most cellular functions utilize magnesium ion complexed to ATP, a form known as Mg-ATP. Mg-ATP is also the source of energy for biochemical reactions causing muscle contraction. Every function from blinking of your eye to the beating of your heart uses chemical energy in the form of Mg-ATP to power contraction. Mg-ATP also functions in cell signaling, a process that regulates turning on and off biochemical pathways. Mg-ATP is converted to cyclic AMP (cAMP) that is the cell's second messenger. Second messengers are responsible for triggering proliferation, growth, and degradation within cells. For example, the hormone PTH is essential to the regulation of calcium in the body. As a peptide, PTH has little ability to enter bone cells (osteoclasts). Instead, it binds to a receptor that results in production of cAMP that activates of the process of dissolving bone calcium. Magnesium deficiency is rare among individuals consuming a well-balanced diet. However, individuals with diseases, such as alcoholism or diabetes, may be at risk for magnesium deficiency. Magnesium salts, such as magnesium oxide, magnesium chloride, and magnesium citrate, are common forms used as supplements. Magnesium sulfate salt is often used for its laxative effect. Magnesium is likely safe when taken as a dietary supplement. Elevated blood level of magnesium can affect blood pressure. The upper limit for magnesium intake is set by the Food and Nutrition Board of the Institute of Medicine, at 350 mg per day.

Sodium (Na) and chlorine (Cl)

Sodium and chloride (Table 5.2) are essential minerals that work together in several physiological processes. The largest store (about 60%) of the body's sodium and chloride is located in bone. The remainder is found in blood plasma and fluids within and outside of cells. The dietary intake of sodium and its salt, sodium chloride, typically exceeds the Adequate Intake (AI) level for most individuals. Sodium intake requirement is about 1.5 g per day for both males and females older than one year. The required intake for table salt is about 3.0—3.8 g per day for both male and female individuals. Sodium supplied as sodium chloride salt is ubiquitous in food. The largest contribution of sodium in the diet comes from processed food most where it is added as a flavor enhancer and/or to prevent microbial spoilage. It is well known that salt enhances sweet taste perception and thus is sometimes included in chocolate and other sweets. Processed meats and cheeses typically contain 1 g or more sodium per serving. Soups typically contain 2g of sodium per serving. Most soups do not have the same level of salty taste as they would at the same concentration in water. The lack of salt perception in soup, gravy, cheese and other foods occurs because sodium ions are complexed to proteins in the food matrix and unavailable to activate their taste receptors.

Biological importance and supplementation

Sodium chloride salt completely dissociates to sodium and chloride ions in water. These mineral ions are important as electrolytes that maintain sodium and potassium ion concentrate across cell membranes. Sodium and potassium ions are at different concentrations on the inside versus the outside of a cell and create an electrical gradient called the membrane potential. If left alone, differences in ion concentrations would be equalized on both sides by simple diffusion. However, an ion pump located in the membrane maintains the differences in sodium and potassium ion concentration (and membrane potential) using energy derived from the hydrolysis of ATP. Control of membrane potential is important to several physiological functions, including skeletal and cardiac muscle contraction, nerve impulse transmission, and the control of blood pressure. Sodium supplementation is generally not warranted given its over-abundance in the diet. Survey data from large nutritional studies indicate that the average sodium intake is above the recommended upper intake level for almost all groups. High levels of sodium intake are linked to high blood pressure and increased risk of cardiovascular and kidney disease.

Salty questions
Where does most sodium come from?

Most sodium comes from the salt (NaCl) used in making processed foods and food additives: MSG (monosodium glutamate), sodium nitrate, sodium phosphate, baking soda (sodium bicarbonate).

What foods contain the most salt?

Foods that contain the most salt include, processed meats and cheese, pizza, soup, gravy, and fried chicken.

How is sea salt different from ordinary table salt?

Sea salt differs from ordinary table salt (a refined product) because it is obtained by evaporating sea water. Thus, it contains additional minerals. Both salts contain additives to prevent caking/clumping. Sea salt typically contains less iodine (an essential mineral), compared to table salt. However, contrary to popular opinion, sea salt is not lower in sodium content.

Potassium (K)

Potassium functions (Table 5.2) in concert with sodium as an electrolyte. The distribution of sodium and potassium ions between the inside and outside of cells establishes an electrical gradient known as the membrane potential (see the description under sodium). The intake level for potassium was established by the Food and Nutrition Board of the Institute of Medicine based on data regarding the level required to lower blood pressure. The required intake for potassium varies with gender and age. The daily level for infants under one year of age is 400–700 mg and increases with age to 3,500 mg (3.5 g) for adults. Vegetables and fruits are good sources of potassium. Baked potatoes are among the most significant source and contain approximately 900 mg per serving. Bananas, raisins, plums, prunes, lima beans, acorn squash, and spinach are also good sources of potassium. All of these contain 400 mg or more per serving.

Biological importance and supplementation

Potassium and sodium ions function together to maintain cell membrane potential that is important to skeletal and cardiac muscle contraction, nerve impulse transmission, and the regulation of blood pressure. Potassium and sodium ions bound to ATP are essential to the enzymatically driven pump that maintains cell membrane potential. Potassium is also important in metabolism. It is a cofactor required for the activity of pyruvate kinase. It is an enzyme in the pathway that generates energy from glucose and other carbohydrates. Low blood potassium level is called hypokalemia and results in a number of symptoms including muscle weakness, abnormal heart rate (arrhythmia), and neurological dysfunction. While modest benefit in lowering blood pressure has been reported for potassium supplements of 500 mg per day, there is serious risk from excessive potassium intake. High potassium blood level (hyperkalemia) can increase the heart rate to dangerous levels. Cardiac arrest and death can also result. For this reason, most multivitamin and mineral supplements contain less than 100 mg per tablet or serving.

Phosphorous (P)

Phosphorous (Table 5.2) is an essential mineral and component of cell membranes in the form of a phospholipid complex. Phosphorous is integral to the structure of both RNA and DNA nucleic acids. Phosphorous is ubiquitous in plant and animal tissues. The daily intake for phosphorous is based on age. Intake requirements for infants under one year are 100–275 mg per day. For children from 1 to 18 years of age, this progressively increases from 460 to 1250 mg per day. The level for adults 19 and older is a lower level (1000 mg per day) presumably because the need for bone growth is decreased. It should be mentioned that phosphorous intake is likely higher than recommended levels for most individuals because of its abundance in plant and animal food. Lentils, almonds, peanuts, and bread are sources of phosphate. Additionally, phosphate is unavoidable in processed foods such as meat, cheese and other dairy products. Most carbonated soft drinks also contain phosphorous. There is concern about exceeding the upper limit of safe intake by those with a diet high in processed food.

Biological importance and supplementation

In addition to its role in the structure of cell membranes and nucleic acids, phosphorous is an essential part of Adenosine Tri-Phosphate (ATP) which is the energy currency of life. The biochemical conversion of ATP to ADP (Adenosine Di-Phosphate) results in the liberation of a phosphate group and energy essential to all cells, tissues, and organisms. The activity of enzymes and other proteins is regulated (turned off or on) by the reversible attachment of phosphate groups to proteins. Phosphorous is essential to strong bones and teeth. The ubiquitous presence of phosphorous in food makes its deficiency rare and supplementation generally not required. Instances of low blood phosphorous are quickly compensated for in the kidney. Phosphorous deficiency is usually limited to individuals with conditions such as diabetes, alcoholism, or chronic kidney disease. Additionally, elevated blood phosphorous level is associated with increased incidence of cardiovascular disease. For these reasons, multivitamin and mineral supplements provide a limited (about 100 mg) amount of phosphorous per tablet or serving.

Chromium (Cr)

Chromium (Table 5.2) is abundant in soil and contained in many foods. Vegetables and fruits represent the greatest sources of chromium. Broccoli, potatoes, whole grains, tomatoes, grape and orange juice are examples that contain between 1 and 10 μg per serving. The daily value is similar for males and females and is approximate 30 μg. While the recommendation for that level of intake is being questioned by the European Food Safety Authority, the cause for concern is low due to its low requirement and wide range of availability in food. The incidence of deficiency is rare (Higdon and Drake, 2012).

Biological importance and supplementation

Biologically, chromium is important for activating insulin receptors in the pancreas and the subsequent effects on glucose uptake and metabolism of carbohydrates and lipids. A deficiency in chromium may result in impaired glucose tolerance. However, there is no credible evidence to suggest a benefit from taking chromium supplements.

Copper (Cu)

Copper (Table 5.2) is found in a variety of foods, but those with the highest level are of animal origin (liver and shellfish). These foods provide between 1000 and 4000 μg (1—4 g) per serving. Good sources of copper can also be found in plant foods, such as cashews, almonds, peanuts, lentils, mushrooms, and whole grain wheat products. The amount of copper in plant foods ranges from 100 to 500 μg per serving. The recommended intake for infants is given as the Adequate Intake (AI) level i.e., 200—220 μg per day. An RDA has been established in older children and adults. The RDA for children of both genders is 340 (1—3 years), 440 (4—8 years), and 700 (9—13 years) μg per day. The RDA for adults is approximately 900 μg per day with a slight increase to 900 μg per day for pregnant women. The recommended intake given as a daily value is 1000 μg 1 g.

Biological importance and supplementation

Copper is a cofactor for numerous proteins and enzymes. It participates in cellular functions of skeletal muscle, heart, bone, connective tissue, and neural transmission in the brain. Cytochrome oxidase is a metabolic enzyme for which copper is an essential cofactor. This enzyme functions in the glycolysis/oxidative phosphorylation pathway producing energy (ATP) from carbohydrate. Copper is also a cofactor for the enzyme lysyl oxidase that cross links and strengthens collagen molecules. Superoxide dismutase is a copper-containing enzyme that provides a defense against reactive oxygen species (ROS). Copper is abundant in foods and thus the incidence of deficiency is low. The body has an effective homeostasis mechanism for maintaining a balance of copper and dietary supplements are needed only under conditions of disease or abnormal diet.

Fluorine (F)

Fluorine (Table 5.2) is a trace element primarily found in bones and teeth. Fluorine increases the strength of bone and tooth enamel through interaction with calcium. The mineral hydroxyapatite, a complex of calcium and phosphate, is responsible for the strength of these hard tissues. Fluorine forms a complex with these two minerals that increases its strength. It can be obtained in the diet and from water to which the mineral has been added. The AI for infants under one year is quite low, at 0.01—0.5 mg per day. The AI for children from one to nine years ranges from 1 to 2 mg per day and that for older children and adults is 2—3 mg per day. Fluoridation of water has been used in most municipal water supplies in the United State for the past 70 years. Fluoridation of water (typically in the range of 1—2 mg per liter) lowers the incidence of tooth decay. Fluorine can also be obtained from foods, such as fruit juices, fish (especially sardines), rice, and processed chicken or turkey. The established intake level (AI) for fluoride varies with age and is the same for both genders. Mechanically deboned poultry (chicken and turkey) in the form of nuggets, sausage, and lunch meat contains calcium and fluorine. Minerals are present in these products because the process used to make them reduces bone to small particles and makes it unnoticeable. The level of fluorine in these products is approximately 2—5 mg per kg (2.2 lbs). While this level is low, there is concern for excessive intake in the diet of children. The upper limit of intake in infants and children is 1 mg and 10 mg per day, respectively.

Biological importance and supplementation

Fluorine hardens tooth enamel that increases strength and resistance to decay. Microorganisms inhabiting the mouth produce acids that can dissolve enamel. Fluorine works by inhibiting the metabolism of mouth bacteria. Specifically, fluorine inhibits metabolic enzymes that convert sugars into energy. In this regard, brushing with fluorine-containing tooth paste is more effective in preventing cavities than eating food or drinking water. Fluorine together with vitamin D is used for the treatment of osteoporosis (Vestergaard et al., 2008). Higher doses of fluorine, such as those given for clinical treatment of osteoporosis, have shown increase in bone density but may not increase bone strength. This may result in more serious change in bone structure that occurs over time. The data regarding dietary fluorine and the incidence of bone (hip) fracture are inconclusive regarding its preventative ability. The recommended fluorine intake given as daily value is 4 mg for adults. The Upper Intake Level (UL) is 10 mg per day for adolescents and adults of both genders. The UL for infants and children (1–8 years) is substantially lower, approximately 1–2 mg per day. Given the low UL for infants and children, it is recommended that their intake from all sources, including toothpaste, dental treatments, food and water, be monitored.

Iodine (I)

Iodine (Table 5.2) is essential to proper functioning of the thyroid gland. Hormones produced by the thyroid have roles in metabolism, growth, and neurological development (Higdon and Drake, 2012). Iodine can be obtained from seafood (fish and shellfish), cow's milk, potatoes, and beans. Seafood and seaweed (kelp) are highest in iodine because they absorb it from sea water. In the United States, table salt is fortified to contain 45 μg of iodine per g of salt. Sea salt, which is obtained by drying salt water, contains only about half that amount per g. It is difficult to know the level of iodine in processed food because salt used in these foods is not required to contain iodine. The dietary intake for iodine given as RDA for children (older than one year) and adults 90 to 150 μg per day. The intake for pregnant and breast-feeding women is higher with the RDA of 220–250 μg per day. The intake for infants less than one year is given as the AI is 110–130 μg per day.

Biological importance and supplementation

Iodine's principle biological role is in the production of thyroid hormones, thyroxine and triidothyroxine that are stored in the thyroid gland in a protein called thyroglobulin. The thyroid gland adsorbs iodine from the blood to use in making thyroxine and triidothyroxine. The release of these hormones is under control of the pituitary gland in the brain. The level of thyroid hormones strongly influences human rate of metabolism and the proper function of the brain and reproductive system. Iodine deficiency, therefore, can have multiple consequences. An enlarged thyroid gland (goiter) can result from a slight iodine deficiency. More serious consequences are caused by severe iodine deficiency in pregnant women. Thyroid deficiency during gestation affects development of the fetal nervous system and impairs cognitive ability in the child. Most of the population receive adequate amounts of iodine from food and salt. It is recommended that pregnant and breast-feeding women increase their intake during these periods with an additional supplement of 150 μg per day (Becker et al., 2006). Most multivitamin-mineral supplements contain 100% of the daily value (as potassium iodide). The Upper Intake Level (UL) for children varies with age and specific recommendations are given by the American Thyroid Association (Leung et al., 2015).

Iron (Fe)

Iron (Table 5.2) has several essential functions in the body. It exists in two major forms: protein bound (iron-dependent proteins) and a complex with heme (heme iron). Iron-dependent proteins, like transferrin, are important in regulating the level of free iron in biological fluids. Heme iron is important for carrying oxygen in blood (hemoglobin) and muscle (myoglobin). The absorption of iron by the body is greater for heme iron compared to non-heme iron. While heme iron accounts for only 15% of the iron in the diet, it provides about a third of the iron absorbed. Food sources of heme iron include meat, poultry, and fish. Plants are a greater source of non-heme iron. Animal sources, such as liver, contain the most iron per serving and plants sources, such as beans, potatoes and spinach contain half or less than form animal sources (Table 5.2). The amount of iron in the diet given as the RDA is approximately 7—11 mg per day for infants of both genders. The RDA for children ranges from 7 to 10 mg per day from age one to thirteen. Males age 14—50 years have an RDA of 8—11 mg per day. The RDA for females age 14—50 years is higher (i.e., 15—18 mg per day) (NIH government fact sheet: https://ods.od.nih.gov/factsheets/Iron-HealthProfessional/).

Biological importance and supplementation

The most important biological function of iron involves oxygen transport. Heme is an iron-containing component of hemoglobin and myoglobin that is essential to their oxygen carrying roles. Hemoglobin carries oxygen from the lungs to muscle where it is transferred to myoglobin. Oxygenated myoglobin in muscle is ultimately used in the metabolism of carbohydrate into energy by cytochromes. Cytochromes are heme-containing enzymes and perform electron transfers essential for production of adenosine triphosphate (ATP). ATP provides muscle with the energy required for contraction. Heme iron is used enzymes, such as catalase and peroxidase that provide protection against reactive oxygen species (ROS). Free (non-heme) iron is essential to proteins such as, transferrin, ovotransferrin, and lactoferrin, that have antimicrobial properties in egg and milk.

Iron deficiency is common to many in the United States and other regions of the world. The deficiency is a form of anemia with symptoms including muscle weakness and shortness of breath. Iron deficiency results in a low level of red blood cells and hemoglobin. Iron is required for the synthesis of red blood cells in bone. There are several causes for anemia, but dietary deficiency (iron-deficiency anemia) can be corrected using dietary supplements. It is suggested that pregnant women take iron supplements as a protection against anemia. The requirement for iron is elevated in the second and third trimesters of pregnancy (RDA 27 mg per day). Some components of food can negatively affect non-heme iron absorption from the gut. Tannins and other polyphenolics (in tea, coffee, and wine) form complexes with iron and inhibit its absorption. Phytic acid, a component of legumes and nuts, is a potent inhibitor of iron absorption from the gut because the complex is resistant dissociation. For example, Vitamin C and organic acids increase intestinal absorption of non-heme iron. Most citrus fruit contains both vitamin C and organic acids and could boost its absorption.

Manganese (Mn)

Manganese (Table 5.2) is a mineral that the body uses as an enzyme cofactor in the metabolism of glucose. Manganese can be obtained for plant foods such as whole grains, nuts, pineapple, and green leafy vegetables. Almonds, peanuts, and pecans contain approximately 0.5 mg per serving. Navy, pinto, and lima beans also contain approximately the same level of manganese. The intake of manganese in the diet is given by the AI level and varies primarily with age. Infants from zero to six months have a very low requirement for manganese, about 0.003 mg (3 μg) per day, and this doubles to 6 μg per day from seven to twelve months. The AI for children for age 1—8 years is 1.2—1.5 mg per day. The AI increases to 2.2 mg (males) and 1.6 (females) in adolescents and adults. The maximum AI level is 2.6 mg per day for breast-feeding women. However, it should be noted that magnesium can be toxic at very low levels and its upper limit of intake is 11 mg per day.

Biological importance and supplementation

Manganese is a cofactor for enzymes found predominately found in liver, kidney, and bones. In addition to its role in the metabolism of carbohydrates, it is a cofactor of superoxide dismutase (SOD). SOD is an enzyme that catalyzes to the conversion of reactive oxygen species (ROS) to non-toxic products. Manganese is also essential in the formation of bone and connective tissue. Manganese is a cofactor for the prolidase enzyme in the production of amino acid proline. Proline is needed for the synthesis of collagen that is made into connective tissue and bone. Manganese deficiency has been difficult to demonstrate and the consequences from excessive intake are a greater concern. Supplementation may be needed only in the case of individuals who are under long term total parenteral nutrition.

Molybdenum (Mo)

Molybdenum (Table 5.2) is an element that does not occur as free form. It exists in soil and biological materials as complex with molybdenum and oxygen (MoO_4^{-2}). Molybdenum ion is a water-soluble compound that is taken up from the soil by nitrogen fixing plants. Legumes like beans, lentils and peas, are thus good sources of molybdenum. Kidney beans, black beans, and soybeans are also good sources of molybdenum. The dietary intake (RDA) for molybdenum (established by Food and Nutrition Board, Institute of Medicine, 2001) is in the low μg per day range. Molybdenum requirement for infants of both genders is specified as the AI level of 2—3 μg per day. The RDA for children of both genders age 1—14 years, and ranges from 17 to 43 μg per day. Adolescents and adults have an RDA of 43—45 μg per day. The level of intake for adults given as daily value is 75 μg per day.

Biological importance and supplementation

In some plants, molybdenum functions as a cofactor for enzymes that fix atmospheric nitrogen (i.e., nitrogen fixation). Molybdenum is also a cofactor essential to the activity of xanthine oxidase that is an enzyme functioning in the catabolism of nucleotides. Molybdenum's role as a cofactor includes enzymes, such as aldehyde oxidase and sulfite oxidase. Sulfite oxidase is essential in the process of metabolism. It catalyzes the breakdown of sulfur containing amino acids, cysteine and methionine. Supplements in the form sodium or ammonium salts of molybdenum are available, but cases of molybdenum deficiency are rare in Western cultures.

Selenium (Se)

There is considerable variation in the selenium content of foods due to differences in the soil content of selenium (Table 5.2). Typical values reported for the selenium content of plant and animal foods in the USDA Food Composition Data base are as follows. Sunflower seeds and Brazil nuts contain the highest level of selenium, approximately 78 and 56 μg per 100 g. Members of the *Brassica* genus of plants, such as broccoli, cabbages, mustards, kale, cauliflower, turnip, and kohlrabi, contain selenium in the range of 1—5 μg per 100 g. Similarly, members of the *Allium* genus, such as garlic and onion contain selenium at approximately 14 and 1 μg per 100 g, respectively. Whole wheat and rice contain approximately 34 and 15 μg per 100 g, respectively. Selenium-containing animal foods include shellfish, tuna, salmon (40—70 μg per 100 g) and beef, pork and chicken, (20—35 μg per 100 g). The recommended dietary intake (RDA) for selenium is the same for both genders, but varies with age. The RDA for infants under one year of age is 15—20 μg per day. The RDA for children from one to thirteen years is 20—40 μg per day and that for adolescents and adults is 55 μg per day. Pregnant and breast-feeding woman have a slightly higher RDA (60 and 70 μg per day). The recommended level of intake give as daily value, is 70 μg per day. The UL for selenium is 400 μg per day.

Biological importance and supplementation

Selenium is a mineral with important biological functions. It has antioxidant properties providing molecules like DNA and proteins from damage by free radicals. Selenium complexed with cysteine and methionine amino acids become selenoproteins with important roles in cell maintenance, skeletal muscle regeneration, calcium homeostasis, thyroid hormone production and immune response. Selenium deficiency or mutation in genes encoding for selenoproteins synthesis is linked to a variety of diseases, including cardiomyopathy (Keshan disease), impaired immunity, and cancer (Bellinger et al., 2009). Sodium salts of selenium (sodium selenite and sodium selenate) are inorganic forms commonly used as dietary supplements. These forms have low levels of absorption compared to foods (Bermingham et al., 2014). It should be cautioned that selenium is potentially toxic. The UL for selenium in adults is 400 μg per day. Intakes of selenium at the gram level have caused fatal reactions.

Zinc (Zn)

Zinc (Table 5.2) is a mineral essential for normal growth, reproduction, and the immune system. Zinc can be obtained from a variety of foods, including shellfish, meat, nuts and legumes. Typical values reported for the zinc content of plant and animal foods as given by the USDA Food Composition Data base are as follows. Oysters have the highest level of zinc and contain approximately 38 per 100 g. The content of zinc of sesame and pumpkin seeds is 5—7.6 mg per 100 g, beans and legumes (peanuts) 1.5 to 6 per 100 g, and meat (beef, pork and poultry) contain about 1.7—12 per 100 g (poultry being lowest). Plant sources, like legumes, contain a substantial level of phytic acid that limits zinc bioavailability due to formation of a tightly bound complex between zinc and phytic acid. Zinc bioavailability is improved in fermented foods because yeasts are able to breakdown zinc-phytic acid complex. The recommended dietary intake (RDA) for zinc varies primarily by age (Food and Nutrition Board Institute of Medicine. 2001). The RDA for infants is 4—5 μg per day and that for children from 1 to 8 years is 7—12 mg per day. The RDA children from 9 to 13 years is 13 mg per day, adolescence for 14—18 years is 34 mg per day and adults over 19 years is 40 mg per day.

Biological importance and supplementation

Zinc is a cofactor for numerous enzymes functioning in a wide spectrum of biological processes. Zinc is especially important in prenatal and postnatal development. Infants, children, adolescents and pregnant woman are at highest risk for zinc deficiency condition that can affect the immune and reproductive systems (Roohani et al., 2013). Zinc deficiency increases the risk of infection such as pneumonia. Zinc supplementation is effective in correcting deficiency related conditions resulting from chronic kidney disease, sickle cell anemia, alcoholism, and inflammatory bowel disease. Zinc supplements are available in several forms, including its salts, zinc acetate, gluconate, and sulfate. The UL dose for zinc is 40 mg per day. Doses in excess of 225 mg per day can cause acute gastro-intestinal reactions and long-term intakes of 60 mg per may also impair copper absorption.

Can supplements be harmful?

Vitamin supplements are highly promoted as a means to improve health. Food and beverage products touting health claims of a product's vitamin content are ubiquitous. However, there is little evidence to substantiate the advantage of taking vitamin supplements by normal, healthy individuals. Conversely, over-consumption of supplements can be harmful. Vitamin A, a fat-soluble vitamin, can have adverse effects on bone. Extra amounts of vitamin A are stored in the liver and fat depots and are slow to be used or eliminated. Vitamin E, also a fat-soluble vitamin, provides antioxidant properties, but over-consumption increases the risk of heart attack. Supplemental intake of iron, while necessary for those with anemia, can also increase the risk of heart attack in normal individuals. Vitamin supplements have been shown to interact with other medications and cause adverse effects. Reliance on supplements, instead of nutrient rich foods, also lacks other components such as dietary fiber, phytochemicals, and essential fatty acids that can affect the gut microbiome and the body as a whole (Dietary Supplement Fact sheets NIH, 2018).

> ## Summary

This chapter provides an overview of vitamins and minerals (micronutrients) their biological importance, and sources in food. They are well established components of diets that support growth and maintain health. This function results from concerted interaction of micronutrients within the body. Micronutrients in food are susceptible to loss and degradation in the transition from farm to fork. Their loss can occur naturally from senescence, exposure to the environment, and/or from processing. The means of loss is described for each micronutrient included in this chapter. In general, however, heat, light, exposure to oxidizing conditions, and leaching are major causes of loss, especially for water soluble micronutrients. Fat soluble vitamins (A,D,E,K) are especially susceptible to loss from free radical oxidation reactions. Limiting micronutrient loss through better understanding of the chemistry of food, represents an important approach to preserve nutrient content and meet the nutritional needs of a growing world population

Glossary

Adequate Intake (AI): The recommended average daily intake based on experientially determined estimates of nutrient intake by a group of healthy people.

Anabolism: Biological process resulting in the synthesis of complex molecules from smaller ones. For example, amino acids are linked together to make proteins.

Catabolism: Biological process resulting in breakdown molecules for energy. For example, carbohydrates are a major source of energy for cells, tissues and organisms.

Cofactor: Metal ion or organic molecule that assist enzymes in preforming their function.

Dietary reference intake (DRI): A system of dietary guidelines that iThe ncludes several different types of reference values, such as RDA, EAR, and UL.

Estimated Average Requirement (EAR): The average daily nutrient intake that is estimated to meet the requirements of 50% of the healthy individuals in a life stage and gender group.

Food Enrichment: Addition of nutrients to foods, that have been lost in processing (e.g., B vitamins in wheat flour).

Food Fortification: Addition of nutrients that may not have occurred naturally in the food (e.g., vitamin D in milk).

Free Radical: An uncharged atom or molecule having an unpaired valence electron. Glucose tolerance and its description is the next item in this list of definition and not part of the definition for Free Radical Glucose Tolerance: Ability of the body to maintain normal glucose levels when challenged with a carbohydrate load

Homeostasis: Ability to maintain an equilibrium absorption (gain) and excretion (loss) of a substance.

Homeostasis (calcium): Regulation of calcium concentration in blood and bone. The process is controlled by hormones and vitamin D.

International Unit (IU): A unit of measurement for the amount of a substance (vitamins in this case) based on its biological activity. The amount of vitamin meeting that activity level is defined through international collaborative study by the World Health Organization (WHO).

Osteoporosis: A condition in which bone density and strength are lost as a result of inadequate calcium intake. Lack of adequate estrogen level in postmenopausal women is a contributing factor to greater risk of bone fracture

Osteopenia: A condition in which bone mineral density is lower than normal and may result in osteoporosis

Percent Daily Value (%DV): The amount each nutrient listed on a food label, per serving.

Sarcopenia: Degenerative loss of muscle mass and strength occurring most often with age.

Tolerable Upper Intake Level (UL): The highest average daily nutrient intake that is likely to pose no risk of adverse effects to almost all individuals in the general population.

Recommended Dietary Allowance (RDA): The nutrient intake level that is sufficient to meet the nutrient requirements of nearly all (97%) healthy individuals by gender and life stage.

Reduction-oxidation (REDOX) reaction: Chemical reactions involving the gain or loss of electrons from molecules or atoms.

Retinol Activity Equivalents (RAE): Difference between the amount of retinol in the food and that actually absorbed

References

Becker, D.V., Braverman, L.E., Delange, F., 2006. Iodine supplementation for pregnancy and lactation-United States and Canada: recommendations of the American Thyroid Association. Thyroid 16, 949—951, 2006.

Bellinger, F.P., Raman, A.V., Reeves, M.A., Berry, M., 2009. Regulation and function of selenoproteins in human disease. Biochem. J. 422, 11—22.

Berdanier, C.D., Dwyer, J.T., Heber, D., 2013. Handbook of Nutrition and Food, third ed. CRC Press. ISBN 978-1-4665-0572-8.

Bermingham, E.N., Hesketh, J.E., Sinclair, B.R., Koolaard, J.P., Roy, N.C., 2014. Selenium-enriched foods are more effective at increasing glutathione peroxidase (GPx) activity compared with selenomethionine: a meta-analysis. Nutrients 6, 4002—4031.

Bikle, D.D., 2014. Vitamin D metabolism, mechanism of action, and clinical applications. Chem. Biol. 21 (3), 319—329.

Boehnke, C., Reuter, U., Flach, U., Schuh-Hofer, S., Einhaupl, K.M., Arnold, G., 2004. High-dose riboflavin treatment is efficacious in migraine prophylaxis: an open study in a tertiary care centre. Eur. J. Neurol. 11 (7), 475—477.

Clayton, P.T., 2006. B6-responsive disorders: a model of vitamin dependency. J. Inherit. Metab. Dis. 29, 317—326, 2006.

Dror, D.K., Allen, L.H., 2012. Interventions with vitamins B6, B12 and C in pregnancy. Paediatr. Perinat. Epidemiol. 26 (Suppl. 1), 55—74, 2012.

Erdman, J.W., MacDonald, I., Zeisel, S.H., 2012. International Life Sciences Institute. Present Knowledge in Nutrition, tenth ed. International Life Sciences Institute, Ames, Iowa.

Food and Nutrition Board (FNB), Institute of Medicine, 2001. Vitamin A. Dietary Reference Intakes for Vitamin A, Vitamin K, Arsenic, Boron, Chromium, Copper, Iodine, Iron, Manganese, Molybdenum, Nickel, Silicon, Vanadium, and Zinc. National Academy Press, Washington, D.C., pp. 65—126 (National Academy Press).

Higdon, J., Drake, V.J., 2012. An Evidence-Based Approach to Vitamins and Minerals: Health Benefits and Intake Recommendations, second ed. ISBN 978-3131324528.

Higdon, J., Drake, V.J., 2013. An Evidence-Based Approach to Phytochemicals and Other Dietary Factors, second ed. ISBN 978-313141842.

Holick, M.F., 2003. Vitamin D: a millennium perspective. J. Cell. Biochem. 88 (2), 296—307.

Leung, A.M., Avram, A.M., Brenner, A.V., Duntas, L.H., Ehrenkanz, J., Hennesy, J.V., Lee, S.L., Pearce, E.N., Roman, S.A., Stagnaro-Green, A., et al., 2015. Potential risks of excess iodine ingestion and exposure: statement by the American thyroid association public health committee. Thyroid 25, 145—146.

Moshfegh, A., Goldman, J., Ahuja, J., Rhodes, D., LaComb, R., 2009. What We Eat in America, NHANES 2005-2006: Usual Nutrient Intakes from Food and Water Compared to 1997 Dietary Reference Intakes for Vitamin D, Calcium, Phosphorus, and Magnesium. US Department of Agriculture, Agricultural Research Service.

Murphy, S.P., Yates, A., Atkinson, S.A., Barr, S.I., Dwyer, J., 2016. History of nutrition: the long road leading to the dietary reference intakes for the United States and Canada. Adv. in Nutr. 7, 157—168.

Roohani, N., Hurrell, R., Kelishadi, R., Schulin, R., 2013. Zinc and its importance for human health: an integrative review. J. Res. Med. Sci. 18 (2), 144—157.

Ross, A.C., 2014. Vitamin A. In: Ross, A.C., Caballero, B., Cousins, R., Tucker, K., Ziegler, T. (Eds.), Modern Nutrition in Health and Disease, eleventh ed. Lippincott Williams and Wilkins, pp. 260—277.

Selhub, J., Bagley, L.C., Miller, J., Rosenberg, I.H., 2000. B vitamins, homocysteine, and neurocognitive function in the elderly. Am. J. Clin. Nutr. 71 (2), 614S—620S.

Sutton, A.L., MacDonald, P.N., 2003. Vitamin D: more than a "bone-a-fide" hormone. Mol. Endocrinol. 17, 777—791.

Vestergaard, P., Jorgensen, N.R., Schwarz, P., Mosekilde, L., 2008. Effects of treatment with fluoride on bone mineral density and fracture risk—a meta-analysis. Osteoporos. Int. 19, 257—268.

Weber, D., Grune, T., 2012. The contribution of beta carotene to vitamin A supply of humans. Mol. Nutr. Food Res. 56, 251—258.

Internet resources

https://www.accessdata.fda.gov/scripts/interactivenutritionfactslabel/factsheets/vitamin_
    and_mineral_chart.pdf [Accessed Dec 2018].

Dietary supplement fact sheets, 2018. NIH National Institutes of Health. https://ods.od.nih.
    gov/factsheets/list-all/.

Food Labeling Guide FDA, 2018. Food and Drug Administration. https://www.fda.gov/
    food/guidanceregulation/guidancedocumentsregulatoryinformation/labelingnutrition/
    ucm2006828.htm Revised 9-16-2018.

Nutrient recommendations, 2018. NIH National Institutes of Health. https://ods.od.nih.
    gov/Health_Information/Dietary_Reference_Intakes.aspx.

Vitamin and Mineral Summary Chart. FDA Food and Drug Administration.

Vitamins and Minerals. OSU MIC (Oregon State University Linus Pauling Institute Micro-
    nutrient Information Center, 2018. http://lpi.oregonstate.edu/mic.

Further reading

Berdanier, C.D., Dwyer, J.T., Heber, D., 2013. Handbook of Nutrition and Food, third ed.
    CRC Press. ISBN 978-1-4665-0572-8.

Food and Nutrition Board (FNB), Institute of Medicine, 2001. Vitamin A. Dietary Refer-
    ence Intakes for Vitamin A, Vitamin K, Arsenic, Boron, Chromium, Copper, Iodine,
    Iron, Manganese, Molybdenum, Nickel, Silicon, Vanadium, and Zinc. National Acad-
    emy Press, Washington, D.C., pp. 65−126 (National Academy Press).

Murphy, S.P., Yates, A.A., Atkinson, S.A., Barr, S.I., Dwyer, J., 2016. History of nutrition:
    the long road leading to the dietary reference intakes for the United States and Canada.
    Advances in nutrition (Bethesda, Md.) 7 (1), 157−168.

Rx for Health, 2018. OSU MIC. Oregon State University Linus Pauling Institute Micronu-
    trient Information Center. https://lpi.oregonstate.edu/publications/rx-health.

USDA Food Composition Database, 2018. https://ndb.nal.usda.gov/ndb/.

Review questions

1. Define the term vitamin.
2. Define the terms DRI, RDA, DV, AI, and IU.
3. What information is provided Nutrition Facts labels concerning vitamins and minerals?
4. What is the primary cause of loss for water soluble vitamins?
5. Which form of vitamin B6 is used in fortification and enrichment?
6. Define the terms vitamin fortification and enrichment.
7. What biological function is common to the B group of vitamins?
8. Name a biological function of vitamin C.
9. What disease is associated with vitamin C deficiency?
10. What foods represent a good source of vitamin C?
11. Why is it important for pregnant women to take folic acid supplements?
12. Which vitamin is destroyed when sulfite is added to food?
13. Which vitamins are good antioxidants?

14. What chemistry is responsible for major losses of fat-soluble vitamins?
15. What foods are good sources of Vitamin A?
16. Name a biological function supported by vitamin A.
17. Is a pumpkin with faded color a poor source of vitamin A? Explain your answer.
18. Name a biological function supported by vitamin D.
19. What foods represent good sources of vitamin D?
20. What food are a good source of vitamin E?
21. What vitamins are essential to metabolizing carbohydrates, fats, and proteins?
22. Which vitamins are likely to be destroyed by heating?
23. What vitamins are lost when vegetables (peas & beans) are blanched?
24. Explain why vitamin D is important to bone strength.
25. Can vitamin supplements be harmful?
26. Name the minerals important to bone strength.
27. Which mineral is essential to thyroid function?
28. Name a biological function for iron.
29. Which minerals have antioxidant activity.
30. What is the greatest source of sodium in the diet? The salt shaker or food? Explain your answer. Hint, the cheese label shows a high level of sodium, why doesn't it taste salty?

CHAPTER SIX

Flavors

Learning objectives

This chapter will help you describe or explain:

- How flavor is created.
- How taste and smell are perceived.
- The chemistry of taste perception
- Sugar substitutes and questions about their contribution to health
- The importance of non-oral taste receptors

Introduction

While it may be counterintuitive, flavor is not in food, rather it is created by the brain from food. The perception we know as flavor is a multi-modal response to chemical and physical inputs that are transmitted to and interpreted by the brain. Flavor inputs are generated when molecules bind to receptors in the mouth and/or nasal cavity. The sense of taste is different from that of smell. It is limited to five basic modalities: sweet, bitter, umami, sour and salty. The perception of taste is also less sensitive than smell perception because it requires higher concentrations of tastant molecules to achieve the threshold for detection. Contrary to earlier models, receptors for sweet, bitter, umami, sour, and salty are not separately clustered into like-sensing

Introduction to the Chemistry of Food
ISBN: 978-0-12-809434-1
https://doi.org/10.1016/B978-0-12-809434-1.00006-2

regions of the tongue. Rather, each taste bud contains receptors for all modalities and are widely distributed on the tongue and palate.

The sense of smell differs from that of taste in its greater discriminatory ability and sensitivity. Smell is created by olfactory receptors capable of detecting millions of compounds at a very low level, typically at part per million (ppm) and lower. To illustrate this, you can think of 1 ppm as 1 inch in 16 miles of road, 1s in 12 days, or 4 drops of ink in a 55 gallon drum. Smell is derived from volatile chemicals in the food. The act of chewing disrupts the food matrix and releases compounds that enter the nasal cavity where they bind to and stimulate olfactory receptors. Olfactory inputs dominate the sense of flavor. You can easily test this idea by pinching your nose while eating something. Almost any food will do for this test, but an apple can provide a good example. Bite into the apple and note its flavor while holding your nose; then notice what happens when you release the hold on your nose. If you also closed your eyes during this exercise, you may not recognize the food as an apple until you stop holding your nose.

Physical inputs, such as texture and color, also influence flavor perception. Food texture contributes to flavor perception through the act of chewing. Force generated by muscles involved in chewing sends signals to the brain that further contribute to flavor perception. Hard or gritty textures are not as pleasant as soft, smooth ones and not likely to be favored. The smoothness of a food, referred to as mouthfeel, is associated with high fat content. This is a desirable attribute because humans are hardwired for evolutionary reasons to prefer foods with high nutrient density. Color is a physical attribute that has a significant influence on flavor perception. In fact, color is the major factor used in selecting food items. The red, yellow, and orange colors of fruits and vegetables indicates ripeness and flavor, whereas a green or brown color is associated with unripe or spoiled food. Ultimately, the phenomena we know as flavor results from the combined effects of food molecules, texture, and color, as perceived by the brain.

This chapter includes questions that will that will help you explore and better understand how taste and flavor perception occur.
- What is a gustatory taste map?
- What is the mechanism of sweet taste perception?
- Do sugar substitutes contribute to health?
- Why is beer bitter?
- What is umami taste?
- What is hidden salt?
- How is flavor created in the brain?

Taste buds and receptors

Taste buds contain molecular mechanisms for the perception of the five basic taste modalities: sweet, bitter, umami, salty, and sour. There are likely taste receptors for lipids and complex carbohydrates (starch), but evidence regarding their mechanism of recognition is not fully established. Taste buds are organelles containing receptor cells specific for various tastants. Taste buds for salt, umami, sweet, bitter, and sour tastants are not segregated in discrete regions of the tongue. Recent studies have shown that multiple taste receptors reside within each taste bud. Taste buds are distributed in the front, side, and back regions of the tongue on structures called papillae. These are visible to the eye as small red dots. The tongue is populated with three types of papillae, including circumvallate, foliate, and fungiform. They are organized as shown in Fig. 6.1. Taste buds have tiny pores containing microvilli that are exposed to the surface of the tongue.

Each taste bud is composed of multiple cell types and precursor cells arranged in a garlic bulb-like structure (Fig. 6.1). Division of the cell types is based on their structure and function. Type I cells have a narrow shape and comprise about 50% of the cells in a taste bud. They operate in a supporting role to other taste bud cell types and functionally detect salty or sour tastants. Other cell types are slightly larger than supporting cells and are important because they are populated by receptors for sweet, bitter, and umami tastants. Lastly, taste buds contain precursor cells that can develop

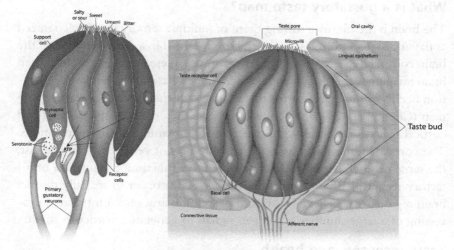

Fig. 6.1 Taste bud structure.

into other cell types (Roper and Chaudhari, 2017). Detection of specific tastants, also known as gustatory stimuli, occurs by routes known as receptors and ion channels. Receptors are protein molecules that participate in a biochemical process responsible for generating an electrical signal to the brain upon binding of a specific molecule. Taste receptors belong to a family of proteins called G-protein coupled receptors (GPCR). They are specialized for binding nutrients, tastants, and poisonous compounds. A portion of the receptor protein is anchored to the cell membrane of taste bud cells, while other regions are exposed to the exterior environment. Slight variations in the amino acid sequence of the receptor protein determine its specificity for sweet, bitter, and umami tastants. Sweet and umami tastants are recognized by type 1 taste bud receptors and given the shorthand notation T1R. Bitter tastants are recognized by type 2 taste receptors and are given the shorthand notation T2R. Ion channels are located on taste bud cell membranes and responsible for detecting salty and sour tastants. Salty taste perception depends on binding positively charged Na^+ ions to a negatively charged component located in pores of the cell membrane. Binding Na^+ ions depolarizes the membrane, resulting in an electrical signal sent to the brain. Sour taste perception involves entry of weak acid molecules into type 1 taste bud cell membranes through its pores. Acetic acid, for example, enters through pores in the cell membrane. The higher pH (6–7) inside the cell causes ionization of acetic acid releasing hydrogen ions and triggers a signal the brain interprets as sour.

What is a gustatory taste map?

The brain is the ultimate sorting agent of multiple sensory inputs recognized as flavors. Signals from taste receptors are received in a distinct region of the brain cortex referred to as the gustatory cortex. Research using sophisticated brain imaging technologies with animal models has shown that taste perception occurs in the brain, cortex this enables construction of a gustatory taste map (Chen et al., 2011). The gustatory taste map contains spatially defined cortical fields corresponding to sweet, bitter, umami, and salty perceptions. The cortical location of sour taste perception is not yet known. In contrast, the sense of smell is a more complex cognitive function. Stimulation of olfactory receptors by odorant molecules does not occur in spatially distinct brain regions. It is believed that odor perception arises from integrated processing of a large number of neuron signals to generate a particular smell.

Taste receptors and health

Discoveries regarding the location of taste receptors have been made in the past decade using biochemical and DNA technologies. GPCR receptors have been found in unexpected locations, such as the gut, pancreas, heart, and upper airway. These receptors, termed extra-oral, do not function in taste perception, rather they have physiological roles (Layden et al., 2010). In the gut, for example, GPCR receptors serve as chemosensors in the absorption of nutrients, such as glucose, and regulation of metabolic processes. When glucose enters the gut, sweet receptors (T1R) in the pancreas are stimulated and respond by releasing a hormone called glucagon-like peptide 1 (GLP-1, for short). GLP-1 stimulates the pancreas to secrete insulin that results in lowering glucose blood level. GLP-1 also slows emptying of the stomach and reduces glycemic load. Similarly, bitter tastants can stimulate type (T2R) receptors in the gut that subsequently cause release of gherlin. Gherlin hormone stimulates the feeling of hunger that causes a short-term increase in food intake followed by a long-term decrease in food intake as the gastric contents are slowly released. Overall, this process reduces food intake. Recently, pharmaceutical companies have explored development of drugs targeting these receptors to mediate obesity and control type II diabetes (Janssen and Depoortere, 2013).

Sweet

Sweet taste perception is generated by a range of molecules, such as sugars (glucose, fructose, sucrose), polyols, glycosides, some D amino acids, and a few proteins. Humans are hardwired for taste, especially sweet tastes, because of the preference for energy dense foods (Acree and Lindley, 2008; Trivedi, 2012). The attribute of sweetness is closely associated with sugars, like fructose and sucrose, but the intensity of sweetness varies widely among other substances (Table 6.1). Fructose is approximately twice as sweet as sucrose. It can be used in place of sucrose to reduce calorie content while providing the same level of sweetness. Galactose and maltose are about one-third as sweet as sucrose. Lactose (milk sugar) is less than one-fifth as sweet. Carbohydrates polymers (oligosaccharides), such as inulin, are about 10% as sweet as sucrose.

What is the mechanism of sweet perception?

The mechanism of sweet taste perception was described by Shallenberger and Acree in 1967. Their model is called the AH/B theory of saporous (agreeable) taste and is illustrated in Fig. 6.2. The model holds that sweet perception involves a multipoint attachment between the tastant molecule and the receptor. Tastant molecules must have two types of chemical groups, AH and B, positioned in a precise geometry to interact via hydrogen bonding with complimentary groups in the receptor (Fig. 6.2). AH groups

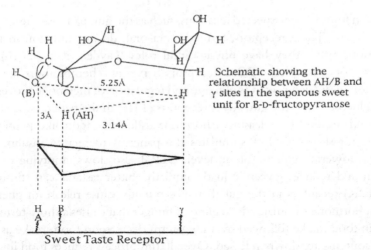

Schematic showing the relationship between AH/B and γ sites in the saporous sweet unit for β-D-fructopyranose

Sweet Taste Receptor

Fig. 6.2 Sweet perception model. *Multipoint Model of Sweet Perception Shallenberger and Acree, 1967.*

provide hydrogen atoms to hydrogen bond with a corresponding electronegative atoms (B). A third site, gamma (γ), contributes a hydrophobic interaction point between tastant molecule and receptor. While not required for sweet taste perception, hydrophobic groups serve to increase the level of sweetness (Nofre and Tinti, 1996; Acree and Lindley, 2008).

Other sweet tastants

While sweet perception is perceived as restricted to sugars, other substances such as polyols, amino acids, sugar substitutes and some proteins are also sweet tasting (Table 6.1).

Table 6.1 Relative sweetness of various compounds.

Name	Composition	Relative sweetness[a]	Calories per g
Fructose	Monosaccharide	1.75	4
Sucrose	Glucose, Fructose Disaccharide	1.00	4
Inulin	Fructose Oligosaccharide	0.1	1.5
Lactose	Glucose, Galactose Disaccharide	0.16	4
Maltose	Glucose Disaccharide	0.4	4
Glucitol (Sorbitol)	Polyol	0.6	2.6
Xylitol	Polyol	1.0	2.4
Glucose	Monosaccharide	0.75	4
D Amino Acids	Glycine, Alanine, and Serine	0.8	4

[a]Normalized to that of sucrose (Joesten et al., 2007).

Amino acids

Several amino acids have a sweet taste that depends on their enantiomeric (D or L) configuration. Some D amino acids are sweet tasting, but L amino acids are not. Enantiomers are stereo isomers (mirror images) of the same molecule. They are chemically equivalent, but biologically different. For example, only L amino acids are found in the proteins of plants and animals. Bacteria are the principal source D amino acids. Sweet tasting amino acids include the D forms of glycine, alanine, and serine. D Glycine is the sweetest of the amino acids, but it is not as sweet as sucrose. Glycine also has interesting tastant properties. It enhances the sweetness of glucose and masks the taste of bitter compounds. Peptides derived from hydrolyzed milk proteins (caseins) and some pharmaceuticals are noted for their bitterness. It is likely that glycine's effect results from its ability to bind sweet receptors and/or block structurally similar bitter receptors. This modifies the intensity of bitter tastants.

Polyols

Polyols are monosaccharide and disaccharide molecules in which the aldehyde group is replaced by a hydroxyl. For example, D glucose and D glucitol (also called sorbitol) share very similar structures. The aldehyde group located on carbon number 1 in glucose (C1, circled in Fig. 6.3) is replaced by a hydroxyl group in glucitol. Polyols, such as glucitol, xylitol, and maltitol, are used to provide sweet taste in food. A major advantage of polyols is the functional ability to provide sweetness with fewer calories per gram.

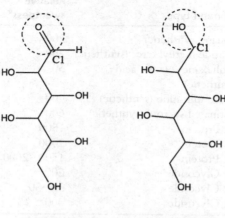

D Glucose   D Glucoitol

Fig. 6.3 Glucose and glucitol.

The absence of an aldehyde group in polyols prevents them from participating in Maillard reaction. Foods in which sugar is replaced by polyol do not brown or develop Maillard flavor during cooking. On a positive note, polyols are not utilized by mouth bacteria and do not contribute to tooth decay. They have a much lower impact on glycemic index. Xylitol and glucitol, for example, have glycemic index values of 4 and 12 respectively, relative to the value of 100 for glucose. Glucitol, xylitol, and other polyols occur naturally in fruits, such as cherries, strawberries, peaches, and plums. They contribute a pleasant sweet taste with a slight cooling sensation in fruits. Polyols are not metabolized as quickly as the corresponding sugars, but gut bacteria are able to ferment them into gas and acids. Unpleasant symptoms, such as bloating and diarrhea, can result from excessive consumption of polyols. For this reason, polyols are classified as FODMAPs (Fermentable, Oligo, Di, Mono, And, Polyols) and capable of causing irritable bowl-like symptoms in some individuals.

Sugar substitutes (synthetic substances)

Sugar substitutes (Table 6.2) are substances that are typically 100 times sweeter than sucrose. They provide a sweet taste with little or no caloric value. There is also little or no impact on glycemic index. Sugar substitutes are obtained through chemical synthesis or extraction from natural sources.

Table 6.2 Sugar substitutes.

Name	Compound type	Relative sweetness[a]	Glycemic index
Acesulfame K	Sulfonate (synthetic)	200	0
Aspartame	Dipeptide methyl ester (synthetic)	200	0
Cyclamate	Cyclohexanesulfamic acid (synthetic)	30	0
Saccharin	Benzoic sulfimide (synthetic)	400	0
Sucralose	Chlorinated sucrose (synthetic)	600	0
Brazzein	Plant Protein	1500	0
Thaumatin	Plant Protein	2000	0
Moellin	Plant Protein	1500—2000	0
Steviol	Plant Glycoside	200	0
Glycyrrhizin	Plant Glycoside	30—50	0
Mogroside	Plant Glycoside	300	0

[a]Normalized to sucrose.

Synthetic sugar substitutes are subject to extensive investigation by regulatory agencies, such as FDA in United States and EFSA in Europe, before they are deemed safe for consumption. The most widely used synthetic sugar substitutes are aspartame, acesulfame K, and saccharin. Sugar substitutes derived from natural sources (plants) represent a growing trend among industry producers in response to public preference for avoiding synthetic substances. They represent an alternative to synthetic substances, but many of the substances, especially sweet tasting proteins, are not as widely used because of their higher cost.

Acesulfame K

Acesulfame K is a synthetic substance and the potassium salt of a sulfonate compound. It is approximately 200 times sweeter than sucrose, but can give a bitter aftertaste, especially when used at high concentration. Acesulfame K is often used in combination with sucralose or aspartame to provide a more sucrose-like taste in soft drinks, shakes, and smoothies. This combination is also used to make pharmaceuticals more palatable. Acesulfame is moderately heat stable. This enables its use in baked foods. The ADI of acesulfame K is 15 mg per kg of body weight which is equivalent to about 1000 mg for a person weighing 165 lbs.

Aspartame

Aspartame is a sweet-tasting synthetic dipeptide composed of aspartic acid and phenylalanine. The free carboxyl group of phenylalanine contains an esterified hydroxyl group. It is approximately 200 times sweeter than sucrose and its taste is very close to that of sucrose. Aspartame is used at very low levels and this makes it essentially calorie free in food applications. Aspartame's chemical stability is best under acidic conditions (pH 3–5) and is well suited as a sweetener for carbonated beverages. Its shelf life is about 1 year at acidic pH, but this decreases to less than one week at pH 7. A slow release of methanol occurs from this degradation over this time, but the level of methanol is too low to be of concern. The ADI for aspartame is 50 mg per kg of body weight, corresponding to about 22 cans of diet soda for a 175 lb person. Aspartame is not suitable for use in cooked foods. Heating causes its decomposition to diketopiperazine and loss of sweet taste occurs. When eaten as part of food, aspartame is hydrolyzed in the gut to aspartic acid, phenylalanine, and methanol. These products are utilized or excreted without ill effect for most individuals. However, some individuals are unable to metabolize phenylalanine and can experience a toxic build-up of this amino acid. The condition known as phenylketonuria (PKU) is caused by a gene defect in protein metabolism. PKU can cause irreversible brain damage if the condition is undetected. Newborns having this genetic defect are most at risk. Fortunately, PKU can be detected at birth with a simple blood test.

Cyclamate

Cyclamate, the sodium salt of cyclamic acid, is approximately 30 to 50 times sweeter than sucrose. Like many synthetic sweeteners, cyclamate has an unpleasant after-taste and this limits its use in food. It was originally purposed as an additive to improve the taste of pharmaceuticals. A mixture of cyclamate and saccharin (10:1) is often used because the combination masks the off taste of both compounds. Cyclamate is heat stable and can be used in baking applications. It was widely used in the United States until 1969 when research studies suggested a link between the compound and bladder cancer in mice. FDA subsequently removed its GRAS status in 1970 and banned its use in the United States. The findings regarding a link between cyclamate and cancer in rats have been contradicted by more recent studies. While cyclamate is banned in the United States, it is approved for use in over 50 countries. It should be cautioned that cyclamate may have toxic effects and more extensive studies have been recommended (Chattopadhyay et al., 2014).

Saccharin

Saccharin (benzoic sulfimide) is a synthetic sweetener approximately 300–400 times sweeter than sucrose. The name saccharin implies sugary taste, but in fact it has a strong metallic after-taste. This taste defect is masked by combining it with cyclamate, as mentioned previously. Saccharin is a heat stable compound with limited water solubility. This property is greatly improved when converted to its sodium saccharin salt form, sodium. Saccharin was discovered in the late nineteenth century and its use continues today. The compound, however, is not without controversy. In 1977, FDA attempted to ban saccharine after studies found that high doses could cause bladder cancer in rats. This effort to ban saccharin or place warning labels on food products was unsuccessful as a result of political pressure. In fairness, subsequent studies found that differences in the physiology of rats, compared to humans, was likely the cause of malignant tumors in rats. In 1995, the

European Scientific Committee for Food concluded that saccharin does not pose a cancer risk to humans. Saccharin is approved for use today in the United States, Canada, and the European Union. The world-wide use of saccharin is third, behind sucralose and aspartame. The ADI for saccharin is 15 mg per kg of body weight, which equates to approximately 1,100 mg for a 175 lb person.

Sucralose

Sucralose is a synthetic sweetener that is approximately 600 times sweeter than sucrose. It is made by partial chlorination (replacing OH groups with Cl) of sucrose. Sucralose is very heat stable and can be obtained in powdered form for use in baking applications. It has numerous food applications, including candy, soft drinks, and sweet syrups. Sucralose is available in granulated and powdered form. Granulated sucralose is mixed with fillers to provide a measure for measure substitution with table sugar (sucrose). The powdered form of sucralose contains 90% bulking agents such as maltodextrin, that are a metabolizable form of carbohydrate. A 50/50 mixture of sucrose and sucralose, plus a bulking agent, is available for baking applications. This mixture reduces caloric content and enables browning reactions in baking applications. The safety of sucralose has been extensively investigated and is considered safe by FDA, the Joint FAO/WHO Expert Committee Report on Food Additives, and the European Union's Scientific Committee on Food. The ADI for sucralose is 5 mg per kg of body weight which equates to approximately 390 mg for a 175 lb person.

Natural sugar substitutes

Natural sugar substitutes, such as glycoside compounds and proteins originating in plants, have potent sweet-tasting properties. Most of these substances are native to tropical regions of the world. Brazzein, thaumatin, and monellin are examples of a growing list of plant proteins that provide an intense sweet perception. These are an insignificant source of calories in food and do not raise blood sugar levels (Kant, 2005).

Brazzein

Brazzein is a sweet-tasting protein found in the fruit of the native West African Oubli plant (*Pentadiplandra brazzeana*). It is a soluble protein with a sweetness that is approximately 1500 times greater than sucrose.

The protein is small in size, containing only 54 amino acids. It is stable over a wide range of pH (2.5—8) and survives heating at 80°C for 4 h. Brazzein can be extracted from berries of the plant, but large-scale harvesting is harmful to the ecology and not economically viable. Alternatively, it is being produced using fermentation technology by at least one company. Brazzein taste is similar to sucrose with no metallic after taste. It effectively reduces the after-taste of other sugar substitutes when used in combination. Brazzein is being studied by FDA for its potential use in food, but it is not currently approved for use. The potential to cause food allergy of brazzein and other novel proteins represents a concern that needs to be examined before it can be deemed safe (Barre et al., 2015).

Thaumatin

Thaumatin is a sweet tasting protein with an intensity approximately 2,000 times greater than sucrose. However, its sweetness is perceived more slowly than sucrose and lingers longer with a licorice-like after-taste. It must be blended with other sweeteners or sucrose to achieve a desirable taste. Thaumatin protein is relatively small and contains 207 amino acids. It is found in the fruit of the katemfe plant (*Thaumatococcus daniellii*) native to West Africa. The protein is heat stable to 70 °C, but losses its sweetness at higher temperature. Solutions of thaumatin are stable throughout a wide range of pH (2.5—10). Thaumatin has been approved for use a sweetener in the United States and the European Union, but no ADI level has been established. While thaumatin is considered to be safe for food use, the protein shares structural similarities with allergens found in apple and kiwi fruit and is resistant to gastric digestion (Bublin et al., 2008). These properties raise the potential that thaumatin could also be allergenic.

Monellin

Monellin is a protein containing 94 amino acids. Its sweetness is approximately 1500—2000 times sweeter than sucrose (Kant, 2005). Monellin is found in the serendipity berry of the *Dioscoreophyllum cumminsii* plant. Like thaumatin, monellin has a slow onset of sweetness response and a licorice-like after taste. Blending monellin with other sweeteners reduces the after-taste effect and can synergistically affect the intensity of sweet perception. Monellin is stable to a wide range of pH (2—9), but it is destroyed by heating above 50 °C. The lack of heat stability limits its potential use in processed foods. The safety of monellin has not been investigated and thus it is not approved for use as a sweetener.

Miraculin

Miraculin is a protein found in the fruit of the miracle fruit plant (*Synsepalum dulcificum*). The protein is not sweet tasting by itself. It has a novel taste-modifying action. It can change the taste perception of a sour substance to sweet. Its modifying taste action results from binding sweet taste receptors when acids are simultaneously present. Miraculin is not approved for use in the United States, but has been given novel food status in the European Union. A novel food is defined as a type of food that does not have a significant history of consumption or is produced by a method that has not been previously used for food.

Steviol

Steviol is a widely used sugar substitute derived from the Stevia plant originating in South America. Steviol is a glycoside type compound occurring in plant leaves. As the term glycoside implies, the steviol molecule contains attached carbohydrates. The stevia plant has long been used by indigenous peoples to sweeten food. Simply chewing the leaves results in a sweet sensation 10 times greater than that of sucrose. The sweetener (steviol) is extracted from the plant's leaves, purified, and sold as the sugar substitute product called Stevia™. It is approximately 200 times sweeter than glucose. Stevia's sweetness tends to linger longer than sugar and can result in a liquorice-like flavor if used at high level. Stevia is heat stabile, pH tolerant, and not fermentable. These properties make it attractive for use in food products. Purified stevia is recognized by FDA as a GRAS (Generally Recognized as Safe) substance. The ADI for stevia is 4 mg per kg of body weight that is equivalent to about 300 mg for a 175 lb person. Stevia does not impact glycemic index. It is the major sweetener used in diet soft drink products marketed as Truvia™ (Coca Cola), and Purvia™ (Pepsi).

Glycyrrhizin

Glycyrrhizin is a plant glycoside extracted from roots of the liquorice plant. It is about 30–50 times sweeter than sucrose, but is not metabolized and has no effect on glycemic index. Its use as a sugar substitute is limited by its intense liquorice flavor. Traditionally, glycyrrhizin extract use is related to its medicinal properties. For example, glycyrrhizin has been used in traditional medicine as an expectorant in cough syrups and as herbal remedy for treating stomach ulcers and constipation. Excess glycyrrhizin intake can elevate blood sodium level and lower potassium level, resulting in high blood pressure and irregular heartbeat. No ADI has been established for glycyrrhizin, but an upper limit of 100 mg per day has been suggested by the European Union.

Mogroside

Mogroside is the glycoside of cucurbitane, a compound with antioxidant activity. This compound is extracted from Monk fruit (s. *Grosvenorii*) and is a native of China and Thailand where it has been used for centuries as a traditional medicine and sweetener. While mogroside exists in several forms in monk fruit, purified mogroside V is the best sweetener. It is approximately 300 times sweeter than sucrose and has been granted GRAS status in the United States. However, mogroside is listed as a novel food in the United Kingdom. Like the other glycosides, mogroside has little or no effect on glycemic index and may be of benefit in managing diabetes.

Do sugar substitutes contribute to health?

Sugar substitutes have been marketed as healthful because of their potential for reducing sugar (sucrose and fructose) consumption and concomitantly lowering the risk of developing type-2 diabetes. However, that has not been the case. Correlational data suggest a link between sugar substitute consumption, weight gain, and type-2 diabetes. While these substitutes produce a sweet taste, some suggest they can also increase appetite. Recent reviews of the literature offer conflicting opinions on this topic (Pepino, 2015; Fernstrom, 2015). The epidemic rise in obesity and paralleled increase in sugar substitute consumption provide a compelling reason to thoroughly examine health effects of sugar substitutes. The use of substitutes is also questioned in light of their potential effects on the gut microbiome (Pearlman et al., 2017).

Bitter

Sensitivity to bitter tastants is an evolutionary trait that has contributed to survival because many poisonous compounds taste bitter. The innate sensitivity to bitter partially explains why bitter tasting foods are less desired and typically an acquired taste. Bitter taste is stimulated by a variety of molecules perceived through T2R type GPCR receptors. There are 20 or more variants of receptor proteins located in oral and extra-oral regions of the body. Extra-oral T2R receptors in gut and pancreas do not create taste perception, but function as chemosensors for nutrient utilization and appetite control. The physiological role of extra-oral taste receptors was demonstrated in mice that were tube-fed bitter tastants. Bitter substances initially caused mice to eat more. However, they quickly returned to eating less (Calvo and Egan, 2015).

Bitter tastants

While avoidance of bitter tasting foods is linked to the fear of being poisoned, some foods are actually prized for their unique, bitter flavors. Arugula, kale, broccoli, coffee, cocoa, citrus peel, olives, and wine are just a few food examples in which bitter components contribute desirable taste attributes. Several bitter tastants and their approximate threshold values for bitterness perception are listed in Table 6.3. Threshold values, given as ppm (parts per million), represent minimum concentration required to produce a bitter taste. These values should be considered to be an approximation because there is a wide range of variation in sensitivity to bitter taste in the human population. For example, PTU (phenylthiourea) is a very bitter compound used as a standard to assess sensitivity to bitter taste. About one-fourth of the population does not detect PTU bitterness. This is proposed to be a result of a genetic mutation in taste receptor protein(s).

Table 6.3 Bitter tastants.

Compound	Bitterness threshold[a]	Source
Denatonium	0.05	Synthetic, most bitter compound known
Quinine	3	Cinchona tree Malaria treatment, tonic water drinks
Phenylthiourea (PTU) or Phenylthiocarbamide (PTC)	3	Synthetic, (genetically linked bitterness)
Isohumulone	3–5	Hops, Beer
Naringin	20	Citrus fruit
Oleuropein	80	Olives
Amino Acids and Peptides	5–50	Hydrophobic L-amino acids
Caffeine	60–100	Xanthine alkaloid/coffee, tea, chocolate
Tannins	50–500	Wine

[a]Approximate threshold of perception in ppm.

Denatonium

Denatonium is the most bitter compound known. Its bitterness is readily detected at 0.05 ppm. This is approximately equal to 1 g of compound in 500 gallons of water. Solutions of denatonium diluted to 10 ppm are extremely bitter to most people. Denatonium is used as an aversive agent. It is a substance added to prevent accidental ingestion in products like denatured alcohol or antifreeze. Denatonium has a medicinal application in treating asthma and allergy. It binds to bitter receptors (T2Rs) in the lungs and causes relaxation of smooth muscles in bronchial tubes. Its action alleviates the restriction and restores normal breathing.

Quinine

Quinine is a very bitter alkaloid compound derived from the bark of the native South American cinchona tree. Extracts of this tree's bark contain appreciable amounts of quinine that have been used to treat malaria since the 1600s. A solution of quinine in carbonated water was often used to ward off malaria symptoms. British troops stationed in tropical regions of the world developed the practice of mixing large amounts of quinine and gin in carbonated water as their regime for malaria treatment. This practice was so common that a quinine containing carbonated beverage (tonic water) became commercially available in the 1850s. More recently, tonic water soft drinks (with or without sugar) containing significantly less quinine are still popular as a base for mixed drinks (e.g., gin and tonic). An interesting property of quinine is that it fluoresces when exposed to UV light. Even present day tonic water (limited in the US to about 80 ppm of quinine) gives off blue–green light under florescent lights.

Why is beer bitter?

Isohumulone is a major contributor of bitter taste in beer. Beer is defined as a beverage produced by fermentation of starch and flavored by isohumulone derived from hops. Dried hops contain as much as 25% humulone (also known as alpha acid) which is the precursor of isohumulone. The beer making process uses barley malt as a source of enzymes to break down starches and proteins into sugars, peptides, and amino acids. The malting process is started when water is added to barley, causing the cereal grain to produce these enzymes. Activated malt is then mixed with crushed grain (rice, corn, or wheat) and heated to about 60 °C to maximize the enzyme catalyzed conversion of starch to sugars. After filtering the mixture to remove solids, the fermentable carbohydrate-rich fraction (wort) is boiled to stop the enzymatic activity. Hops are added to the hot mixture causing conversion of humulone to its bitter form, isohumulone. The bitterness threshold for isohumulone is similar to that of quinine, about 3 ppm. After cooling, yeast is added to ferment sugars into alcohol and carbon dioxide. Isohumulone provides much of beer's unique flavor, but it can also produce off flavors. A major defect in beer is the off flavor caused by the action of sunlight (UV) on sensitizers that initiate formation of free radicals. Subsequent reaction of free radicals with isohumulone is responsible for off flavors described as stale, cardboard-like, and skunky. Thus, most beer is protected from light by packaging in brown or green bottles or cans. It is interesting to note that isohumulone is a bioactive agent with medicinal applications. Hop extracts in which humulone is converted to isohumulone may reduce chronic inflammation through a reduction in pro-inflammatory cellular factors. It has also been suggested that dietary supplementation of isohumulone improves insulin resistance (Bland et, al., 2015). However, it is doubtful that the advertising slogan "Guinness is good for you", has medicinal merit.

Proteins

Bitterness in proteins is rare, but some amino acids and peptides can have bitter properties. Bitter tasting amino acids include the L enantiomer of proline, isoleucine, and valine. Amino acid bitterness is associated with the shape, size, and hydrophobic character of the R group. Bitter amino acids lose their taste when incorporated into proteins because they are buried in the interior of folded proteins and prevented from binding to T2R receptors. In contrast, peptides of 15 or fewer amino acids exist as random coils with little hindrance to bind taste receptors. The difference in the taste of milk casein as a whole protein or peptides is a well-known example. Casein proteins are a very bland tasting. However, when casein is hydrolyzed into peptide fragments, a strong bitter taste occurs (Maehashi et al., 2008).

Caffeine

Caffeine is the most widely consumed psychoactive drug in the world. It stimulates the brain, increases cognitive function, blood pressure, and gastric acid secretion. Caffeine is principally responsible for the bitterness of coffee and tea. Caffeine is also a natural component of cocoa beans and provides chocolate, especially the dark variety, with bitter taste. Caffeine is an additive in many foods (i.e., energy drinks, snack foods) and in pain relief medications. The taste of caffeine pairs well with sweet. Adding sugar to coffee or having coffee with your dessert are common examples of tastes that pair well. However, the combination of sweet and bitter can also reduce the perception of sweetness and increase the desire or craving for sweet tasting foods.

Flavonoid bitterness

Hundreds of flavonoid compounds are known in plants where they provide color (red, blue, yellow and more) to many flowers, and fruits. Biologically, flavonoid compounds also serve as a plant defense mechanism against microbes and insects. Flavonoids are bioactive molecules important to health. They are effective antioxidants that control damage caused by free radical reactions. Flavonoids inhibit the blood aggregation and reduce the risk of clot formation that can result in stroke or heart attack. Flavonoids molecules share a common structural backbone consisting of three rings two phenyl (A and B) and one heterocyclic ring (C) (Fig. 6.4). Additional diversity in flavonoid structure occurs by adding hydroxyl (OH) groups to the structure (Circled group in Fig. 6.4). These structures can also include carbohydrates such as glucose, galactose, or mannose, added via a glycosidic, link to hydroxyl (OH) groups. Flavonoids readily form dimers and larger polymers (polyphenols and tannins) that are more intensely bitter and astringent. Flavonoids create bitterness by binding bitter taste receptors (T2R) in oral and extra-oral locations. More than 20 T2R bitter taste receptors are known and the pattern of receptor binding varies with each flavonoid. Flavonoid monomers and dimers are responsible bitterness of foods such as wine, beer, tea, citrus fruit, and chocolate. A low or moderate level of tannin in wine provides desirable bitterness and astringency. High tannin

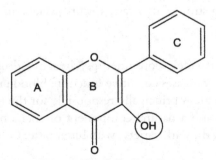

Fig. 6.4 Generalized Flavonoid structure.

level, however, causes a puckering sensation that is considered to be a defect. Wine makers use an old method of adding egg white or gelatin to reduce the tannin level in wine. The method works because tannins bind strongly to proteins and form a precipitate that is easily removed by filtration. Bitterness in grapefruit is due to naringin. This flavonoid is quite bitter (detection threshold 20 ppm) and represents a major problem in citrus juices. Naringin bitterness can be removed on a commercial scale by treatment with the naringinase enzyme. This enzyme cleaves the carbohydrate from the molecule and yields the non-bitter flavonoid known as naringenin.

Oleuropein

Oleuropein is a phenolic compound found in olive leaves and the oil of its fruit. Oleuropein occurs in glycosylated and non-glycosylated forms that alters its solubility. The non-glycosylated form oleuropein predominates in the lipid fraction (olive oil). Like many phenolic compounds, oleuropein serves as protection against microbial and insect damage to the plant. Oleur-opein has significant health promoting properties that include antioxidant activity and protection against cardiovascular disease and metabolic disorders such as diabetes. Its strong antioxidant activity may also contribute to the prevention of cancer through its ability to neutralize reactive oxygen species (ROS) (Barbaro et al., 2014). Oleuropein is a moderately bitter tastant with a threshold for perception of about 80 ppm. It contributes to the desirable bitterness of finished olive oil that is eaten with foods or used in cooking. Raw or unprocessed olives are objectionably bitter.

Processing technologies are used to remove or destroy oleuropein in olives. Oleuropein can be substantially reduced by soaking olives in a lye and salt brine prior to pressing. Lye in the soaking solution creates an alkaline environment that effectively destroys oleuropein. Lactic acid fermentation is another method used to reduce oleuropein bitterness. Fermentation con-verts oleuropein into non-bitter products and adds flavors to olives. The taste profile of oleuropein includes bitterness, astringency, and some peppery or pungent notes. The level of these tastants in olive oil varies with the press-ing fraction resulting in the various grades of olive oil sold. The first pressing (extra virgin olive oil) has the highest content of oleuropein and the greatest bitterness and dark color. The next pressing fraction (virgin olive oil) contains less oleuropein, is lower in bitterness, and is lighter in color. Finally, the pressing fraction labeled simply as olive oil contains the lowest level of oleuropein, is least bitter, and has the lightest color.

Umami

Umami taste is characterized as a meaty and/or savory taste perception. Meat and meat broth, shellfish, seaweed, mushroom, cheese, soy, and tomato are all examples of foods rich in umami taste. Fermented foods like soy and fish sauce are known for their umami taste. Umami is characterized as a less intense taste perception, but it persists for a longer period time on the tongue. Umami tastants were previously considered flavor enhancers. Its recognition as a basic taste modality did not occur until the late in the twentieth century (Kurihara, 2015). The work of a Japanese scientist, Kikunae Ikeda, provided the first evidence in 2002 of a substance responsible for umami taste. His work identified a compound in seaweed (*komku*) that provided umami taste in soup made from this plant. The compound L-glutamic acid has the interesting quality of tasting sour in the acid form, but changing to a uniquely pleasant taste when neutralized with sodium hydroxide (NaOH) (Ikeda, 2002). Neutralization of glutamic acid with NaOH converts the acid to monosodium glutamate (MSG) salt. MSG is now one of the most widely used substances for seasoning and enhancing the taste of food. Glutamate creates umami taste by binding T1R receptors in taste buds.

Understanding the physiology of glutamate action led to the establishment umami as the fifth taste perception. Biologically, glutamate acts as a strong neurotransmitter that is released by nerve cells in the brain. It plays a role in cognitive activities, such as memory and learning. In food, glutamate-containing di- and tri-peptides are potent stimulators of umami taste perception. Glutamate also occurs naturally as the enzymatic breakdown product of proteins in plant and animal foods (Table 6.4). For example, ripe tomatoes have a strong umami taste because of their high

Table 6.4 Glutamate in foods. (mg/100g).

Plant		Animal	
Kombu	1200–3400	Anchovies	630–1440
Nori	1380	Cheese	300–1680
Tomato	150–250	Fish Sauce	620–1380
Garlic	110	Soy Sauce	410–1260
Potato	30–100	Green Tea	220–670
Chinese Cabbage	40–90	Dry-Aged cured ham	340
Onion	20–50	Scallop	140
		Kuruma shrimp	120
		Crab	20–80

Adapted from Kurihara, K., 2015.

glutamate level, approximately 150—250 mg/100 g. The rise in glutamate level in tomato results from ripening induced breakdown of its proteins into peptides and amino acids. Cow's milk protein has a high content of glutamic acid (about 20% of its total amino acids). Cheese containing protein breakdown products are exceptionally high in glutamate (approximately 1200 mg/100 g) and have a strong umami taste. It should be mentioned that there has been controversy over the potential side effects of MSG. A condition known as Chinse restaurant syndrome was proposed following anecdotal reports of symptoms (i.e., headache, flushing, and sweating) that occurred after eating food seasoned with MSG. Chinese restaurant syndrome been the subject of popular press articles and published reports over the 50 years since its rise to attention. However, controlled studies have not found evidence to substantiate a link between MSG and the reported symptoms (Obayashi and Naganura, 2016). Food safety agencies (FDA in the United States and EFSA in the European Union) consider MSG safe for human consumption and without a toxicological concern.

Nucleotides

Nucleotides also have a role in umami taste perception. Biologically, nucleotides are the building blocks of ribonucleic acids (RNA) and deoxyribonucleic acids (DNA). In foods, the nucleotides, 5′-inosine monophosphate (IMP) and 5′-guanosine monophosphate (GMP) are produced by the enzymatic breakdown of corresponding nucleic acids (Table 6.5). While IMP and GMP nucleotides do not directly stimulate umami taste, they act synergistically with glutamate to enhance its perception. Glutamate initially stimulates umami taste on binding to T1R receptors. Taste perception increases two-fold or more when nucleotides also bind the receptor. The practice of aging meat before cooking facilitates enzymatic breakdown of proteins and

Table 6.5 Nucleotides in foods. (mg/100g).

5′-Insoinate		5′ Guanylate	
Dried Bonito	470—800	Dried Shittake	150
Dried Sardine	350—800	Enoki	50
Yellowtail	410—470	Dried Morel	40
Sardine	420	Dried Porcini	10
Tuna	250		
Chicken	230		
Pork	230		
Beef	80		

Adapted from Kurihara, K., 2015.

nucleic acids and enhances its taste. Similar enzymatic breakdown processes in fish, mushrooms, and potatoes are responsible for enhanced umami taste by the release nucleotides and glutamate.

Salt

The mechanism of salty taste perception does not involve specific protein receptors like those for sensing sweet, bitter, and umami tastants. Rather, a salty taste is created when sodium ions (Na^+) enter channels in epithelial cell membranes of type I taste bud cells. Positively charged sodium (Na^+) ions bind to negatively charged molecules in the channel and neutralize its charge. A short-term depolarization of the membrane causes release of calcium ions and triggers a cascade of chemical reactions that signal gustatory regions of the brain. Contrary to previous models, saltiness is not perceived in isolated regions of the tongue. Instead, sodium ion channels are distributed throughout the oral cavity. There are two types of ion channels. One specific for low levels of sodium and the other for higher sodium levels. The second type can also be triggered by potassium (K^+) ions when they are present at the same time. Potassium's presence reduces salty taste perception and creates an aversive bitter taste. Sodium ion (from NaCl) enhances other taste perceptions, such as sweetness. Salt added to sweet in a food increases its sweetness. Chocolate manufactures now offer products with added salt in recognition of sodium's effect. Added salt inhibits bitter taste perception in food. The bitter taste of grapefruit and other citrus fruits due to the flavonoid naringin is reduced because sodium ions inhibit the response of bitter taste receptors.

What is hidden salt?

Foods such as, cheese, soup, bread, processed meat, and many other processed food contain added salt, but these are only moderate salty. Salt is added to processed food for several reasons. It is used to inhibit the growth of spoilage microorganisms in cheese and processed meat. Salt inhibits the bitter flavor of amino acids and peptides liberated by protein breakdown. It is also added to solubilize meat proteins essential to making processed products, such as sausage, ham, and deli meats. A 4-ounce serving (about 1/2 cup) of processed meat contains about 800 mg of NaCl, but its taste is only moderately salty. The same amount of salt in a cup of water is almost too salty to drink. The explanation for lower salty taste in meat, cheese, and other foods is that sodium ions are bound to food proteins and not available to activate its taste perception.

> ## Sour

Sour taste perception is stimulated by weak acids in food. Direct acid-ification of foods is a time-honored method to improve taste and prevent spoilage. Added acid lowers the pH of a food matrix and inhibits the growth of spoilage organisms by disrupting ion channel functions in their outer membranes. Sour taste perception results from the properties of weak acids, such as lactic or citric acid. The taste of these acids is described as tart and this is a desirable quality in many foods. The perception of sourness is linked to the action of hydrogen ions (H^+). However, the mechanism is not explained by the chemical property of acidity alone. Weak organic acids, such as acetic, lactic, and citric, have stronger sour taste than hydrochloric acid (HCl), even when adjusted for molar or hydrogen ion concentration. The intensity of sour taste among weak acids is not correlated with their acid strength (i.e., pKa of their carboxyl groups). At equal concentration, the order of sour intensity in weak acids is citric > tartaric > succinic > lactic > acetic. Previously, several mechanisms have been proposed to explain how weak acids cause sourness. It is suggested that sour taste perception is due to the total amount of acid present as measured by titratable acidity. Another model suggests that sourness of weak acids is influenced by the corresponding negatively charged conjugate of the acid (e.g., acetate from acetic acid or citrate from citric acid). Neither of these models adequate to explain differences in weak acid sourness. Recently, it has been found that sour taste perception occurs via two pathways (Frings, 2010). In the first pathway, protonated (uncharged) acids introduced on the tongue's surface diffuse through the membrane of taste cells. Once inside the cell, the acid is ionized and liberated hydrogen lowers the intracellular pH. This causes a signal to be sent to the brain's gustatory center. The second pathway of sour perception involves an ion channel mechanism. In this case hydrogen ions from disso-ciated acids enter the channel and bind to PKD2L1, a protein receptor. This molecule is a member of a receptor family whose primary function is detection of pain stimuli.

Food acids and sour taste

Yogurt and cheese have a pleasant sourness due to lactic acid created from lactose during fermentation. The tartness of wine results from its content of three weak acids: tartaric, malic and citric. Malic acid has the strongest sour taste of the three and a second bacterial fermentation is often used to convert it to lactic acid and reduce sourness. The process in wine-making is called malolactic fermentation and is typically used with red varieties that would otherwise be too tart. Citrus fruit, such as lemons, grapefruit, and oranges, contain substantial amounts of citric, malic, and ascorbic acids, but their sourness is varied by competing tastants. The intensity of sourness in lemons, for example, can be mouth puckering due to a high level of citric acid (48 g/L) and a pH of about 2.5 (Table 6.6). Lemon sourness can be made palatable by adding sugar as the competing tastant in lemonade, pie, and other desserts. Similarly, oranges contain a moderately high level of citric acid (about 17 g/L and a pH of 3.3) (Table 6.6). The principal acid in oranges is citric acid. Oranges also contain a high level of sweetness in the form of fructose that suppresses the sour taste of citric acid. Additionally, a compound called limonin stimulates a bitter perception that further competes with and modifies perception of sourness. Overall, the balance of sweet, bitter, and tart provides the orange with a pleasant, sweet, and tangy taste, despite the level of acid.

Grapefruit is another example of a moderately high acid food (about 25 g/ L with a pH of approximately 3.3) (Table 6.6). The acid level of grapefruit is expected to result in a very sour taste, but it contains two tastants, fructose and naringin, that modify its sourness. The fructose level in grapefruit is about 16 g per cup and naringin is present at about 120-400 mg/L of juice. Soda pop is a carbonated beverage containing phosphoric and carbonic acids, the latter provided by dissolved CO_2. The pH of soft drinks and lemons is about the same (2.3-2.5), but their sourness is greatly different. This again illustrates that multiple factors influence sour perception.

Table 6.6 Foods, acids, and sourness.

Food	Acid(s)	Amount of acid (g/L)	pH	Sour taste
Soda Pop	phosphoric, carbonic, citric	9-10	2.3-2.8	+
Lemon	citric, ascorbic, malic	48	2.3-2.5	++++
Vinegar	acetic	9.6	3.0	++++
Grapefruit	citric, ascorbic	25	3.3	+
Orange	citric, ascorbic, malic	17	3.3	+
Wine (red)	tartaric, malic, lactic	5-6	3-4	+
Grapes	malic, tartaric, citric	1.5-5	2.8-3.0	++
Tomato	citric	5	4.5	++
Yogurt	lactic	0.7-1.0	4.7-4.9	+
Coffee	citric, malic, acetic	0.5	4.6	+

Pungency

The taste of pungent foods is described as a hot or stinging sensation in the mouth. Pungent foods are often referred to as spicy. However, the term pungent is preferred over hot because it avoids possible confusion with the temperature of food. Additionally, in culinary usage the term spicy is reserved for flavors derived from spices like cinnamon and nutmeg. Foods with a pungent character are important to the cuisine in many parts of the world. Interestingly, pungent foods are often favored dishes in populations that live in hot climates. The word piquant is sometimes used to describe a less intense pungent sensation in foods like mustard or curry. Pungent tastants can cause sensations that are painful or pleasurable. These sensations include coolness, tingly feelings, or numbness in the mouth and lips. The mechanism responsible for pungent sensations is called chemesthesis. This chemical reaction is initiated by molecules that activate somatosensory nerves in skin, nasal, or oral tissues. Chemesthesis is also the cause of expulsive reactions such as sneezing, tearing, and nasal drip. Sneezing after vigorously shaking a pepper shaker is a chemesthesis reaction. Biologically, pungency perception involves ion channels residing in the membranes of cells that line the mouth and throat. Ion channels in these membranes belong to the transient receptor potential (TRP) channel family. Common chemesthesis activating agents like capsaicin, menthol, piperine, and weak acids activate the TRP channel pathway. Pungency has been said to be the sixth taste, but technically this is not the case. Unlike true taste receptors, molecules causing the sensation are not required to have a specific configuration or functional group.

Capsaicin

Capsaicin is the active molecule responsible for the pungency in chili peppers and other members of the capsicum family (e.g., paprika) (Fig. 6.5). Capsaicin causes its pungent response in the mucous membranes of the oral and nasal cavities and eyes. Biologically, capsaicin deters animals from using the plant for food. It has antimicrobial properties and inhibits the growth of fungi. Chili peppers are perhaps the most widely used source of capsaicin for pungent foods. It is not uniformly found in all parts of chili peppers. The highest content of capsaicin is found in the fleshy tissue that holds the seeds. Chili pepper seeds do not contain capsaicin. The pungency of chili peppers varies widely and the Scoville heat unit is to rate their hotness. The range for Scoville heat units varies from zero for bell peppers to greater than

Pungent Molecules

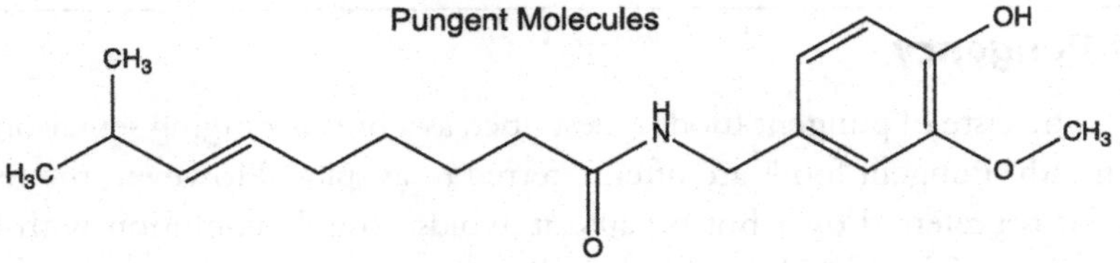

Capsaicin

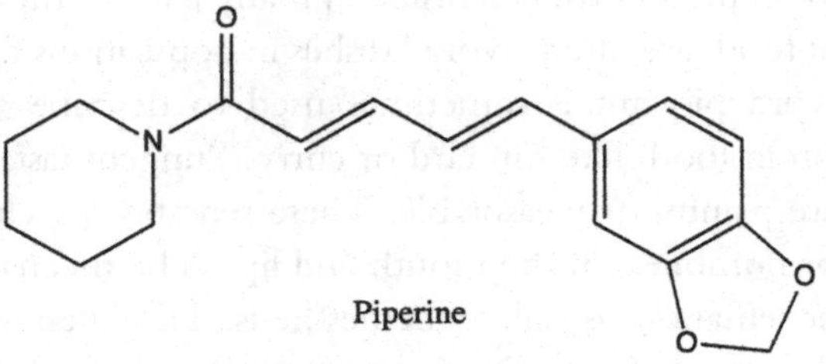

Piperine

Fig. 6.5 Capsaicin and piperine.

one million for habanero and komodo dragon peppers. Capsaicin is a some–
what hydrophobic molecule with limited solubility in water. It is consider–
ably more soluble in alcohol and oils and these liquids are used to make
liquid hot sauce preparations. In addition to its pungency, capsaicin also
activates secretion of stomach acid. Nausea, vomiting, and burning diarrhea
can be a consequence of consuming too much capsaicin. Non–food applica–
tions of capsaicin include its use as an analgesic in topical treatments for relief
of muscle pain. Topical application can relieve the symptoms of arthritis. It is
also used in pepper spray for personal defense.

Piperine

Black pepper is the most widely used pungency enhancer. Like salt, its use in
cooking and its presence on the table is ubiquitous. The pungent portion of
the plant (*Piper nigrum*) is the fruit body that becomes black after maturation
and drying. The pungency of black and white pepper is provided by
piperine, an alkaloid type compound (Fig. 6.5). Black and white pepper
are derived from the same plant. These peppers differ in the ways in which
they are processed. The distinctive color and aromatic flavors of black
pepper come from enzymatic and Maillard browning reactions occurring
in the skin during drying. Piperine is responsible for the sharp, pungent taste
of black pepper. Chavicine, isopiperine, and isochavicine isomers of piperine
are also present in the seeds of *P. nigrum,* but they lack pungency. Freshly

ground black pepper has the greatest pungent and aromatic quality because molecules responsible for these attributes are volatilized by the grinding action. As black peppercorns age, they lose flavor and pungency due to volatilization. Piperine is light-sensitive, losing its pungency due to isomerization of piperine into other forms. Like most pungent molecules, piperine creates its sharp taste through the pain sensing TRP cation ion channel pathway. Black pepper has a long history of use in herbal medicine. Recent investigation has shown that piperine has bioactive properties as an antioxidant and anti-inflammatory agent. Its antioxidant activity also inhibits lipid oxidation in foods. However, the effectiveness of piperine in these roles is limited due to its poor solubility in water (about 0.14 g/L). White pepper is made from the peppercorn seeds, but is processed differently. Mature berries (typically red) are soaked in water and undergo bacterial fermentation. This step softens the outer skin making them easy to be remove from the peppercorn seeds. The seeds are then dried and finely ground to make white pepper. These processing differences between black and white pepper affects their pungency and other taste qualities. It thought that removal of skins in making white pepper takes away some of its piperine and volatile compounds that contribute to the taste of black pepper. The fermentation process used to soften the skin also changes its overall flavor. The pungency in white pepper is desirable in soup and stir-fry dishes. The lighter color of white versus black pepper makes it the better choice for adding pungency and avoiding the appearance of black specks.

Glucosinolates

Glucosinolates are sulfur-containing compounds found in the cruciferous family of plants. The family includes broccoli, cabbage, kohlrabi, mustard, cabbage, cauliflower, radish, horseradish, kale, and bok choy. Biologically, glucosinolates are present in these plants for defense against microbial pathogens (fungi) and to ward off herbivores. The glucosinolate molecule itself is not responsible for pungency. Rather, pungency results from enzymatic action on glucosinolate. Glucosinolates are composed of glucose (dashed circle) and an organosulfur compound, connected by a glycosidic link to the sulfur atom (Fig. 6.6). Pungency is caused when myrosinase enzyme hydrolysizes of the glycosidic link. The multi-step reaction results in several products, most notable are isothiocyanate and nitriles. The active molecule responsible for pungency is isothiocyanate. It is a highly reactive molecule that initiates a pain sensation through the TRP ion channel and activates a pain sensation

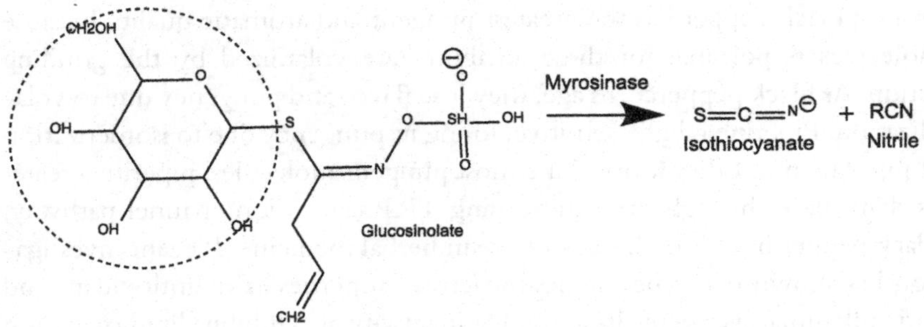

Fig. 6.6 Glucosinolate.

through the trigeminal nerve. In plant tissues, myrosinase and glucosinolate are spatially separated. Isothiocyanate is produced when the plant material is chopped, chewed, or crushed. Isothiocyanate is toxic to many insects and this prevents damage to the plant. Isothiocyanate is also toxic to herbivores. Rape seed, for example, is the commercial source of canola oil. This oil is extracted by pressing rape seed. The remaining solid matter is dried and used as animal feed. It is problematic for animals to eat this by-product because it contains significant amounts of isothiocynate generated from the pressing operation. Isothiocyanate is toxic to animals because it reduces iodine uptake and affects thyroid gland function. Lack of iodine can cause a deficiency in the thyroxine hormone. Therefore, glucosinolate is an anti-nutritional for animals, including humans. Boiling destroys myrosinase activity and controls the level of isothiocyanate produced.

Carbonated beverages

Bubbles in carbonated beverages like sparkling water, soft drinks, and beer contribute a pleasant taste perception. Bubbles of carbon dioxide (CO_2) create a pungent sensation that is both pleasant and tingling though a low-level stimulation of pain sensing nerves. Carbon dioxide (CO_2) readily dissolves in water to form carbonic acid. The acid dissociates into hydrogen ions (H^+) and bicarbonate ions (HCO_3^-) and makes the solution acidic. Soft drinks often contain additional acids like phosphoric and this contributes additional hydrogen ions and further lower its pH. Hydrogen ions stimulate a burning or tingly sensation through the process of chemesthesis. Increased hydrogen ion concentration causes the trigeminal nerve to send a mild pain signal to the brain. Because the pain response is small, it is generally referred to as a pleasurable, tingly sensation.

$$CO_2 + H_2O \rightarrow HCO_3^- + H^+$$

The presence of the taste receptor mechanism specific for carbon dioxide is interesting because CO_2 is not a nutrient. One hypothesis suggests that this system for detection of carbonation is an evolutionary protection against spoiled foods.

Olfactory perception (is flavor created in the brain?)

The influence of flavor on the brain is much more than a physiological response to chemical stimuli. The sense of smell is linked to cognition, memory, and emotion. Humans are poor smellers compared to other animal species. Based on relative body mass cats, for example, have approximately 10 times more olfactory epithelium than humans. The sense of smell is keenest in the young and declines with age. An unfortunate fact of life is that the sense of smell typically declines by over half as people age from 20 to 70. The loss of smell can be a signal of serious change in brain function. It has been suggested that a test based on the sense of smell can be used to identify individuals at risk for developing Alzheimer's disease. Tests using familiar odors like coffee, peppermint, and peanut butter are being investigated as a means for detection of the disease (Albers et al., 2016). One of the most interesting aspects of human sense of smell is its linkage to memory. Olfactory perception occurs in the brain's limbic system that contains the regions involved in emotion, behavior, and long-term memory. The smell of baking bread, for example, can bring back a vivid memory of its first encounter. Feelings described as craving a food are also associated with food flavor. Craving chocolate is an often-cited example of this linkage. In the extreme, emotions connected to food flavor can be linked to eating disorders and their associated health consequences. A term coined by Gordon Shepherd (Shepherd, 2006) in the field of psychology has recognized the relationship between the brain and flavor in the new field of study called neurogastronomy. This new field of science examines how the brain creates flavor and how it matters to human culture. In addition to traditional psychology research, the findings of neurogastronomy research are being used by chefs and food companies to study how food flavor can enhance the eating experience (Humphries, 2012).

Smell

The sense of smell, like that of taste, depends on receptors in specialized cells that communicate with the brain. The sense of smell together with that of taste comprise the chemosensory system responsible for flavor.

It differs from that of taste because it has a greater number of receptors, greater sensitivity, and ability to differentiate between a large number of odorant molecules. The sense of smell is used by animals to find food, avoid spoiled food, and influence mate selection for both male and female species. It is also biologically important in protection from toxic compounds. Hydrogen sulfide, ammonia, and mercaptans are examples of compounds that can be detected in the low ppm (part per million) range. Mercaptans are incorporated into natural gas supplies for safety. Methane is odorless, but the smell of mercaptan provides a warning that a combustible gas is present. Smell perception is carried out by specialized olfactory neuron receptors, that are located in epithelium at the back of the nasal cavity. When stimulated by an odorant molecule, olfactory neurons transmit signals to the brain. Odorants are volatile molecules that enter the nasal cavity via the nose or from the back of the throat when food is chewed. Humans have approximately 400 types of odor receptors. It is thought that they can differentiate between millions of odorants. This odorant differentiation is achieved by the pattern of neuron signals sent to the brain. Olfactory data is further processed in regions of the brain that controls cognition, memory, and emotion.

The sense of smell is the single most important factor affecting food flavor. Odor perception typically depends on the concentration of molecules available to stimulate olfactory receptors. However, some molecules are better olfactory stimulators than others. The lowest concentration of an odorant required for human smell detection is called the limit of detection. Green peppers have a very low limit of detection, about 4 ppb (parts per billion), that can disproportionately affect its impact on smell. The structure of an odorant molecule is also important to the sense of smell. Carvone, for example, is a terpenoid type of molecule found in seeds and essential oils. Interestingly, carvone is responsible for two distinctly different aromas. Carvone is a chiral molecule existing in two enantiomeric forms (R and S) (Fig. 6.7). The two carvones enantiomers are structurally identical, except for being mirror images of one another. However, the R enantiomer of carvone produces the aroma of spearmint and the S enantiomer produces the aroma of caraway seeds. Carvone illustrates the discriminatory ability of olfactory receptors.

Foods like coffee and wine are extremely complex mixtures of odorants containing of hundreds or perhaps thousands of molecules that combine to produce a unique smell or aroma. The word aroma is often used when describing the smell of wine and is the primary means through which it is

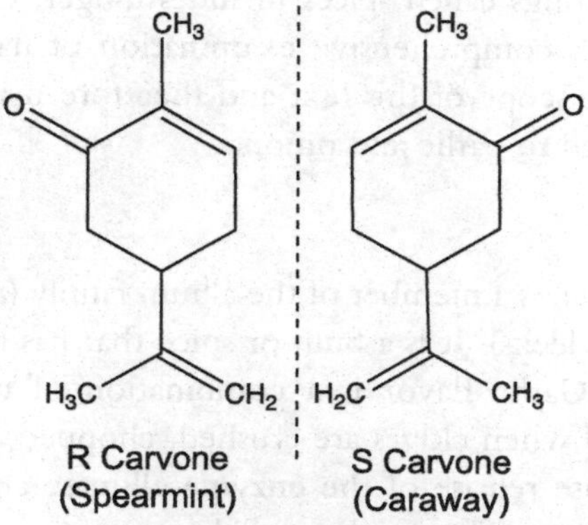

Fig. 6.7 Carvone enantiomers.

evaluated. Wine aroma is composed of distinctly different notes. Notes described as fruity (e.g., blackberry, strawberry, and cherry) are typical of young red wines. As the wine ages, more complex notes described as earthy or chocolate are developed due to ongoing chemistry within wine. It is possible to train your nose to discern desirable and undesirable compounds in wine and other foods. The technique of sensory analysis is widely used by the food industry to evaluate the taste and smell of foods. A sensory analysis typically consists of a panel of 8–10 people who smell and/or taste the food under controlled conditions. Trained panels are especially important for detection off flavors, many of which result from lipid oxidation (see Chapter 4 for a description of lipid oxidation chemistry). Volatile aldehyde and ketone reaction products of this chemistry are responsible for undesirable odors described as grassy or beany. Similarly, off flavors from lipid oxidation reactions are a major problem for meat products.

Herbs and spices

Herbs and spices are plant-derived materials with a long history of use in flavor enhancement of food. More recently, many have been found to contain bioactive components with health promoting properties. The words herb and spice are often used as synonyms, but they are technically differentiated by the part of the plant from which they are derived. The word herb refers to flavors derived from the green leafy tissues of plants. Herbs include, but are not limited to, basil, bay leaf, rosemary, and sage. In contrast, the word spice refers to flavors derived from the seed, root, or bark of plants.

The family of flavorings called spices includes ginger, cinnamon, pepper-corns, and cloves. A comprehensive examination of flavors in herbs and spices is beyond the scope of this text and therefore material presented in this chapter is limited to garlic and onion.

Garlic

Garlic (*Allium sativum*) is a member of the allium family (also represented by onions, shallots and leeks). It is a bulb or spice that has long been used for culinary purposes. Garlic flavor is a combination of olfactory and taste sensations produced when cloves are crushed, chopped, or chewed. These physical actions cause release of the enzyme alliinase and allow it to mix with its alliin substrate. The product of the reaction L-cysteine sulfoxide (allicin) is responsible for the hot stinging (pungent) sensation encountered when chewing raw garlic (Fig. 6.8). The pungency of raw garlic is short lived. Allicin is a very reactive molecule and is spontaneously broken down to other compounds such as diallyl sulfide and diallyl disulfide. As anyone who has cooked with garlic knows, its flavor changes with heat. The hot taste contributed by allicin is lost, but garlic flavors contributed by allicin decomposition compounds remain. These compounds are hydro-phobic and volatile. They are responsible for potent aromas and flavors that diffuse through the fat of food. Freshly chopped garlic mixed with a little oil is an excellent way to infuse food with its flavors. Garlic's cysteine content containing dipeptides, such as γ-glutamyl-L-cysteine, contribute to the kokumi taste effect. Kokumi, while not a new taste in itself, has been shown to enhance sweet, salty, and umami tastes.

Health benefits of garlic

The dipeptide γ-glutamyl-L-cysteine is a precursor of glutathione. This strong antioxidant provides protection against cellular damage caused by reactive oxygen species (ROS). In food, gamma-glutamyl peptides are water soluble molecules that contribute to kokumi and umami taste perception. Garlic and garlic-derived supplements are suggested to have health

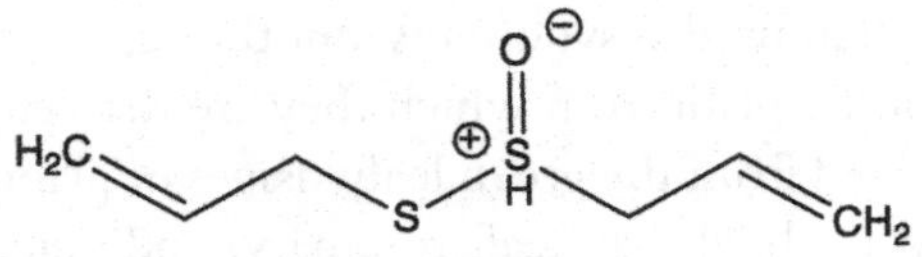

Fig. 6.8 Allicin.

promoting activities that include anti-inflammatory, antioxidant, cardio-protective, and cancer preventative properties (Stabler et al., 2012; Trio et al., 2014). However, data supporting the health benefits of garlic-derived compounds in supplement form are not conclusive (Garlic, 2019; Zong and Martirosyan, 2018).

Onion

Onion is another member of the allium family that derives its characteristic odor and flavor from organosulfur compounds produced by action of the alliinase enzyme. The ultimate source of sulfur for these compounds is related to its level in the soil. Low sulfur soils result in lower levels of organo-sulfur compounds and less pungency. Intact onions have little or no odor. However, the action of chopping or crushing breaks open cells and mixes S-propenyl cysteine sulfoxide (the substrate) with alliinase (the enzyme). This produces sulphenic acid and other products. Unlike garlic, onions contain an additional enzyme called lachrymatory factor synthase (LFS) that converts sulphenic acid into a volatile molecule called thiopropionalde-hyde. This volatile molecule is rapidly released into the air when onions are cut. While not a contributor to onion flavor, it is a potent eye irritant that causes a stinging sensation and immediate tearing. It also responsible for the hot, pungent sensation found in eating raw onion. Onion odors are derived from spontaneous reactions which convert 1-propenylsulphenic acid into thiosulphinate. This is also a precursor of compounds associated with health promoting properties.

S-propenyl cysteine sulfoxide + alliinase → sulfenic acid + LFS → thio-propionaldehyde (Irritant)

Options are available for those who wish to avoid the unpleasant stinging and tearing reactions to chopping onions. Keeping the onion cold slows the action of alliinase and LFS enzymes and reduces the amount of the lachry-matory irritant propanethial S-oxide. Sweet onion cultivars have been developed that contain little LFS enzyme or the lachrymator irritant. Onions have been recently recognized as a source of organosulfur compounds with potential health benefits, including anti-thrombosis (prevent blood clot for-mation) and antimicrobial activity. Onions are a good source of vitamin C. Red onions, in particular, owe their color to high levels of anthocyanin flavonoid compounds that have antioxidant activity. Cooking onions dramatically changes their flavor. Heat destroys the pungency of raw onion and replaces it with savory, meaty flavors such as those developed upon sautéing. Maillard reactions between onion sugars (glucose and fructose) and proteins also contribute to cooked onion flavor.

Summary

Food flavor is a fusion of taste and smell inputs with lesser, but important, contributions made by texture and color. The sense of taste (sweet, bitter, sour, salty, and umami) is generated by a relatively small group of compounds that stimulate specialized cells located in the mouth and tongue. Taste perception is influenced by inputs from temperature, pungency, and smoothness (mouth feel). The sense of smell involves specialized cells in the nasal cavity that detect a wide range of volatile compounds. Sense of smell is more sensitive and discriminatory than is the sense of taste. Flavor is created in the brain by stimulation of multiple centers activating feelings of emotion, memory, and reward. Knowing the flavor of an apple from simply seeing it demonstrates the brain's role in creating flavor. Neuro-gastronomy is a new area of science that studies the creation of flavor by the brain and its effects on behavior. This interdisciplinary field, involving neuroscience, psychology, and culinology, contributes to understanding of flavor's influence on emotion, food preference, cravings, and eating disorders. This new science is important to foodies, wine connoisseurs, chefs, and those wanting to know more about flavor and food.

Glossary

Alkaloid: A group of naturally occurring molecules found in plants, animals, and fungi. They are composed of multiple ringed structures containing one or more nitrogen atom.

Anion: An atom or molecule with a negative charge. Anions are attracted to, and form, ionic bonds with positively charged species (cations).

Astringency: Substance that causes a dry and or puckering sensation in the mouth. The feeling is commonly encountered in wine and some berries that contain high levels of tannins (i.e., polyphenols).

Cation: An atom or molecule with a positive charge. Cations are attracted to, and form, ionic bonds with negatively charged species (anions).

Chirality: Molecules that are identical except they are mirror images of each other and thus are not superimposable. The mirror images of chiral molecules are called enantiomers.

Chemesthesis: Chemically initiated sensations of the skin or mucus membranes caused by molecules that activate receptors associated with pain or touch. Substances that cause chemesthesis are responsible for the pungent, cooling, and tingly sensation of foods.

Chemosensor: Specialized sensory receptor (e.g., GPCR) which responds to a chemical substance and generates a biological response.

Electronegative: A measure of the tendency of an atom to attract electrons.

Enantiomers: A pair of molecules with the same chemical composition that are mirror images of each another. The difference can be seen when superimposing the structures on each other.

Extra-oral taste receptor: Receptors found in the gut and other locations that function in sensing nutrients.

G protein coupled receptor (GPCR): Cell surface receptors that function in detecting the presence of a substance e.g., sugar, and send a signal to the brain.

Glutamate: A form of glutamic acid. Free glutamic acid combines with sodium ions to form glutamate, a substance described as having umami flavor.

Gustatory taste map: Region of the brain responsible for interpreting sensory input signals corresponding to the taste perceptions of sweet, bitter, umami, salty and sour. Also referred to as the gustatory cortex.

Glycosidic bond: Covalent bond that links a carbohydrate (e.g., glucose etc) to other molecules, using one of its oxygen atoms.

Glycoside: A carbohydrate molecule to which another molecule is attached through a glycosidic bond using one of its hydroxyl groups.

Hydrogen Bond: An attractive force between a hydrogen atom and an electronegative atom.

Hydrophobic: Molecules are defined as water hating or hydrophobic because of their nonpolar composition. An attractive force is created between non-polar molecules by association of their hydrophobic groups.

Incretin: A hormone that stimulates insulin secretion in response to eating. Glucagon-Like Peptide (GLP-1) and glucose-dependent insulinotropic polypeptide (GIP).

Ion channel: A cell membrane channel that is selectively permeable to specific ions, e.g., sodium, calcium, potassium, or hydrogen.

Neurogastronomy: A field of psychology devoted to the study of how the brain creates flavor and how it matters to human culture.

Organosulfur compounds: Organic molecules containing a sulfur atom. Examples include essential amino acids like cysteine and methionine.

Papillae: Small, rounded protuberances on the surface of the tongue.

pKa: A measure of the dissociation of a weak acid into hydrogen ions and its conjugate base. The lower the pKa value, the stronger the acid.

Receptors: A protein molecule that receives chemical signal from outside the cell which results in chemical or electrical response.

Scoville heat unit (SHU): A measure of pungency based on the concentration of capsaicin in the food material. SHU are used to rate the degree of "hotness" in peppers.

Terpenoids: Lipid-based substances that typically have a strong odor. They protect plants and insects from predation. Butterflies secrete terpenoids to lessen the risk of being eaten.

Titratable acidity: A method used to determine the total amount of weak acid in foods.

Weak acids: An acid that is only partially dissociated into hydrogen ions and its conjugate base. In contrast, strong acids like HCl dissociate completely in water.

References

Acree, T.E., Lindley, M., 2008. Structure-activity relationship and AH-B after 40 years. Sweetness and Sweeteners, American Chemical Society 979, 96—108.

Albers, A.D., Asafu-Adjei, J., Delaney, M.K., Kelly, K.E., Gomez-Isla, T., Blacker, D., Johnson, K.A., Sperling, R.A., Hyman, B.T., et al., 2016. Episodic memory of odors stratifies Alzheimer biomarkers in normal elderly. Ann. Neurol. 80, 846—857.

Barbaro, B., Toietta, G., Maggio, R., Aricello, M., Tarocchi, M., Galli, A., Balsano, C., 2014. Effects of the olive derived polyphenol oleurorein on human health. Int. J. Mol. Sci. 15, 18508—18524.

Barre, A.S., Caze-Subra, C., Gironde, 2015. What about allergenicity of sweet-tasting proteins? Rev. Fr. Allergol. 55, 363—371.

Bland, J., Minich, D., Lerman, R., Darland, G., Lamb, R., Tripp, M., Grayson, N., 2015. Isohumulones from hops (Humulus lupulus) and their potential role in medical nutrition therapy. PharmaNutrition 3 (2), 6—52.

Bublin, M., Radauer, C., Knulst, A., Wagner, S., Scheiner, O., Mackie, A.R., Mills, E.N.C., Breiteneder, H., 2008. Effects of gastrointestinal digestion and heating on the allergenicity of the kiwi allergens Act d 1, actinidin, and Act d 2, a thaumatin-like protein. Mol. Nutr. Food Res. 52 (10), 1130—1139.

Calvo, S.S., Egan, J.M., 2015. The endocrinology of taste receptors. Nat. Rev. Endocrinol. 11 (4), 213—227.

Chattopadhyay, S., Raychaudhuri, U., Chakraborty, R., 2014. Artificial sweeteners - a review. J. Food Sci. Technol. 51, 611—621.

Chen, X., Gabitto, M., Peng, Y., Ryba, N.J., Zuker, C.S., 2011. A gustotopic map of taste qualities in the mammalian brain. Science 333, 1262—1266.

Fernstrom, J.D., 2015. Non-nutritive sweeteners and obesity. Annu. Rev. Food Sci. Technol. 6, 119—136.

Frings, S., 2010. The sour taste of a proton current. P.N.A.S. 107, 21955—21956.

Garlic, 2019. OSU Micronutrient Information Center. https://lpi.oregonstate.edu/mic/food-beverages/garlic.

Humphries, C., 2012. Cooking: delicious science. Nature 486, S10—S11.

Ikeda, K., 2002. New seasonings. Chem. Senses 27 (9), 847—849.

Janssen, S.I., Depoortere, I., 2013. Nutrient sensing in the gut: new roads to therapeutics? Trends Endocrinol. Metab. 24, 92—100.

Joesten, M.D., Hogg, J.L., Castellion, M.E., 2007. Sweetness relative to sucrose, fourth ed. The World of Chemistry: Essentials, Belmont, California.

Kant, R., 2005. Sweet proteins- Potential replacement for artificial low-calorie sweeteners. Nutr. J. 4, 1—6.

Kurihara, K., 2015. Umami: the fifth basic taste: history of studies on receptor mechanisms and role as a food flavor. Biomed. Res. Intl. https://doi.org/10.1155/2015/189402.

Layden, B.T., Durai, V., Lowe, W.L., 2010. G-Protein-Coupled receptors, pancreatic islets, and diabetes. Nat. Educ. 3 (9), 13.

Maehashi, K.M., Matano, H., Wang, L.A., Vo, Y., Yamamoto, L., Huang, 2008. Bitter peptides activate hTAS2Rs, the human bitter receptors. Biochem. Biophys. Res. Commun. 365, 851—855.

Nofre, C., Tinti, J.M., 1996. Sweetness reception in man: the multipoint attachment theory. Food Chem. 56 (3), 263—274.

Obayashi, Y., Naganura, Y., 2016. Does monosodium glutamate really cause headache?: a systematic review of human studies. J. Headache Pain 17, 54.

Pearlman, M., Obert, J., Casey, L., 2017. The association between artificial sweeteners and obesity. Curr. Gastroenterol. Rep. 19 (12), 64.

Pepino, M.Y., 2015. Metabolic effects of non-nutritive sweeteners. Physiol. Behav. 152, 450–455.

Reddy, A., Norris, D.F., Momeni, S.S., Waldo, B., Ruby, J.D., 2016. The pH of beverages in the United States. J. Am. D. Asoc. 147, 255–263.

Roper, S.D., Chaudhari, N., 2017. Taste buds: cells, signals and synapses. Nat. Rev. Neurosci. 18, 485–497.

Shallenberger, S.H., Acree, T., 1967. Molecular theory of sweet taste. Nature 480–482.

Shepherd, G.M., 2006. Smell images and the flavor system in the brain. Nature 444 (7117), 316–321.

Stabler, S.N., Tejani, A.M., Huynh, F., Fowkes, C., 2012. Garlic for the prevention of cardiovascular morbidity and mortality in hypertensive patients. Cochrane Database Syst. Rev. 2012 (8), CD007653.

Trio, P.Z., You, S., He, X., He, J., Sakao, K., Hou, D.X., 2014. Chemopreventive functions and molecular mechanisms of garlic organosulfur compounds. Food Funct 5 (5), 833–844, 2014.

Trivedi, B.P., 2012. Hardwired for taste. Nature 486, S7–S9.

Zarka, M.H., Bridge, W.J., 2017. Oral administration of γ-glutamylcysteine increases intracellular glutathione levels above homeostasis in a randomized human trial pilot study. Redox Biology 11, 631–636.

Zong, J., Martirosyan, D.M., 2018. Anticancer Effects Pf Garlic and Garlic Derived Bioactive Compounds and its Potential Status as Functional Food. Bioactive Compounds Health Disease: Online ISSN 2574-0334.

Further reading

Acree, T.E., Lindley, M., 2008. Structure-activity relationship and AH-B after 40 years. Sweetness and Sweeteners, American Chemical Society 979, 96–108.

Fernstrom, J.D., 2015. Non-nutritive sweeteners and obesity. Annu. Rev. Food Sci. Technol. 6, 119–136.

Shepherd, G.M., 2012. Neurogastronomy. How the Brain Creates Flavor and Why it Matters. Columbia University Press. Retrieved from. http://www.jstor.org/stable/10.7312/shep15910.

Trivedi, B.P., 2012. Hardwired for taste. Nature 486, S7–S9.

Review questions

1. Name the 5 basic taste modalities.
2. Where are sweet, sour, and salty taste receptors located on the tongue?
3. What is a taste receptor and how does create a perception (sweet for example)?
4. Define the term gustatory map.
5. Describe how the sensations of taste and smell differ?
6. What is an extra-oral taste receptor and what is its function?
7. What is a polyol? Give an example and describe the flavor and sensation it causes.

8. What are some advantages and disadvantages for using polyols as a sweetener?
9. Describe the synthetic sweeteners, aspartame and sucralose and their food uses.
10. Give an example of a natural sugar substitute that can be used in baked foods.
11. What is the evolutionary importance of bitter taste?
12. Name two substances valued for their bitter flavor. What foods are they found in?
13. How is the bitterness in olives controlled?
14. What does "umami" taste like?
15. Give two examples of substances with strong umami taste.
16. What is MSG? Is it safe to eat?
17. Is MSG an example of hidden salt?
18. Explain why grapefruit's high level of acidity dose not cause a strong sour taste.
19. What taste perception is caused by chemesthesis? Give a food example that has this taste sensation.
20. What enzyme is responsible the taste of garlic and onion?
21. Are the terms pungent and spicy, synonymous? Explain your answer.
22. What does the field of neurogastronomy study?
23. What benefits does neurogastronomy provide?

CHAPTER SEVEN

Food additives

Learning objectives

This chapter will help you to:
- Describe FDAs role in regulation of food additives
- Define the terms natural and synthetic additives
- Describe uses for food additives
- Define the term hydrocolloid and describe uses in creating novel foods
- Explain the difference between toxins and toxicants

Introduction

Food additives are employed for a variety of purposes. They are used to improve nutrition, enhance safety, add color and flavor, modify texture, and preserve food. Some food additives contribute more than one function. Acids, for example, give food a tangy flavor, but also inhibit the growth of microorganisms that cause spoilage. Nitrite also has multiple functions. It is responsible for the distinctive flavor and color of cured meats and inhibits the growth of the microbial pathogen responsible for botulism. While food additives are generally chemical substances, biologicals (enzymes) can also be additives. For example, the enzyme chymosin is added to milk in cheese making. This chapter contains an overview of substances added to food directly and some of those that find their way into food indirectly. However, the number of substances used as food additives is quite large (3000+), thus this chapter focuses on the most commonly used ones. Specifically, it covers additives used as acidulants, antioxidants, antimicrobials, hydrocolloids, emulsifiers, and toxins. Additives used to provide flavor and color are covered separately in Chapters 6 and 8, respectively.

This chapter includes questions that will help you explore and better understand the role of additives in food.
- What is alkaline water and does it have health benefits?
- Why is bacon considered an unhealthy food?

Introduction to the Chemistry of Food
ISBN: 978-0-12-809434-1
https://doi.org/10.1016/B978-0-12-809434-1.00007-4

- Does the sulfite in wine cause headache?
- What is antioxidant and how does it work?
- What is the difference between food poisoning and foodborne illness?

Regulation of food additives in the United States

In the United States, food additives are regulated by the United States Food and Drug Administration (FDA). Food additives are defined by FDA as "any substance the intended use of which results, or may reasonably be expected to result, directly or indirectly, in its becoming a component or otherwise affecting the characteristics of any food". The history of governmental regulation of food began in the late nineteenth century with Dr. Harvey Wiley, chief chemist for the Bureau of Chemistry. Dr. Wiley and others led the fight against the practice of adulteration and use of hazardous substances in food and drugs. These efforts established the first nationwide regulation known as the "Pure Foods and Drug Act of 1906". Under this act, interstate transport of adulterated or misbranded food and drugs was prohibited, adulteration was defined, and misleading or false statements were prohibited by law. For the first time, if a food or drug was in adulterated or mislabeled, it could be seized and the seller could be fined and/or jailed. The United States Congress passed additional legislation called the The Federal Food, Drug, and Cosmetic Act of 1938. This act gave the Federal government authority to closely oversee food, drugs, and cosmetics. Manufacturers were required to provide proof of a drug's safety and efficacy before putting it on the market. The act required that safe tolerances be met for unavoidable poisonous substances in food. The act established standards of identity for foods and authorized factory inspections. A standard of identity means that specific foods (e.g., milk) must meet the definition set by FDA if the word "Milk" is used on its label. If a product labeled as milk contained added water or other substances, it is considered to be an act of adulteration, a punishable offense.

Additional major legislation

The Food Additives Amendment of 1958, was passed in response to public concern about the growing use of chemicals added to food, especially those used in processed foods. There was particular concern over pesticide substances occurring indirectly in foods. Two major outcomes of this legislation were establishment of substances classified as "Generally Recognized As

Safe," or GRAS, and adoption of the Delaney Clause. GRAS status was given to those substances with a history of use without harm as determined by review from experts in the field. Initially, 200 substances were listed as GRAS and grandfathered for use without further testing. The number of GRAS substances has increased to more than 800 as a result of FDA's petition and review process. Additives introduced after the 1958 act are required to be scientifically tested before being allowed in foods. The Delaney Clause was initially intended to address the issue of pesticides in foods. It broadly prohibited the use of any food additive shown to cause cancer in animals. At the time of the regulation, no provision was made for risk factors such as carcinogenic potency of the substance or its level in food. It is now known that substances vary greatly in their ability to cause cancer. Additionally, advances in analytical technology have made detection of these substances possible at extremely low levels. FDA has subsequently adopted a *de minimis* (Latin phrase meaning "about minimal things") exception to the Delaney Clause. This rule allows FDA to grant an exception to Delaney Clause if the substance occurs at very low levels. Currently, if a potentially carcinogenic substance is present in a food at 1 ppm (part per million) or lower, FDA can invoke the exception (Overview of Food Ingredients, FDA 2018).

Regulation of additives in other countries

The European Food Safety Authority defines food additives as "any substance added intentionally to foodstuffs to perform certain technological functions, for example to color, to sweeten or to help preserve foods." In the European Union, food additives are given an E-number (e.g., E-951 aspartame, sweetener). All additives in food sold in the EU must be listed on the food product label. E-number labeling has been adopted by other countries, e.g., Australia, and New Zealand. A summary of food additives grouped by their function and corresponding E-number range is given in Table 7.1 (European Food Additive Identifier Numbers 2018).

Types of food additives

A food additive is defined as a substance not normally found in food. Additives can occur in food as the result of indirect or direct addition. Indirect additives are substances that migrate into food from the environment and include production, processing, and packaging. They do not result from an intentional act. Examples of indirect additives include compounds

Table 7.1 E number classification of food additives.

E number range	Description of function
E100–E199	Colors
E200–E299	Preservatives
E300–E399	Antioxidants, Acidity Regulators
E400–E499	Thickeners, Stabilizers, Emulsifiers
E500–E599	pH Regulators, Anticaking Agents
E600–E699	Flavor Enhancers
E700–E799	Antibiotics
E900–E999	Glazing Agents, Gasses, Sweeteners

from packaging materials, toxicants such as heavy metals, aflatoxin from mold, and pesticide residues. Direct additives are substances added to a food to provide a desired effect as a colorant, preservative, antioxidant, or sweetener. It is important to note that use of an additive to conceal damage or spoilage is illegal and considered to be an act of adulteration. Food additives are also classified as natural or synthetic. Natural additives are defined as substances occurring in foods or extracted from a natural source. For example, ascorbic acid (vitamin C) is considered to be a natural additive even when obtained as an extract. Conversely, synthetic additives are person–made substances. The distinction between a natural substance and its synthetic counterpart can be blurred when they are demonstrated to be equivalent. Organic foods produced in the United States and Europe must contain at least 95% organic ingredients. They are, however, permitted to contain additives of natural origin. Organic foods produced under the European Food Safety Authority (EFSA) also require that both the additive's name and its function must be stated on the food label. Most additives do not contribute to the food's nutritive value because they are used at very low levels. Vitamin additives are an exception to this statement because they are used for fortification or restoration of a food (Food Additives EFSA, 2018, Food Additives WHO, 2018).

Food acids and acidity regulators

Several weak acids (e.g., acetic, propionic, malic, lactic, citric, and tartaric) are common food additives used to create flavor, preserve food, provide functionality (e.g., leavening systems) and improve safety. Acetic acid, a short chain fatty acid, is the predominant component of vinegar and is responsible for its characteristically pungent taste. Vinegar is used as a tastant and preservative in pickled foods, salad dressings, marinades, and as a condiment. Other short chain fatty acids such as butyric and propionic, are volatile compounds with distinctive tastes that contribute to the overall flavor of cheese and other dairy products. Lactic acid produced by bacterial fermentation of lactose in milk is responsible for the sour taste of yogurt. Lactic acid is also the taste of spoiled milk and the principle tastant in sausage and sourdough bread. Fruits such as apples, oranges, and grapes derive much of their tart taste from malic, tartaric, and citric acids. Phosphoric acid is an inorganic acid commonly used as an acidulant in carbonated soft drinks. Phosphoric acid in combination with carbonic acid (produced when CO_2 is dissolved in water) is responsible for the tangy taste of these beverages.

Weak acids preserve food by inhibiting the growth of spoilage micro-organisms. The mechanism of microbial inhibition involves diffusion of the acid through pores in the bacteria's outer membrane, causing disruption of essential biochemical pathways. Weak acids contribute to food safety by preventing botulism poisoning. *Clostridium botulinum* is a pathogenic and ubiquitous microorganism that grows well in food environments containing little oxygen. Spores of this organism survive normal heat treatments, such as boiling. This enables them to germinate and release botulinum neurotoxin that causes muscle paralysis and death. Foods, such as sausages, represent an optimum environment for botulism poisoning. Sausages were a major cause of illness and death during the 18th and 19th centuries. Fortunately, it was discovered that spore germination and toxin production could be prevented by keeping the pH at 4.5 or lower. Sausage is made safe from botulism poisoning by directly adding lactic acid or indirectly as a fermentation by-product. However, acidification is not desirable for foods such as green beans and other vegetables. Low pH causes chlorophyll-containing foods to turn brown. The risk of botulism in low acid canned foods (e.g., vegetables) is avoided by using a rigorous canning process that inactivates or kills *C. botulinum* spores.

Chemical leavening

Acids are used in chemical leaving systems for bread dough applications when yeast fermentation producing carbon dioxide (CO_2) is not practical. Chemical leavening systems produce CO_2 through a reaction between a carbonate salt and an acid. An example of this chemical reaction using sodium bicarbonate ($NaHCO_3$ - baking soda) as the source of carbon dioxide and an acid (R—COOH) is shown below. In this reaction, a hydrogen ion from the acid adds to bicarbonate producing the ionized acid (ROO^-), sodium ion, water, and carbon dioxide.

$$NaHCO_3 + R - COOH \leftrightarrow ROO^- + Na^+ + H_2O + CO_2$$

While this chemistry works well in producing CO_2 gas, it has a major limitation in food. Specifically, the production of gas begins as soon as the reactants are mixed. Foods such as refrigerated bread doughs, self-rising flour, and cake mixes must rise during cooking. Gas production is delayed in these products by using different acid sources. Compounds called leavening acids provide hydrogen ions slowly or when heated. Examples of leavening acids include monocalcium phosphate hydrate $[Ca(H_2PO4)_2] \cdot H_2O$, and sodium aluminum phosphate hydrate $[NaH_{14}Al_3(PO_4)_8 \cdot 4H_2O]$. Monocalcium phosphate is a more water-soluble acid than is sodium aluminum phosphate and generates acid more quickly. Its rate of CO_2 production is slow compared to acetic acid, but it begins to produce gas as soon as water is added. Sodium aluminum phosphate primarily releases its hydrogen ions in response to elevated temperatures, such as in baking. Monocalcium phosphate and sodium aluminum phosphate are called slow-acting and fast-acting leavening acids, respectively. These leavening acids are commonly used in the household baking ingredient known as double acting baking powder.

Acidity regulators

Excess acid in food can make it unacceptably sour. Acidity regulators are used to reduce the sour taste of a food. For example, sourness in cherries can be reduced by adding the sodium salt of malic acid called sodium malate. This is based on the chemistry of weak acids. Weak acids (HA) are by definition incompletely dissociated into hydrogen ion (H^+) and its corresponding anion (A^-). The amount of H^+ present depends on its dissociation from HA in an equilibrium process. When the sodium salt of A^- is added, an equilibrium is re-established by driving the reaction (shown below) in reverse (to the left) and this decreases level of hydrogen ion in the food.

$$HA \leftrightarrow H^+ + A^-$$

Other examples of weak acids used as food acidity regulators include the sodium salt of citric acid in fruit juice, lactic acid in cheese, and phosphoric acid in soft drinks.

Buffering capacity

Food acids also provide buffering capacity in food systems. Buffering capacity is defined as the ability to resist change in pH when acid or base is added to a system. Milk, for example, contains a substantial amount of phosphate complexed to its proteins and free in solution. Milk phosphate acts a buffer to absorb hydrogen ions and maintain the pH at a normal level. However, as bacteria in milk metabolize its lactose into lactic acid, hydrogen ions are produced. At some point in time the amount of hydrogen ion produced exceeds the buffering capacity of milk phosphates and causes a drop in pH. The casein fraction of milk is destabilized and precipitates when a pH of about 4.9 is reached.

> ### Bases

A base is chemically defined as a substance that donates hydroxide ions (OH⁻). Bases (alkaline substances) have a number of food uses. Sodium bicarbonate is a weak base used for its ability to generate carbon dioxide in the presence of an acid. Sodium bicarbonate also accelerates Maillard browning. Potatoes coated with sodium bicarbonate brown quicker and have a crisp texture. A secret to making French baguettes with a crisp, brown crust is a quick dip in sodium bicarbonate (baking soda) solution before baking. Similarly, sodium bicarbonate accelerates browning of sugar in caramelized products like peanut brittle and other confections. Stronger bases, such as potassium carbonate and sodium hydroxide, are used in olive processing to destroy the bitter compound oleuropein and to give olives a darker color. Potassium carbonate base is used in Dutching of cocoa beans. It neutralizes the acidity of the bean and gives the resulting cocoa powder a darker color and a more intense chocolate flavor. Dutched cocoa is also more soluble in aqueous solution after treatment with base. Strong bases are used in the production of soy protein isolates that are the main ingredient in many soy-based food products. Soy proteins are tightly packed in the soybean and biologically function as a store of energy and amino acids for the plant's growth. Extraction of protein from soybeans is made possible by adjusting the pH to 9 or 10. High pH disrupts protein complexes in the seed and increases the amount of protein extracted. Alkaline pH treatment with sodium pyrophosphate salt is used to make instant pudding. Elevated pH promotes starch gelatinization and has the desirable effect of thickening pudding mix as soon as milk is added. As a result, no cooking is needed.

What is alkaline water? Does it have health benefits?

Alkaline water is a type of bottled water to which bicarbonate salts (sodium, calcium, magnesium, or potassium) are added to raise its pH from 7 to about 9.0–10. Sodium bicarbonate (also known as baking soda) is a safe additive for the purpose of increasing the pH of water. Alkaline water can also be homemade by adding one teaspoon of sodium bicarbonate to a gallon of deionized water. A potential concern for alkaline water, however, is its sodium content. For example, the sodium intake (approximately 1.34g) from drinking a gallon of alkaline water over the course of a day represents about two-thirds of FDA recommended daily sodium intake (2.3 g). Alkaline water has been proposed as a means to mediate the negative effects of diets with a high Potential Renal Acid Load (PRAL). Metabolism of foods such as cheese, meat, and fish elevate the level of acid processed in the kidney and excreted in urine. High PRAL diets are proposed to be linked to higher risk of kidney stone formation and bone mineral loss. Conversely, diets high in vegetables and fruits have a lower PRAL value and reduced risk of stone formation (Remer and Manz, 1995). It should be noted that blood pH is closely regulated (buffered) in the kidney and consuming excess alkaline substances (carbonates) raises potential concern for causing alkalosis. This condition affects the heart's contractile force and may cause neurological symptoms (Al-Abri and Olson, 2013). No health benefit from drinking alkaline water has been demonstrated.

Salts

Sodium chloride (NaCl) is perhaps the most used food additive. Salt enhances flavor, preserves food, and is essential to many processed foods. Saltiness is one the basic five taste modalities with a specific receptor mechanism. Sodium chloride enhances the perception of sweetness (see Chapter 6, Flavors). This effect on sweetness is demonstrated by the popularity of adding salt to chocolate and other sugary foods. Sodium chloride is a preservative used to extend the shelf life of perishable foods like meat. The practice of salting meat and fish has been used for thousands of years to keep it from spoiling. The preservation effect of salt results from its effect on the availability of water to spoilage microorganisms. Water availability in food is measured as water activity (a_w), as described in Chapter 1 (water and pH). The minimum a_w for spoilage bacteria to grow is 0.9 and fresh meat's a_w is typically 0.98. Salt added to pork, beef, and fish reduces the a_w level to 0.88 or less. This extends the period of time these foods can be kept without spoiling. Sodium chloride is also essential in making processed meat. Large amounts of salt are added during the grinding operation to improve protein extraction and binding between meat pieces. Salt extracted proteins (e.g., myosin) enhance the emulsification of fat and increase water holding capacity in processed meat products.

Antimicrobials

Acetic acid

Acids contribute several attributes to food, including flavor (taste), preservation, and safety (Fig. 7.1). Tastes provided by acids are generated by stimulation of pain receptors located in taste buds. The taste of food acids is described as tart, tangy, or crisp. Acetic acid (Fig. 7.1A) is a ubiquitous food flavoring (vinegar) and preservative. Acetic acid and its salts (sodium, potassium, and calcium) are used at low levels in baked goods as dough conditioners. The distinctive tasting condiment known as vinegar is produced by bacterial (oxidative) fermentation of alcohol. In some cases, alcohol derived from wine is used as the starting media for fermentation. *Acetobacter aceti* is commonly used to make acetic acid used as a 3% solution and has excellent bactericidal effects. It may prove to be a suitable antiseptic agent in treatment of burn wounds (Ryssel et al., 2009).

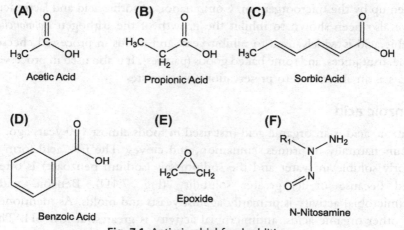

Fig. 7.1 Antimicrobial food additives.

Propionic acid

Propionic acid is a short chain saturated fatty acid containing 3 carbons (no double bonds) and has a characteristic flavor described as cheesey or dairy-like. It is also rather pungent (Fig. 7.1B). It is the predominant acid in Swiss cheese, comprising about 1% (by weight) of the cheese. Propionic acid has antimicrobial activity, primarily against molds in acidic environments. Antimicrobial activity of propionic acid is only effective at low pH because the ionization of its carboxyl group creates a negative charge and limits transport across the microbial cell wall. Sodium and calcium salts of propionic acid (i.e., sodium propionate and calcium propionate) are used as a preservative (mold inhibitor) in baked goods and processed cheese. Calcium propionate occurs naturally in butter and is used in processed cheese and meat for its preservative properties.

Sorbic acid

Sorbic acid is a natural preservative first isolated in 1859 from the berries of an ash tree. Sorbic acid is unsaturated fatty acid containing two carbon-carbon double bonds and both are in the trans configuration (Fig. 7.1C). Investigation of its antimicrobial activity found it to be an effective inhibitor of molds, yeasts, and fungi. Sodium and potassium salts of sorbic acid (i.e., sodium sorbate and potassium sorbate) are not as effective. The free acid form is more effective as an antimicrobial. Antimicrobial activity of sorbic acid increases as the pH of the media decreases. Enhanced activity at low pH results from the acid being fully protonated (unionized) and thus readily taken up by the microorganism. Combination of sorbic acid and lactic acid have also been shown to inhibit the growth of the pathogen *Salmonella*. Sorbic acid is widely used to inhibit yeast and molds in processed cheese, wine, fruit juices, and some baked goods (pastries). It is also used in processed meats as an alternative to preservation using nitrites.

Benzoic acid

Benzoic acid is an organic acid first used in foods almost 100 years ago. It occurs naturally in prunes, cinnamon, and cloves. The free acid form is poorly soluble in water and the sodium salt (sodium benzoate) is often used because of its greater solubility (Fig. 7.1D). Benzoic acid's antimicrobial activity is primarily against yeasts and molds. As mentioned for other organic acids, antimicrobial activity is greatest at low pH. The effect results from greater permeability of the unionized form into microorganisms. Benzoic acid's most common uses are in carbonated beverages, pickles, sauces, and jelly. Non-food applications for benzoic acid's antibacterial function are found in cosmetics. Benzoic acid itself has low toxicity, but there has been concern because of a potential reaction that converts it to benzene. Although benzene is a toxic and carcinogenic compound, the reaction causing this change has a very low chance of occurring in food. Typically, benzene is rapidly converted to hippuric acid in the body and excreted in the urine. Parabens are antimicrobial compounds chemically derived from benzoic acid. Chemically, parabens are esters made by combining benzoic acid and alcohols such as methanol or propanol. Paraben esters have antimicrobial activity against molds and yeasts and are used in beer, soft drinks, and olives. Cosmetics and pharmaceuticals represent the largest use of parabens.

Antibiotics

The peptide antibiotic nisin is used in cheese making. It is composed of 34 amino acids that includes two unusual products of post-translational modification known as lanthionine and methyl-lanthionine. Nisin is more correctly termed a bacteriocin that involves peptides with the ability to kill or inhibit the growth of other bacteria. Nisin is produced by *Streptocpccus lactis*. This organism is found in milk and has little or no activity against yeasts and molds. Nisin's antibacterial activity is very selective. It benefits cheese making by controlling the growth of bacterial strains that cause spoilage. It can also inhibit the growth of pathogens and spore growths, such as *Staphylococcus aureus* and *C. botulinum*. Nisin's antibiotic activity does not present concern for development of antibiotic-resistant strains. Nisin was found to be an effective inhibitor of more than 90% of gram-positive spoilage organisms in beer (Suganthi et al., 2012). In the United States, nisin is approved for use in pasteurized process cheese spreads. In Europe, it is used in processed meats and cheese.

Epoxides

Epoxides, such as ethylene and propylene oxide, are small, highly reactive, and volatile molecules that effectively kill most microorganisms, including bacterial spores (Fig. 7.1E). The effectiveness of epoxides results from the molecule's reaction with the organism's proteins (enzymes) and nucleic acids. As a result, microorganisms are unable to perform essential processes and die off. Epoxides are used in gas form to chemically sterilize processing and food packaging equipment. Ethylene oxide is also used as a sterilant for medical equipment. The treatment of foods with epoxides is limited to applications where the microbial load is typically high and washing with water-based detergents is not applicable. Spices and nuts are examples of food materials that are treated with epoxides. The increased popularity of spicy foods has logically increased the demand for hot peppers, turmeric, and other ground dry spices. These ingredients are often produced in remote regions, transported great distances, and stored for long periods of time. It is no surprise that they often carry high levels of microorganisms, including pathogens.

Nitrite

Nitrite (NO_2) and nitrate (NO_3) salts (typically sodium) are used in food as a preservative and to create cured meat color and flavor. Nitrite is a potent preservative because it inhibits the growth of some microorganisms that cause spoilage and pathogens, such as *C. botulinum*. Nitrite does not inhibit *Escherichia coli* or *Salmonella* pathogens. While instances of poisoning from botulinum toxin in food are rare, can be fatal. Nitrate is converted to nitrite in meats by microorganisms and/or reducing agents such as ascorbic acid. Nitric oxide (NO) derived from nitrite is the source of the pink color and flavor characteristic of cured meats, such as ham and bacon. Nitric oxide is also an antioxidant that prevents free radicals from causing lipid oxidation and off flavors in meat. However, there are health concerns associated with excessive consumption of nitrite. The LD50 value for nitrite is 71 mg per kg of body weight. This translates to an intake of 4.6 g for a 150 lb person. Lesser amounts of nitrite are known to cause methemoglobinemia when it binds to hemoglobin in red blood cells. The nitrite-hemoglobin complex has low affinity for oxygen and reduces the amount delivered to tissues and organs. High nitrite levels have occurred in ground water supplies due to fertilizer runoff. This causes methemoglobinemia in infants (blue baby syndrome) when the water is used to make their formula. The largest source of nitrite in food is from cured meats and these foods contain 10—30 mg per kg of product. The largest cause for concern regarding nitrites comes from N-nitroso compounds. Specifically, nitrosamines are produced by a reaction between nitrites and proteins under acidic conditions found in the stomach. Nitrosamines (Fig. 7.1F) are also found in foods such as bacon (especially when cooked at high temperature), beer, salted fish, and cheese. There is evidence for a link between nitrosamine intake and risk of gastric and esophageal cancer (Jakszyn and González, 2006). A combination of ascorbic acid (water soluble) and alpha tocopherol (fat-soluble) antioxidants have been shown to inhibit nitrosamine formation in cooked products such as bacon.

Why is bacon considered an unhealthy food?

Bacon is one of those foods that many people find very desirable because of its flavor. Despite its popularity, there are reasons to use moderation when eating bacon. It is high in fat and salt. A typical two slice serving of bacon (16g) contains about 90 calories (66% from fat and 24% from protein), 160 mg of sodium, and lesser amounts of other minerals. The fatty acid profile of bacon fat is predominately saturated fatty acids. A more serious concern results from cooking bacon at high temperature to make it crispy. Frying temperatures cause nitrites in bacon to react with amino acids, such as proline, and form N-nitrosopyrrolidine, a carcinogen. While most bacon contains added ascorbate and tocopherol to reduce nitrosamine formation, the amount formed is not zero. Reducing the exposure to nitrosamines and their associated health risk can be achieved by alternative cooking methods. Bacon cooked at lower temperature or in the microwave has a significantly lower nitrosamine content (Park et al., 2015).

Sulfur dioxide and sulfite

Sulfur containing compounds, such as sulfur dioxide (SO_2), sulfite (SO_3^{-2}), bisulfite (HSO_3^-), and meta bisulfite ($S_2O_3^{-2}$), are effective antimicrobial agents. Sulfur dioxide (SO_2) added to water forms the complex ($SO_2 \cdot H_2O$) that has strong antimicrobial activity at low pH. Hydrogen sulfite is a more effective antimicrobial in the pH range of 3 to 7. Wine makers have a long history of using sulfites to inhibit the growth of bacteria, molds and wild type yeasts in grape musts. Without an initial sulfite treatment of the must, these microorganisms would take over fermentation and ruin the wine. The use of sulfite in wine making was first accomplished by burning elemental sulfur to produce sulfur dioxide gas. Current and common practice today is to add purified sulfur dioxide gas or sulfite salts to control the growth of undesirable microbial species. The addition of sulfur dioxide results in acidification of the media and contributes to its antimicrobial activity. Sulfite salts used in wine making are either sodium sulfite (Na_2SO_3) or potassium sulfite (K_2SO_3). The antimicrobial activity of sulfite depends on the pH of the environment. In general, the lower the pH, the more effective is the antimicrobial action. Disulfite, also known as metabisulfite, provides antimicrobial activity over a longer period of time. Disulfite also protects wine must from oxidation and turning brown before and during fermentation. Disulfite added at approximately 2% is used as a sanitizing agent for wine bottles. Most wines produced in the United States contain some sulfite. Any wine that contains 10 ppm sulfite must be labeled as, "contains sulfite". The maximum level permitted by the FDA is 350 ppm. In contrast, the maximum level in the UK is 210 ppm and 160 ppm for white and red wines, respectively. Organic wines made in the United States are not permitted to contain sulfite.

Sulfite reactions affecting color and flavor

In wine, acetaldehyde formed during fermentation provides a pleasant fruity aroma. When hydrogen sulfite (HSO_3^-) is present, a complex is formed that reduces the level of this aldehyde and negatively affects wine flavor. Sulfite is also used as an inhibitor of Maillard reactions responsible for brown pigments in food. Sulfites inhibit Maillard browning by reacting with aldehyde group of reducing sugars. Sulfite addition also affects flavonoid pigments found in a number of fruits. Chemical interaction between sulfite and flavonoid compounds destroys their color and causes a bleaching effect. Added sulfite is an effective agent to control brown color formation in dried fruits and molasses. Sulfites are used to inhibit enzymatic browning in lettuce and freshly cut apples and potatoes. The polyphenol oxidase enzyme responsible for this reaction causes oxidation of phenolic compounds which subsequently become brown pigments. Sulfite interaction with oxidized phenolic compounds prevents pigment formation. Fresh shrimp and lobster can also become brownish in color due the action of the same enzyme. Sulfite treatment is used to stop color development in these shellfish. Sulfites are effective reducing agents that can be used to improve the dough properties of wheat flour. Sulfite addition to wheat flour causes chemical reduction of disulfide bonds in wheat proteins. Overall, new disulfide bonds are formed and these crosslink wheat proteins and improve dough strength.

Physiological effects of sulfite

The body produces about 1000 mg of sulfite per day. Sulfite and other sulfur containing compounds, such as the amino acids cysteine and methionine, are converted to sulfate by the enzyme sulfite oxidase. They are eliminated in the urine without ill effect. However, there is a caution for asthmatics. About 1% of the U.S. population have sulfite reactions that cause symptoms such as shortness of breath, wheezing, and hives. Dried fruits such as apricots, figs, and prunes are of particular concern for those with sulfite sensitivity because they can contain excessive amounts of sulfite. For this reason, sulfite is prohibited from use on fruits and vegetables presented in salad bars. A final negative aspect of sulfite is its destructive reaction with thiamin (vitamin B_1). Sulfites are therefore prohibited from use in processed meat as they are a good source of thiamin.

Does the sulfite in wine cause headache?

Adverse reaction to sulfite is a condition known as sulfite sensitivity. Sulfites are components of all wine as a result of direct addition or an indirect addition from yeasts used in fermentation. It is commonly assumed that sulfite is the cause of adverse reactions experienced from drinking a glass of wine. However, there is another possible culprit. A substance called histamine found in fermented foods like beer, soy sauce, and wine is known to cause headache, flushing of the face, and allergy–like symptoms, including rash and difficulty breathing. These are symptoms of histamine intolerance. Are adverse reactions from drinking wine due to sulfite, histamine, or something else? First and foremost, anyone who has an adverse reaction to any food should seek a medical professional's advice for help. But a clue to the cause of the adverse reaction might be found in its histamine content.

Chelators (sequestering agents)

Chelating agents are compounds that tightly bind and sequester metals like copper, iron, zinc, calcium, and manganese. Chelators, such as EDTA contain an unshared pair of electrons that enable formation of a complex with metal ions. Examples of chelators used in foods are polycarboxylic acid compounds such as; citric, malic, tartaric, and succinic acid. Pyrophosphates and the synthetic compound ethylenediaminetetra acetic acid (EDTA) are also chelating agents used in foods. The chelating effect on metal ions can be illustrated using EDTA as a model. Two nitrogen atoms with an appropriate geometry are able to share unpaired electrons with the metal ion as shown in Fig. 7.2. The pH of the medium affects the metal chelating ability of EDTA and other chelators. Ionization of carboxyl groups

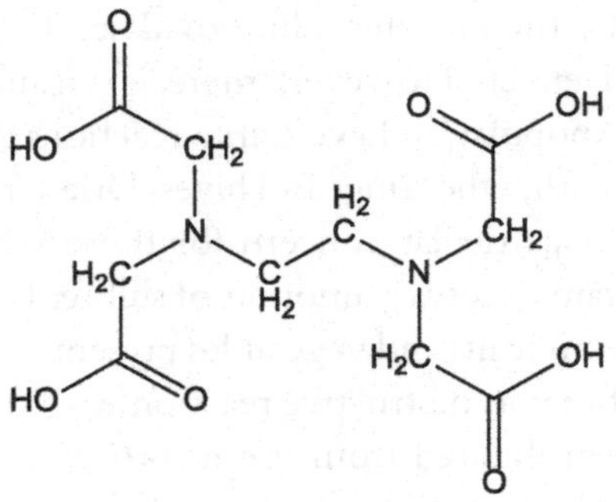

EDTA- Metal Chelator
Fig. 7.2 Metal ion chelator (EDTA).

equates to a more effective metal chelator. Conversely, protonation removes the negative charge and diminishes the strength of the interaction between chelator and metal ion. The relative chelating ability decreases in the following order; EDTA > pyrophosphate > citric acid. A similar chelation of iron occurs in the porphyrin containing proteins, myoglobin and hemoglobin. Iron is essential to the oxygen carrying function of these proteins and is very tightly held. However, iron is released from heme the group during the digestion and is in fact, the most easily absorbable form of iron intake. Chelators can provide antioxidant activity by binding iron (a pro-oxidant) and inhibiting lipid oxidation.

Antioxidants

Oxidation is chemically defined as the loss of electrons from an atom or molecule. More importantly, oxidation is a chemical process causing the breakdown of molecules. The process of oxidation is also accompanied by the release of energy. For example, burning wood is a rapid oxidation process producing energy in the form of heat and light. A much slower oxidation of food materials occurs through the process of digestion and provides the energy required by all forms of life. Oxidation also occurs from chemical species known as free radicals (discussed in Chapter 4, Lipids). Free radicals are reactive atoms or molecules containing unpaired electrons in their outer shell. They can be produced by natural processes, such as metabolism, or can be formed by extrinsic factors, such as ultraviolet light or ionizing radiation. Unchecked free radicals in living systems cause damage by reacting with proteins, lipids, and DNA. DNA modified by reaction with a free radical, such as hydroxide radical (•OH), represents a mutation in the genetic code and a potential cause of cancer. Fortunately, living systems have compensatory mechanisms that repair and restore the damage caused by free radicals. In food, free radical reactions are the cause of several adverse effects most noticeable as loss of color and production of off flavors. A more immediate effect of free radicals is the destruction of essential fatty acids and amino acids, vitamins, and naturally occurring antioxidant molecules. The net result of these oxidations is loss of food quality and reduction of nutritional value. Additives included in processed foods perform their antioxidant function either as free radical scavengers or chelators of pro-oxidant metal ions.

What are antioxidants? How do they work?

Antioxidants are compounds that inhibit or terminate free radical reactions. Antioxidant molecules are free radical scavengers (FRS). They function by donating a hydrogen atom to the radical, taking away the unpaired electron and converting it to a low energy (non-reactive) species (Shahadi et al., 2002). An effective FRS must be able to react with the radical species before it can damage other molecules (e.g., unsaturated fatty acids, vitamins). The pro-oxidant activity of transition metals, such as iron and copper, can be inhibited by adding a chelating agent. Chelators such as EDTA form tight complexes with iron and copper and limit the ability to participate in the chemistry of oxidation reactions. Additionally, the EDTA-iron complex is more soluble that the mineral alone and exists in the aqueous environment where target molecules are concentrated.

Ascorbic acid

Ascorbic acid exists in two forms: L-ascorbic acid (vitamin C) and the inactive optical isomer, L-isoascorbic called erythorbic acid (Fig. 7.3A). Both forms of ascorbate are equally effective antioxidants and control free radical reactions in several ways. Erythorbic acid is considerably less expensive compared to L-ascorbic acid and therefore is commonly used as an antioxidant additive in food. Ascorbates are potent free radical scavengers (FRS). In this process of controlling free radicals, ascorbate is oxidized (a loss of electrons) to a form that can be readily reduced back to L-ascorbic acid by another antioxidant such as, alpha tocopherol (vitamin E). The combined antioxidant effect of ascorbate and tocopherol is greater than the sum of two, thus they are said to act synergistically. Ascorbate and tocopherol are often used in combination as antioxidant additives in food

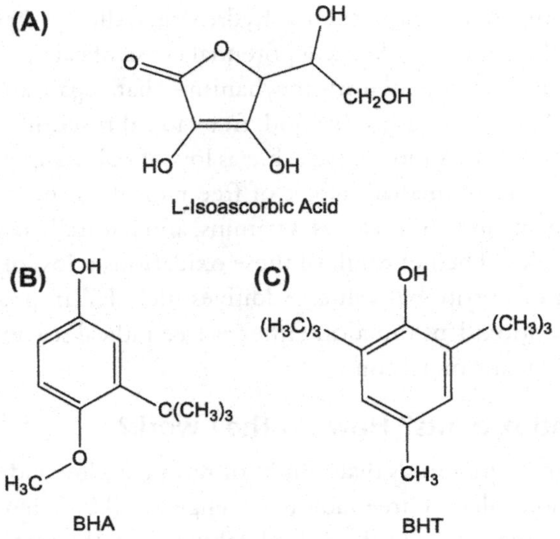

Fig. 7.3 Antioxidants.

because they effectively inhibit free radical reactions in both aqueous and lipid environments. The combination is commonly used to inhibit lipid oxidation in processed meats. Ascorbates are also effective for controlling enzymatic browning in foods such as freshly cut apples or potatoes. Phenolase and other similar enzymes oxidize phenolic compounds in these foods to the corresponding quinone that subsequently changes to a brown color. Ascorbic acid controls enzymatic browning by lowering the pH of the medium and binding pro-oxidant metal ions (Fe and Cu).

BHA and BHT

Butylated Hydroxyanisole (BHA) and Butylated Hydroxytoluene (BHT) are antioxidant compounds that share a similar structure (Fig. 7.3B and C). These function as free radical scavengers and terminate the reaction by donating hydrogen atoms from their hydroxyl (OH) group. Both BHA and BHT are synthetically derived and have long-standing GRAS status. They are effective at very low levels in preventing lipid oxidation in a wide variety of processed foods, such as breakfast cereals, cake mixes, dehydrated meats, and food fats and oils. BHA and BHT are also used in animals feed, pet foods, cosmetics, pharmaceuticals, and plastic packaging. Both BHA and BHT have been in use since the 1950s without major issues. Recent studies have raised concern over their safety. A link between

BHA, but not BHT, and cancer in rats was reported by Ito et al. (2008). BHA's possible carcinogenicity is based on evidence that it caused cancer in the fore-stomachs of tested rats. Humans do not have this organ. Perhaps the health risk of consuming BHA and BHT will be more clearly understood in the future.

> ## Hydrocolloids

The term hydrocolloid is a contraction of the words hydrophilic and colloid. Colloids are technically defined as a homogeneous mixture of one substance dispersed or suspended in another. Polysaccharides (i.e., starch, modified cellulose, pectin, and gums) are the most common form of hydrocolloids used in food. These polymers contain a large proportion of hydroxyl (OH) groups and provide a strong interaction with water via hydrogen bonding. Some proteins, like gelatin (collagen), perform well as hydrocolloids for similar reasons. Hydrocolloids are used in food systems to provide functional properties such as gelation, thickening, and emulsification. Most hydrocolloids are able to provide these properties at low concentration (e.g., < 1%) with little or no impact on flavor. A detailed summary of hydrocolloid agents and their food applications can be found in recent reviews (Saha and Bhattacharya, 2010; Banerjee and Bhattacharya, 2011).

Molecular gastronomy and hydrocolloids

Molecular gastronomy represents a new field combining technical, artistic, and social components in the creation of novel foods. The phrase molecular gastronomy was coined by two scientists, Nicholas Kurti, a physicist, and Herve' This, a physical chemist. Modernist cuisine and culinary physics are other labels used to describe this field. Unlike traditional Food Science, molecular gastronomy is focused on applying scientific principles to enhance understanding of process (cooking) at a molecular level (McGee, 1984). Examples of techniques used in molecular gastronomy include low temperature-immersion (sous-vide) cooking, liquid nitrogen fast freezing and shattering, and dehydrator made fruit jerky. Notably, molecular gastronomy makes extensive use of hydrocolloids (e.g., starch, pectin, and gelatin) in the creation of novel foods. Cola caviar, chocolate noodles, salmon mousse, and red wine jelly are just a few examples that rely upon various types of hydrocolloid to provide these innovative foods.

Gelation

Gels represent one of the most important ways to alter food texture. Hydrocolloid gels are made by treatments resulting in formation of a stable 3-dimensional network. The treatment for many polysaccharides involves a heating and cooling process that enables formation of new hydrogen bonds (i.e., cross-links) between polysaccharide chains. Stability of the network is provided by the newly established cross-links between the polymer molecules. Gel networks contain microscopic cavities that entrap water. Gels can contain as much as 90% water by weight. This provides an obvious advantage in texture, appearance, and economy. Examples of additives using gelling agents include agar, gelatin, pectin, carrageenan, alginate, and starch.

Agar

Agar is a galactose-based heterogenous polysaccharide derived from red algae. It is a heterogenous polysaccharide composed of agarose and agaropectin polymers. A typical agar composition is 70% agarose and 30% agaropectin. Agarose is a linear (no branch points) polysaccharide composed of a repeated galactose disaccharide. The composition of agaropectin is more varied and contains D and L isomers of galactose with sulfate and pyruvate substituents that give the polymer a strong negative charge. Agar is best known as the growth media used in identification and enumeration of microbial organisms. Agar gels are reversible (melts at 85°C and solidifies at 32–40°C) and translucent. Refined grades of agar are used in food applications. Agar's properties are similar to gelatin. It is a good substitute for animal-based gelatin in vegetarian foods. Agar is easier to use in food gels than many other substances. Common food applications of agar include puddings, custards, and soft candies. Agar improves the texture of processed cheese and frozen desserts. It is also added to baked goods to inhibit staling. A creative food application uses agar-based gel cubes that are infused with fruit extract or wine to make a vegetable-based aspic. You may have encountered agar in the dentist's office where it is often the polymer used to make dental impressions.

Gelatin

Gelatin is another name for the animal protein collagen that is obtained as a by-product of the meat processing industry. Collagen is extracted from skin, bones, and connective tissue using a variety of conditions (e.g., salt solutions of weak acid, weak alkali, and/or enzymes). The choice of extraction method is important because it affects gelatin's end use properties. In addition to food, gelatin is used in pharmaceuticals (gelatin capsules) and photography (film emulsions) applications. Gelatin is also made from extracted collagen through enzymatic process. Enzymes break down the protein into smaller fragments called peptides that are better gelling agents. Collagen is an unusual protein because the amino acids proline and hydroxyproline make up about one-third of its total amino acid composition. Proline and hydroxyproline impose structural constraints on the collagen molecule. These amino acids are principally responsible for the ability to make reversible gels. The melting temperature of gelatin ($95°F/35°C$) is just below body temperature ($98°F/37°C$). This provides a desirable mouth feel to the gel. The gelation properties of collagen are used to make aspic, marshmallows, gummy bears, ice cream, yogurt, and more. Gelatin is also used as a clarifying agent in wine and beer. Cloudiness caused by substances such as yeasts, proteins, and tannins bind to gelatin and can be precipitated or filtered out of the solution.

Pectin

Pectin is a polysaccharide found in the cell walls of plant tissues. The native pectin polymer is responsible for the strength of the cell wall and firmness of the fruit as a whole. During ripening, pectin is modified and broken down into shorter segments by enzymes. This natural process makes the fruit softer. Pectin is composed of galacturonic acid monomers, some of which contain an esterified methyl group (Fig. 7.4).

Those galactose monomers without methyl groups have a negative charge when they ionize. The charge inhibits the interaction between pectin molecules and results in a weak gel. Pectin with a high percent of methyl ester groups (high methoxyl pectin) has less negative charge and is better able to form strong gels. The enzyme pectin methyl esterase (PME) is responsible for removing methyl groups during ripening. The extent of

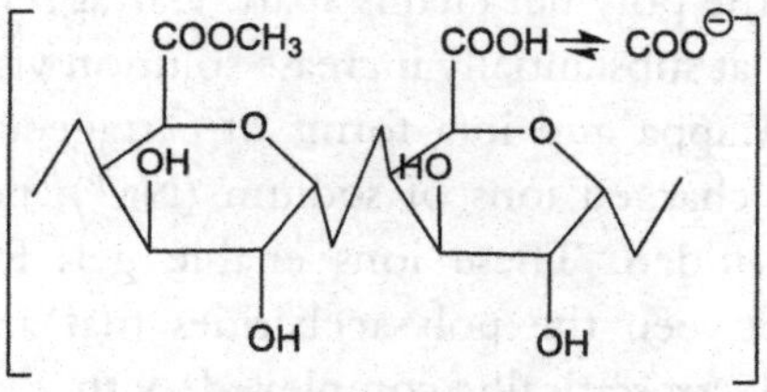

Pectin (Glacturonic Acid Polymer)

Fig. 7.4 Pectin.

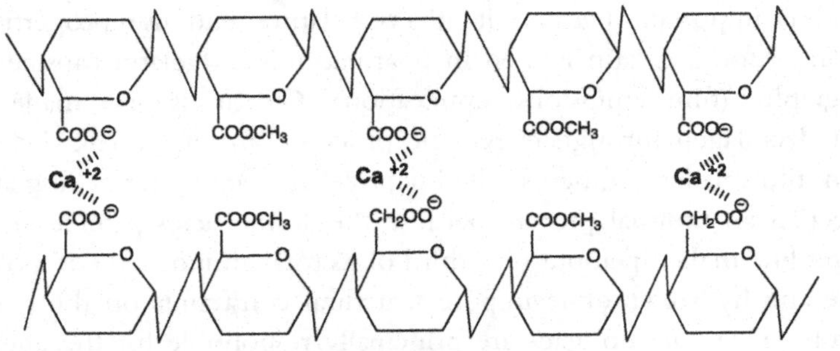

Fig. 7.5 Pectin gel with calcium.

methyl group hydrolysis from pectin results in different commercial grades of pectin, such as low–methoxyl and high methoxyl. High methoxyl pectin is the form of choice for making traditional jams and jellies. A combination of high methoxyl pectin, low pH, and high levels of sugar (50%) is essential to making a firm jelly. It is possible to make firm gels using low methoxyl pectin by the addition of calcium. Calcium (Ca^{+2}) is a positively charged, divalent cation that forms gels with low methoxyl pectin through electrostatic cross-linking between negatively charged pectin molecules. This type of gel has the advantage of requiring lower levels of sugar (10% −15%) and is the choice for making reduced calorie jams and jellies. Added calcium forms a complex between pectin polymers that resembles an egg crate structure (Fig. 7.5). Pectin's ability to form clear gels with calcium has made it a favorite tool for creating novel foods in culinary applications.

Carrageenan

Carrageenan is a polysaccharide gelling and thickening agent obtained from sea weed. The carrageenan polymer is composed of galactose units containing numerous sulfate groups. Carrageenan exists in kappa, iota, and lambda major forms that differ in their content of esterified sulfate groups. The negative charge contributed by sulfate groups creates a mutual repulsion that keeps the polymer chains apart. Carrageenans are thus soluble polysaccharides that substantially increase solution viscosity (thicken) at low concentration. Kappa and iota forms of carrageenan also form gels when the positively charged ions of sodium (Na^+), potassium (K^+), or calcium (Ca^{+2}) are added. These ions enable gels formation through electrostatic links between the polysaccharides that are bridged by the ions. Proteins are electrostatically complexed with carrageenan for the same reason. For example, a weak, pourable gel is formed with milk caseins through this type of interaction. Carrageenan works well in milk-based desserts such as shakes and ice cream. Carrageen addition greatly improves the dispersion of chocolate in milk. Processed meats, such as ham containing carrageenan, have a softer texture and retain 20% to 40% more water.

Alginic acid/alginate

Alginic acid/alginate is a polysaccharide found in brown seaweed. It is the world's most abundant marine polysaccharide. The polymer is linear and composed of glucuronic and mannuronic acid monosaccharides. Because of its carboxylic acid content, it is very acidic and negatively charged at pH 4 and above. The free acid form of this polysaccharide is called alginic acid. When a base such as sodium hydroxide (NaOH) is added, it neutralizes carboxylic acid groups and creates sodium alginate salt. Sodium alginate is soluble in water and forms highly viscous solutions at low concentration. Sodium alginate solutions can form gels when calcium ions are added. Alginate is used to make restructured products such as the pimento in stuffed olives, onion rings, and novel foods. For example, restructured pimento is made by combining its puree with sodium alginate and rolling the mixture into a thin sheet. Spraying a calcium solution on the material causes it to solidify. Gelled pimento is then cut in small pieces and stuffed into olives. Sodium alginate is used to make onion rings. A mixture of minced onion and sodium alginate is extruded through dies of various sizes. Rings formed by the process are solidified by dipping the extruded material in a calcium solution. Rum caviar is a novel application in which alginate-containing liquid is added dropwise to a calcium containing solution (Fig. 7.6). Droplets quickly solidify into small pearl-shaped beads as they hit the

Fig. 7.6 Rum caviar -alginate + calcium.

calcium solution. This technique is used make faux caviar from a variety of liquids, including juices and wine.

Stabilizers and thickeners

Several types of polysaccharide are used in foods to provide viscosity and/or stabilize suspensions. These large polymers contain numerous hydroxyl (OH) groups that provide good interaction with water and increase the viscosity or resistance to flow of solutions. They are also used to suspend poorly soluble substances, such as colorants. Starch is often used as a viscosity enhancer. In general, starches from root plants (e.g., potato and tapioca) have higher ratios of amylopectin to amylose and provide stable thickening properties to gravies and stews. High amylopectin starch develops high viscosity after heating through gelatinization and cooling (Chapter 3 Carbohydrates). High amylopectin starches retain much of their viscosity after a freeze-thaw cycle and are the form of starch used for frozen food products.

Carboxy methyl cellulose (CMC) is a thickener and emulsifying agent made from modified cellulose. Cellulose is a linear polymer of glucose with a starch-like structure with little or no solubility in water. A chemical process is used to add polar carboxymethyl (CH_3-COOH) groups to most glucose units in the cellulose backbone to create a negative charge. This greatly improves compatibility with water. The product of this process carboxymethyl cellulose, has many food applications. Notably, CMC is used in ice cream where it stabilizes the fat and liquid emulsion, prevents ice crystal formation, and improves product texture. It is also used to thicken salad dressing, gravies, and ketchup.

Locust bean and guar gum are polysaccharide gums with a similar composition. These polysaccharide are composed of a repeating galactose and mannose structure referred to as a galactomannan. The ratio of galactose to mannose in guar and locust bean gums is 2:1 and 4:1, respectively. Guar gum does not contain ionizable groups, but hydrates readily in cold water and forms a highly viscous solution at low concentration. Guar gum solutions retain viscosity over the range of pH found in most food systems. Combinations of guar gum with other polysaccharides, notably locust bean gum, result in a synergistic effect that substantially increases solution viscosity. Locust bean gum dissolves in water, but heat is required to develop maximum viscosity. This gum can form gels, but its strength is improved when guar gum is also added. Major applications for locust bean and guar gum include prevention of ice formation in dairy products such as ice cream and frozen novelties. Gums are also used to make spreadable cheese products. Gum arabic is a polysaccharide derived from the sap of the Acacia tree. It is one of the oldest and best-known gums. Gum arabic is composed of several different monosaccharides, including galactose, arabinose, and glucuronic acid. The later monosaccharide contains an ionizable carboxyl group that contributes a negative charge and water solubility, depending on the pH. Gum arabic readily dissolves in cold water and forms highly viscous solutions. It is used to keep poorly soluble substances, like colorant lakes and flavors, in suspension. Kool Aid™ is perhaps the best-known product that uses gum arabic as a stabilizer for its colors and flavors. Xanthan gum is a polysaccharide derived from a bacterial (*xanthomas campestris*) fermentation. It is composed of glucose monosaccharides arranged in a cellulose-like polymer. Xanthan forms highly viscous solutions at low concentration (1% and less). Xanthan gum is commonly used in tooth paste formulations. While xanthan gum is technically not an emulsifier, it works well to keep oil from separating in salad dressings. In gluten free bakery products, xanthan gum provides the texture that would normally be provided by wheat protein (gluten). Xanthan gum also has health benefits. It is a form of natural fiber that binds cholesterol in the gut and thus can lower cholesterol level in the blood.

Emulsifiers

Emulsifiers have long been used to overcome the challenge of combining incompatible substances like oil and water. An emulsion is defined as a stable mixture of two liquids that are normally immiscible. Emulsions exist basically as two types: oil-in-water and water-in-oil. The defining factor is the predominate phase. Oil-in-water represents the most common type of food emulsion. Major examples include homogenized milk, mayonnaise, salad dressing, and vinaigrettes. Water-in-oil emulsions are exemplified by foods such as butter, margarine, and high fat spreads. Emulsifying agents enable formation of a stable dispersion of oil-in-water or water-in-oil. In order for an emulsifying agent to work, it must contain both polar or hydrophilic and non-polar or hydrophobic regions to provide interaction with the corresponding phase. Emulsifying agents used in food systems include natural substances such as lecithin, hydrocolloids, proteins, fatty acids, and other amphiphilic molecules. Several synthetic substances are also available for use as emulsifying agents.

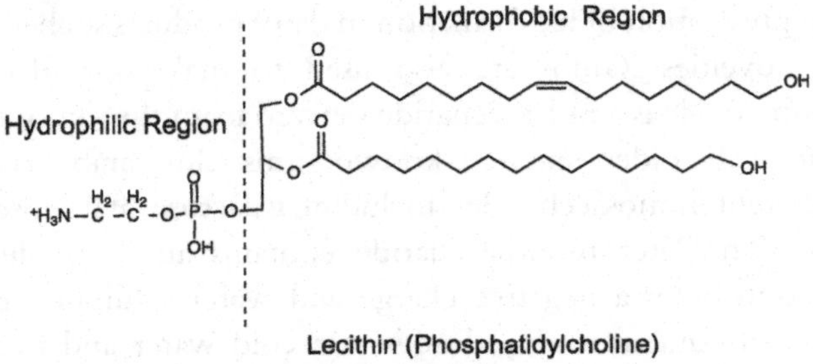

Lecithin (Phosphatidylcholine)
Fig. 7.7 Lecithin.

Lecithin is a phospholipid component of all cell membranes. As a food additive, it is a widely used emulsifying agent for water-in-oil applications. Lecithin is commercially derived from soy beans, egg yolk, and sun flower seeds. It is an amphiphilic molecule with the polar character contributed by phosphate and choline groups attached to the glycerol backbone (Fig. 7.7). Lecithin's hydrophobic character is contributed by two non-polar fatty acid molecules that are also attached to the glycerol backbone. Lecithin is actually a mixture of several phospholipids, including phosphatidyl choline (major form), phosphatidyl inositol, phosphatidyl ethanolamine, and phosphatidic acid. It is widely used in food applications, such as controlling sugar crystallization in chocolate, improving wettability of cocoa powder, increasing loaf volume in bread, and as a lubricant in such things as non-stick pan coating. Lecithin is also sold as a dietary supplement. The phosphatidyl choline component of lecithin is a source of the essential nutrient choline.

Mono- and di-glyceride esters of fatty acids are widely used synthetic emulsifiers composed of fatty acid and glycerol molecules. They are made through a chemical process that links unsaturated and/or saturated fatty acids to glycerol hydroxyl groups (Fig. 7.8). The aliphatic tail of fatty acids provides non-polar character needed to interact with lipid and glycerol's hydroxyl groups provide polar character. Glyceride esters are used in bakery, dairy products, peanut butter, and margarine. In bread dough, mono- and di-acylglycerides are used to inhibit staling and preserve a longer soft texture (Fig. 7.9). They also strengthen the gluten network and provide greater loaf volume. A concern has been voiced that some synthetic acylglycerol emulsifiers contain trans fatty acids. However, a recent re-evaluation of their safety by the European Food Safety Authority found no evidence for adverse effects when these were used as food additives (Younes et al., 2017).

Polysorbate 80 is an emulsifier derived from chemical combination of polyethoxylated sorbitan and oleic acid. It is commonly used to make ice

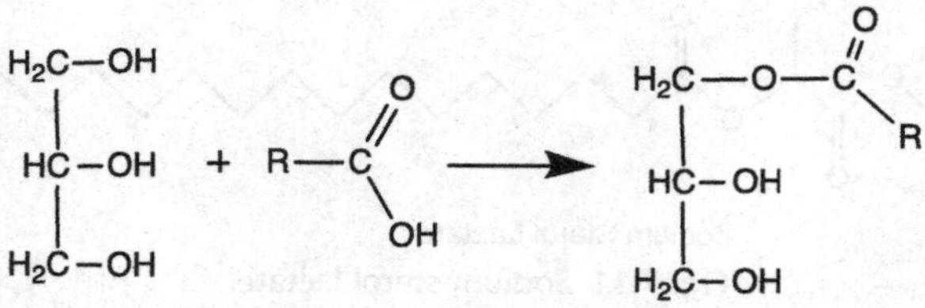

Glycerol Fatty Acid Monoacylglycerol

Fig. 7.8 Monoacylglycerol.

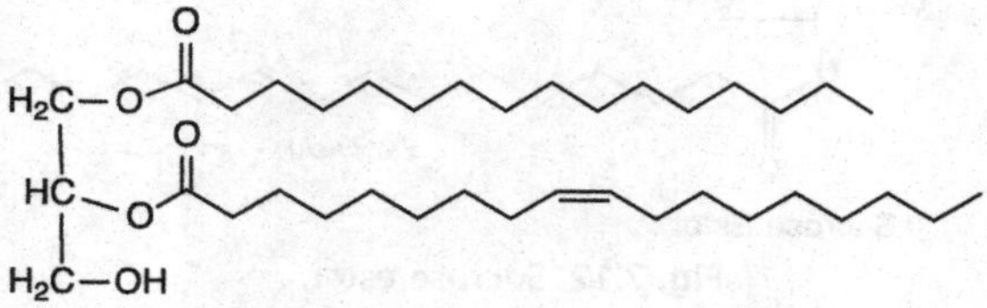

Diacylglycerol (DAG)

Fig. 7.9 Diacylglycerol.

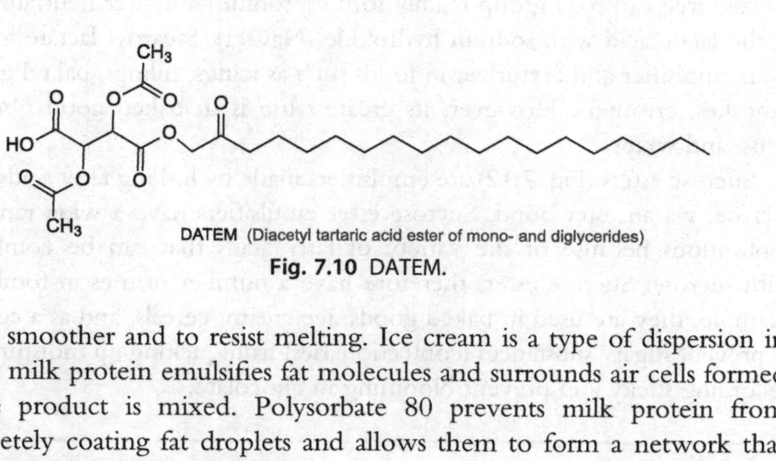

DATEM (Diacetyl tartaric acid ester of mono- and diglycerides)

Fig. 7.10 DATEM.

cream smoother and to resist melting. Ice cream is a type of dispersion in which milk protein emulsifies fat molecules and surrounds air cells formed as the product is mixed. Polysorbate 80 prevents milk protein from completely coating fat droplets and allows them to form a network that holds it shape as the ice cream melts.

Diacetyl Tartaric acid Ester of Monoglyceride (DATEM) (Fig. 7.10) is a synthetic emulsifier composed of tartaric acid to two acetate and one fatty acid molecules. It is widely used in bakery products where it improves bread texture and loaf volume. DATEM strengthens bread dough by interaction between gluten proteins and air. A stronger gluten is able to remain stable as loaf volume increases.

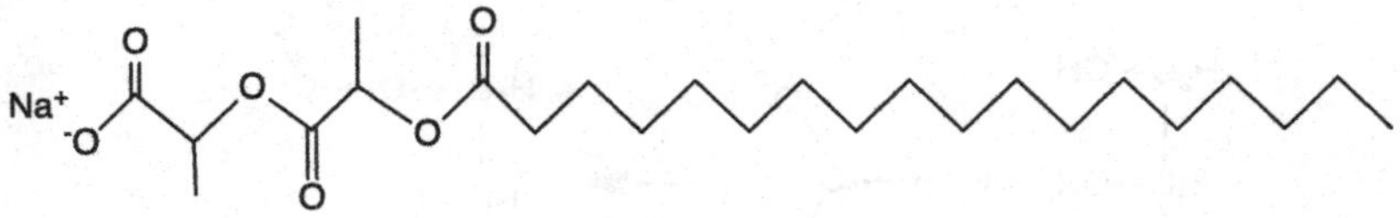

Sodium Sterol Lactate
Fig. 7.11 Sodium sterol lactate.

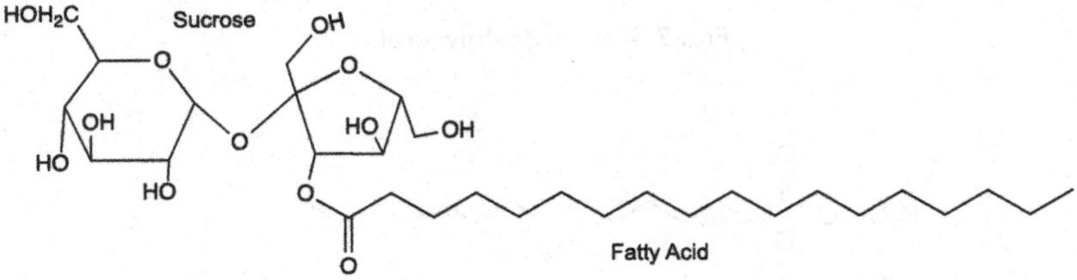

Sucrose Ester
Fig. 7.12 Sucrose ester.

Stearoyl lactate (sodium salt) (Fig. 7.11) is made by combining the saturated fatty acid, stearic acid, with lactic acid through an ester link. The ionized free carboxyl group readily forms a sodium salt after neutralization of the lactic acid with sodium hydroxide (NaOH). Stearoyl lactate is used as an emulsifier and texturizer in foods such as icings, fillings, baked goods, pancakes, creamers. However, its greatest use is in baked goods, breads, buns, and wraps.

Sucrose esters (Fig. 7.12) are emulsifiers made by linking fatty acids with sucrose, via an ester bond. Sucrose ester emulsifiers have a wide range of applications because of the variety of fatty acids that can be combined with sucrose. Sucrose esters therefore have a number of uses in food. For example, they are used in baked goods, ice cream, cereals, and as a coating to prevent sugary substances (confectionaries) from picking up moisture and becoming sticky and prevent blooming in chocolate.

Fat replacers

Fat replacers are substances added to food to mimic the eating qualities of foods that are high in fat. Smooth texture and mouth feel are enhanced. They are used in such products as low-calorie ice cream, dessert novelties, butter-like spreads, and low-fat cheese. Replacement of fat in food potentially benefits health by lowering the calorie content and eliminating undesirable saturated and trans fatty acids from the diet. Several types of fat

Table 7.2 Fat replacers.

Type	Examples	Source	Applications
Protein-based	Simplesse™, Finesse™, Dairy-Lo™	Milk and/or egg protein	Ice cream, butter, margarine, yogurt
Carbohydrate-Based	Modified cellulose, Avicel™, Methocel™	Cellulosic polysaccharides or oligosaccharides	Frozen desserts, dairy products
	Litresse® or Sta-Lite®	Polydextrose	Dietary fiber, Baked goods, chewing gum, gelatins and puddings, gelatins
	Oatrim (Beta-Trim™, TrimChoice)	Hydrolyzed oat flour	Prebiotic fiber, Baked goods, frostings, frozen desserts, processed meat
	Inulin (Raftiline™, Fruitafit™,	Fructose oligosaccharide	Prebiotic fiber, Baked goods, frostings, frozen desserts, processed meat
Lipid-based	Salatrim (Benefat^T)	Tri-acylglyceride mixtures	Bakery and confectionary products
	Olestra (Olean®)	Sucrose fatty acid ester	Snack foods

replacer additives based on unique or modified proteins, carbohydrates, and lipids are available (Table 7.2).

Protein-based fat replacers are made from proteins created by a micro-particulation process to provide a smooth creamy texture. Most protein-based fat replacers are made from milk and egg sources. The process of making a microparticulated fat replacer involves denaturation of the protein, using a combination of heat and homogenization treatments to create small diameter particles with a charged surface. A small particle size with an average diameter of less than 1 μm is key to achieving a texture similar to that of fat. Examples of protein-based fat replacers include Simplesse™, Finesse™, and Dairy-Lo™. Simplesse™ is made from whey milk protein whey. Its applications include ice cream, yogurt, cheese spread, and mayonnaise. Dairy-Lo™ is a similar whey protein-based fat replacer used in low-fat cheeses, ice cream, and frosting. Carbohydrate-based fat replacers are made from polysaccharide materials such as starch, gum, and cellulose that are derived from several plant sources. These substances are extracted from their source by chemical and enzymatic methods, purified, and ground to a fine particle size. Most starch and gum carbohydrate-based fat replacers hydrate well in water and swell to create a smooth, fat-like texture. Cellulose is also used in making fat replacer additives, but it must first be chemically treated to improve its interaction with water. Products such as Avicel™, Methocel™, and Solka-Floc® are cellulose-based and have applications in dairy products, sauces, frozen desserts, and salad dressing. Examples of non-cellulosic polysaccharides used as fat replacers in food include polydextrose, Oatrim™, inulin, and several gums. Polydextrose, commercially available as Litresse® or Sta-Lite®, it is a synthetic polymer of glucose containing 10% sorbitol, and 1% citric acid. It provides only one-fourth the calories of sucrose, but can create a fat-like mouth feel. Additionally, polydextrose is a source of dietary fiber that contributes to a favorable intestinal microflora population and provides beneficial short-chain fatty acids. Polydextrose applications include dairy products and baked goods in which glucose components participate in Maillard browning. Oatrim is composed of the polysaccharide beta-glucan. This water-soluble glucose polymer is derived from enzymatically modified oat flour. It is commercially available under the trade names Beta-Trim and TrimChoice. Oatrim is a good source of soluble fermentable fiber that is beneficial as a substrate for intestinal microflora that produces short chain fatty acids. Oatrim fiber absorbs cholesterol from the gut and aids in the digestive process. Food applications for Oatrim include baked good, fillings, frostings, desserts, dairy beverages, cheese, salad dressing, and processed meats. Inulin is a fructose-based polysaccharide derived from chicory root. It has a slightly sweet taste that is approximately 10% as sweet as sucrose. Inulin is a soluble fiber that can be used as a fat replacer. Its food applications include yogurt, cheese, frozen desserts, whipped cream, and processed meat. A negative aspect of inulin is its potential activity as a FODMAP. FODMAPS are forms of carbohydrate that can be fermented in the gut and produce bloating, gas, and diarrhea.

The major lipid-based fat replacers include Salatrim™ and Olestra™. These fat replacers are chemically modified and/or synthesized lipids. Salatrim (Benefat™) is a reduced calorie fat replacer composed of triacyl glycerides that contains mixtures of hydrogenated short and long chain fatty acids. The name Salatrim is an abbreviation for short and long chain acyl triglyceride molecule. Salatrim is a synthetic fat made by the process of interesterification. Olestra (Olean™) is a synthetic fat replacer made by esterifying fatty acids to hydroxyl groups of sucrose. Olestra is a polyester fat so it is called a sucrose fatty acid ester. Olestra is a very stable fat-like compound that can be used in high heat cooking such as frying chips and fries. It is not metabolizable, has zero calories, and little or no flavor. Olestra has been commercially used in frying snack foods as a means to offset its high calorie content. Major negatives for the use of Olestra involve interference with fat soluble vitamin and carotenoid uptake in the gut. Anal leakage may be caused when olestra fried foods are over consumed.

Food enzymes

Enzymes are biological catalysts. As such, these protein molecules speed up chemical reactions without being transformed in the process. Enzymes are essential to digestion and energy production, vision, the immune system, and every other function needed to sustain life. Enzymes within food materials (endogenous) are responsible for numerous reactions that affect food quality. For example, ripening of fruits and vegetables is a result of endogenous enzymes that bring about changes in color, texture, and flavor. Exogenous enzymes, in contrast, are those added to food systems to bring about change. Exogenous enzymes provided by yeast convert starch to glucose and subsequently turn glucose into ethanol. Fermentation of milk performed by exogenous bacterial enzymes is used to make yogurt and cheese. Microbial enzymes produce many of the compounds that create distinctive cheese flavors and preserve a perishable commodity. Chymosin, also known as rennin, is a proteolytic enzyme widely used in cheese making. It hydrolyzes a specific casein protein and causes precipitation of this fraction from milk. The remaining fraction of protein is left in the whey along with lactose and other soluble constituents. Rennin is a mixture of enzymes containing chymosin. It has traditionally been extracted from the stomach of calves to make cheese. This practice requires sacrifice of the animal, its loss from the milk producing population of cows, and inconsistent cheese quality because of its heterogenous nature. Chymosin is now made using recombinant technology. In a process approved by FDA in 1999, the genes for chymosin were inserted into yeast that produces the enzyme through fermentation. The product of this process is called recombinant chymosin. It provides a consistent enzyme supply without the need for calf stomach.

Toxins and toxicants

There is a subtle distinction between the terms toxin and toxicant. Toxins are the natural products of organisms, such as substances found in snake venom and poisonous mushrooms. Toxins can occur in food as a result of biological processes or food processing. This section of the chapter describes some of the more often encountered toxic substances in food, but should not be considered as comprehensive. Toxicants are person-made substances, such as pesticides and herbicides, with the potential to cause harm if ingested. Toxicant substances can also enter foods through contact with packaging materials. A comprehensive review of naturally occurring toxins in food can be found in the work of Dolan et al. (2010) and also in Toxic Substances (Center for Disease Control 2018).

Food toxins

Plants are essential to life. They use sunlight to convert carbon dioxide and water into sugars in the process known as photosynthesis. In turn, sugars are converted into energy-rich polysaccharides molecules that animals use to sustain life. In the delicate balance between these different forms of life, plants have developed mechanisms that provide a competitive advantage and protect against predation. Some plants produce toxins as a strategy to enhance their survival. The discussion below describes examples of plant toxins and their occurrence in food (Table 7.3 Food Toxins).

Alkaloids

Alkaloids are a large and diverse class of plant-based molecules whose structure is characterized by a nitrogen containing ring. The word alkaloid is a contraction of the term alkali–like. This term refers to nitrogen compounds that form salts when neutralized by acids. Alkaloids are noted for pharmacological effects in animals, ranging from analgesic to poison. Morphine, opium, strychnine, quinine, ephedrine, and nicotine are members of the alkaloid family. Alkaloids are most often bitter tasting, a trait that reinforces the association between bitterness and poison. Pyrrolizidine alkaloids function in plants as a biological defense against insect predation. Variants of pyrrolizidine alkaloids (PA) are common to several plant families such as *Asteraceae, Leguminosae,* and *Orchidaceae.* PA is found in borage, comfrey, and herbs used in traditional Chinese medicine. It is not unexpected that pyrrolizidine alkaloids occur in herbal teas, infusions, and even some

Table 7.3 Food toxins.

Substance	Food origin	Biological effects
Alkaloids		
Solanine	Potatoes, Eggplant	Gastrointestinal
Caffeine/Theobromine	Coffee and cocoa beans	Increased heart rate and blood pressure
Furanocoumarin (Psoralen)	Grapefruit, parsnip, celery	Photodermatitis, inhibition of DNA synthesis
Glycosides		
Amygdalin	Cassava, almonds, apricot pit, lima beans	Cyanide poisoning
Glucosinolate	Cabbage, broccoli, horseradish, mustard, rape seed	Impaired thyroid function
Saponins	Potatoes, beans, tomato seeds, and quinoa	Intestinal mucosal damage
Proteins		
Lectins	Soybeans and other legumes	Anti-nutritional, hemagglutination, diarrhea
Trypsin Inhibitor	Soybean, lentils, and peas	Anti-nutritional (limits protein digestion)
Allergens	Milk, eggs, peanut, tree nuts, wheat, fish, and shellfish	Wide ranging, (skin rash to life-threatening anaphylaxis)
Histamine	Decomposition product of amino acid, histidine	Rash, edema, and gastrointestinal
Other		
Phytoestrogens	Soybeans, nuts	Weakly mimics human estrogen hormone

food supplements. Pyrrolizidine alkaloids also known to contaminate honey and milk as a result of transmission from bees to cows when they feed on silage and hay containing this alkaloid can also contaminate honey. PA represents a significant health concern because they are known to cause liver damage and cancer. There is a risk in processed food because PAs are stable to most cooking procedures, such as boiling. Solanine is a poisonous glycoalkaloid found in potatoes, and eggplants. Tomatoes also contain a solanine-type glycoalkaloid, but it is not as toxic as the compound found in potatoes. The structure of solanine is composed of an alkaloid to which mono- and disaccharides are attached. Solanine can also be found in the flowering night shade plant that is sometimes used as a food. Potatoes produce solanine as a defense against insect and herbivore animal predation. Solanine is present in all parts of the potato plant but tubers contain the highest level. Freshly dug, potatoes will turn green when exposed to sunlight. The green color is due to chlorophyll, but light exposure also increases solanine content. Thus, green colored potatoes are a warning of elevated solanine content and should not be eaten. Bitter tasting potatoes are also a warning of high solanine content. While boiling potatoes does little to destroy solanine, higher temperatures such as in frying can reduce the content by half. Solanine is not a very potent toxin and few fatalities have been attributed to solanine poisoning. Symptoms of solanine poisoning are typically limited to gastrointestinal upset, such as vomiting and diarrhea. Caffeine and theobromine are structurally very similar methylxanthine alkaloids found in the seeds or beans of coffee and cocoa plants. Coffee beans predominantly contain caffeine at about 1%—2%, while cocoa beans contain 2% or more theobromine and 0.2% caffeine. Caffeine is a stimulant that generally improves cognition, reaction time, and coordination, but it also increases blood pressure. Caffeine is the most widely used psychoactive compound in the world. It contributes a pleasant bitter taste to coffee, tea, chocolate, and cola drinks. Caffeine is a weak toxin requiring doses of 10g per day to cause toxic effects in the average adult. A cup of coffee typically contains about 150 mg of caffeine. It would require consumption of over 50 cups of coffee to reach the level of toxic effects.

Furanocoumarins are phytochemicals produced in a variety of plants as part of their defense mechanisms against predation and microbial pathogens. Grapefruit, parsnips, and celery are common food sources of furanocoumarin. Furanocoumarins in grapefruit can have adverse effects on prescription medications. This is known as the grapefruit juice effect. Furanocoumarins inhibits an enzyme in the small intestine that degrades medications such as those used to treat hypertension, infection, and cancer. The inhibiting enzymatic breakdown of the drug causes individual to receive a much larger a dose than intended and this may have adverse side effects. Furanocoumarins also enter the system through skin contact and can cause very unpleasant photodermatitis skin reactions. Cleaning celery or coming into contact with weeds, such as giant hogweed, can cause photodermatitis that presents as redness, rash, and large welts on the skin. In contrast, furanocoumarins in grapefruit are also beneficial as antioxidants with the ability to neutralize free radicals (Hung et al., 2017). Psoralens are compounds belonging to the furanocoumarin group. Plants use psoralens to ward off predation by a variety of animals. Psoralens are found in figs and the leaves of celery. They are mutagenic compounds that modify DNA by intercalating between thymidine residues of DNA molecules. When exposed to UV light, a chemical reaction forms a bond between psoralen and thymidine base. The result of this reaction is inhibition of DNA synthesis, cell replication, and darkening of skin. An interesting application of psoralens involves its use for treatment of serous skin diseases such as psoriasis, vitiligo, and skin cancer. Oral doses of psoralen make psoriatic cells susceptible to UV light and causes them to die off preferentially. The treatment is known as Psoralen UltraViolet A (PUVA) therapy.

Glycosides

Glycosides are compounds consisting of a sugar molecule (typically a monosaccharide) attached to a functional group through a glycosidic bond. A variety of functional group molecules, such as flavonoids, phenolics, steroids, thiols, and nitriles, are present as glycosides in food. The most toxic glycosides contain a nitrile functional group that potentially can be converted to hydrogen cyanide. These substances are thus termed cyanogenic glycosides. Cyanogenic glycosides become poisonous after enzymes act on the molecule and release hydrogen cyanide. The most significant source of toxicity occurs in plant materials that have been ground or crushed. Some gut microorganisms also contain the enzymes acting on cyanogenic glycosides and produce hydrogen cyanide that is absorbed into the blood stream. Cyanide's poisonous properties result from inhibiting energy producing pathways, especially in the heart and respiratory system. A lethal dose for most people is approximately 0.5–3.5 mg per Kg of body weight, which translates to about 50 mg for an adult male. Long term exposure to low levels of cyanide have serious consequences. Conditions such as goiterism, blindness, ataxia (loss of motor control), and paraparesis (progressive weakness of the legs) can result. Amygdalin is a cyanogenic glycoside present in several foods, including cassava, almonds, apricot pits, lima beans, clover, and flax. The molecule is composed of a glucose disaccharide linked through a glycosidic bond to a nitrile group. Cassava and the pits of stone fruits such as apricots contain levels of amygdalin that make them a potential health hazard. Cassava contains sufficient amygdalin to generate approximately 50 mg of hydrogen cyanide per 100g of fresh root. An equally hazardous dose of hydrogen cyanide can also be obtained from the seeds taken from 10 apricot pits. Amygdalin is the source of synthetically derived substance called laetrile. Laetrile is purified from amygdalin after hydrolyzing one of its glucose units. It is important to note that laetrile is not a vitamin and is ineffective as a cancer treatment (Wade, 1977). Several individuals have been incarcerated in the United States for promoting it as such.

Glucosinolate is a glycoside found in cabbage, broccoli, horseradish, mustard, rape seed. The oil extracted from rape seed in known as canola oil. Glucosinolate is converted by the enzyme myrosinase to the active components, thiocyanate and a nitrile, when the plant material is sliced or crushed. They are a plant defense mechanism and a natural pesticide against predation by insects. However, excessive consumption of glucosinolate is known to impair thyroid production (hypothyroidism) and may result in enlargement of the gland in a condition called goiterism. Saponins are glycosidic compounds with surface-active properties. They are found in a variety of legumes such as soybeans, peanuts, garbanzo beans. They are also present in potatoes, tomato seeds, and quinoa. The surface-active property of saponin is due to the presence of both non-polar and polar groups that provide the ability to mix with both lipid and aqueous fractions. Saponin molecules are composed of a polycyclic ring structure and a carbohydrate group. The polycyclic ring is lipophilic and responsible for the molecule's structural diversity. Saponin ring structures occur in two chemical types: terpenoids and steroids. Conversely, monosaccharide carbohydrates such as glucose or galactose, provide aqueous solubility to saponins. Saponins are bitter tasting substances that have long been considered anti-nutritional factors in animal feed. They damage the membranes of intestinal mucosal cells and cause a condition known as leaky gut. Saponins bind cholesterol and reduces the amount of this essential molecule. This contributes to poor growth in animals fed materials high in saponins. The damaging effects of saponins on cell membranes can also cause hemolysis of red blood cells and result in anemia. More recently, triterpenoid saponins are suggested to have human health benefits including lowering serum cholesterol level and providing antioxidant and anticancer activity (Bishayee et al., 2011). Saponins are not destroyed by typical cooking methods. A frothy foam formed during boiling of beans is due to solubilized saponins and should be discarded. Fermentation used in making products such as soy sauce is the only method known to eliminate saponins. A review of saponins in foods can be found in the work by Price et al. (2009). Fava beans contain the toxic vicine and convicine pyrimidine glycosides that are normally broken down in the gut. The absorption of vicine and convicine at sufficient levels triggers destruction of red blood cells and can cause a condition known as hemolytic anemia. The adverse effects most often occur in individuals with an inherited deficiency in the enzyme glucose 6 phosphate dehydrogenase.

Proteins

Most food proteins are provide a source of energy and essential amino acids. There are few examples of proteins encountered in food which have toxic side effects. Some food proteins, especially those from plants, are well known sources of allergens that cause a variety of symptoms ranging from discomfort to anaphalxisis and death. Proteins, such as trypsin inhibitor and lectins (heamagglutinins), are examples of proteins with toxic and/or anti-nutritional properties. Ricin is a lectin and very toxic protein found in castor bean. Two mg (a dose about the size of several grains of salt) inhaled, ingested, or absorbed through the eyes is sufficient to kill an adult. Consequently, ricin has been used as a biological weapon.

Lectins are proteins that recognize and bind to specific carbohydrate-containing molecules. They have important roles in biological processes, including the immune system and metabolism. Lectins are not enzymes or antibodies, but have bioactive properties upon binding to receptor carbohydrates located on the surface of cells. For example, influenza virus contains a hemagglutinin that recognizes a glycoprotein on the surface of host cells. This allows it to attach and gain entry. Plants are the predominant source of the lectins often referred to as phytohemagglutinins. Lectins protect plants from microorganism, insect predation, and can pose a problem when eaten. Lectins are found in many plant foods including wheat and legumes (soybeans, peanuts, and beans) (see the review by von Buul and Brouns, 2014). Red kidney bean has the highest lectin level of all legumes. Food lectins are considered to be anti-nutritional factors because they interfere with the uptake of nutrients from the gut. This mechanism protects plants from being eaten by insect predators. Most animals, including human, are unable to breakdown lectins. The toxic effects of lectin in food include nausea, vomiting, and diarrhea. Some lectins, such as soybean agglutinin, can cause damage to the cells forming the wall of the small intestine. Lectin's antinutritional properties can be reduced or eliminated by some processing techniques. Soaking for several hours followed by boiling is a time-honored method for preparing beans. This practice is effective because soaking allows a substantial portion of lectin to diffuse out and boiling destroys most of the remaining content. Fermentation is another effective method to reduce lectin content. Fermented products such as beer, tempeh, and soy sauce retain only small amount of the original lectin content. Wine is the exception because its fermentation process does not destroy lectins. Ricin is found in castor beans and the oil extracted from it (castor bean oil). The biological toxicity of ricin results from a complete shut-down of protein synthesis and eventually causes multiple organ failure. Ricin poisoning in foods is rare because the castor beans and extracted oil are always heated to destroy the lectin's biological activity Trypsin inhibitors are common to many legumes such as soybean lentils and peas. Trypsin inhibitor acts in the small intestine to reduce the activity of the major digestive enzyme, trypsin. The net result is lower protein uptake which can negativelty affect animal growth. Trypsin inhibitors are fairly heat stable proteins and remain active in processed soy protein, which is widely used in food products.

Food allergens are proteins that cause adverse immunological reactions soon after the ingestion or contact with food containing the offending protein. Reactions to food allergens involve the immune system and are mediated by IgE type antibodies. The allergic reaction can be initiated by very small amounts of protein. For example, a few mgs of peanut in 100g of chocolate can produce a serious reaction. Symptoms of food allergy range from mild skin rash, to welts, wheezing, and difficulty in breathing. Unfortunately, reaction to a food allergen can be lethal when it causes anaphylactic shock. It is estimated that 150 to 200 deaths occur each year due to food allergy-related anaphylaxis. Eight foods represent the most common source of food allergy, milk, eggs, peanut, tree nuts (walnuts, almonds, pine nuts, brazil nuts and pecans), soybean, wheat, fish, and shellfish. Food allergy affects 5% to 8% of children and 2% of adults in the United States. Fortunately, most of these reactions are not life-threatening. Proteins known to cause food allergy are diverse in nature, but share some common properties. Most allergens are stabile to thermal processes and resistant to enzymatic digestion. However, it remains unclear regarding how protein structure and their resistance to digestion is involved in development of a food allergy (Breiteneder and Mills, 2005).

Histamine is a decomposition product of the amino acid histidine. Free histidine is converted to histamine by loss of its CO_2 group through the action decarboxylase enzymes. Histamine is found is all cells and present in most foods. Biologically, histamine has several roles. It functions as a neurotransmitter in inflammation processes, and in the immune response to allergens. Its role in allergy is to cause additional release of histidine from its stores in mast cells. This release triggers the symptoms associated with allergies, including runny nose, itchy eyes. and sneezing. However, the reaction to histamine is not a food allergy, rather it is another type of intolerance resulting from lack of enzymes (histamine-methyltransferase and diamine oxidase) necessary to break down the compound. Histamine naturally occurs in many foods such as, smoked meats, aged cheese, fish, shellfish, nuts, chocolate, and other cocoa-based foods. Additionally, foods produced by fermentation processes, such as beer, red wine, and cheese, can contain high levels of histamine. Ingestion of histamine-containing foods can trigger a reaction that resembles food allergy. Scombroid food poisoning is an example of a foodborne illness caused by eating fish high in histamine. Decomposition of fish muscle occurs quickly after fish are caught and results in high levels of histamine. High levels are also caused by storage at elevated temperatures or for extended time. Initial symptoms of scombroid poisoning initially include facial flushing, dizziness, and headache. This can progress to edema, rash, and gastrointestinal illnesses.

Phytoestrogens are substances bearing structural similarity to the female estrogen hormone responsible for fertility. Phytoestrogens, such as isoflavone, genistein, and coumetarol, are found in nuts, beans and other legumes. The structural similarity between phytoestrogens and estradiol is sufficient to cause a weak hormone-like effect in some animal studies. For example, isoflavone has a ring structure containing hydroxyl groups that is overall very similar to estradiol. Structural similarity enables phytoestrogen to weakly mimic the effect of human estrogen and may be beneficial to women approaching menopause. Phytoestrogens show some health benefits, including increased bone density and maintenance of hormone levels (Bacciottini et al., 2007).

Microbial toxins

Toxins produced by bacteria, molds, and viruses are responsible for many foodborne illnesses and food poisonings world-wide. It is estimated that there are 48 million cases of foodborne illness and 3000 deaths each year in the United States alone. In most cases the adverse consequences of encountering microbial toxins in food are short lived. However, some toxins have serious consequences. Aflatoxin from aspergils molds and botulinal toxin from *C. botulinum* cause life-threatening illness. This section describes notable examples of microbial toxins and their properties (Table 7.4). A more comprehensive summary of this important subject can be found in the publication Bad, Bug, Book and Foodborne Illnesses: What You Need to Know (FDA 2018).

Table 7.4 Microbial sources of food poisoning and foodborne Illness[a].

Organism	Sources	Symptoms
Algae (Dinoflagellates)	Mussels, clams, oysters, scallops	Nausea, vomiting, and diarrhea Neurologic symptoms include tingling, numbness, and weakness
Campylobacter jejuni	Water, unpasteurized milk, undercooked meat, poultry, shellfish	Diarrhea, abdominal pain, fever
Clostridium botulinum	Improperly canned or vacuum-packed foods	Slurred speech, muscle weakness, paralysis and death
Clostridium perfringens	Cooked meat and poultry held for long periods of time at room temperature	Nausea, cramps, and diarrhea
Escherichia coli O157:H7	Uncooked meat, unpasteurized milk or apple cider, fruits and vegetables	Severe diarrhea, abdominal cramps, and vomiting. Hemolytic uremic syndrome and kidney failure in children under 5
Listeria monocytogenes	Ready to eat foods, soft cheeses from unpasteurized milk, smoked seafood, deli salads, especially made with ham, chicken or seafood	Fever, muscle ache, nausea, diarrhea, can cause miscarriage or death to newborns.

Table 7.4 Microbial sources of food poisoning and foodborne Illness[a].—cont'd

Organism	Sources	Symptoms
Mycotoxins	Fungal infections of crops, such as corn, sorghum, and wheat	Acute and chronic effects gastroenteritis to life-threatening illness
Norovirus	Food, water, and infected individuals	Vomiting, diarrhea, headache Symptoms usually short-lived, 1—3 days
Salmonella	Contaminated eggs, poultry, unpasteurized milk or juice, cheese, seafood, fruits and vegetables	Diarrhea, fever, abdominal cramps, and death for immuno-compromised individuals
Shigella	Food and water contaminated by workers with poor hygiene e.g., especially salads and other foods	Diarrhea, (bacterial dysentery), abdominal cramps
Staphylococcus aureus	Food contaminated by skin contact from people who carry the organism, typically without symptoms	Severe nausea, abdominal cramps, and diarrhea occurring within 1 h after eating
Vibrio vulnificus	Uncooked or raw seafood; fish shellfish or oysters	Diarrhea, abdominal cramps, can become life-threatening

[a]Adapted from USDA Food Safety and Inspection Service, 2017.

What is the difference between food poisoning and foodborne illness?

The terms and food poisoning and foodborne illness are often used interchangeably, but they are different. Food poisoning refers to adverse reactions such as nausea, vomiting, and diarrhea resulting from eating a toxin-containing food. Symptoms of food poisoning usually occur within a few hours after eating. Botulism is an example of food poisoning caused by Clostridium botulinum toxin produced during its growth in a food, such as sausage. Foodborne illness is caused by eating food contaminated with pathogenic microorganisms. Illness results from toxin produced by the pathogen growing in the gut. Toxin-producing strains of E. coli are examples of microorganisms causing foodborne illness. Table 7.3 contains a summary of microorganisms most often associated with food poisoning and foodborne illness.

Algae

Dinoflagellates are unicellular organisms (members of the *Gambierdiscus* species) that attach to algae and produce neurotoxins such as ciguatoxin. Filter-feeding organisms such as mussels, clams, and oysters concentrate toxins into their flesh. The level of toxin is progressively increased as larger fish concentrate toxin via the food chain. Symptoms of eating contaminated fish or shellfish include both gastrointestinal and neurological disorders. Nausea, vomiting and diarrhea are typical gastrointestinal symptoms. Neurological symptoms are more significant and include numbness, tingling, muscle weakness that often progress to vertigo, irregular heartbeat, and low blood pressure. Neurological symptoms may persist for years. The risk of eating ciguatoxin-contaminated fish or shellfish is much greater in warm water areas. Ciguatoxin is very stable and its toxicity is not affected by cooking, freezing, or other food processing methods.

Campylobacter

Campylobacter jejuni is a pathogenic microorganism common to animal feces, especially poultry. Campylobacter is often found in unpasteurized milk and the cheese made from it. It is estimated to be third leading cause of foodborne illness in the United States and Europe. The illness called campylobacteriosis is an inflammation of the intestinal tract. Symptoms include diarrhea, vomiting, and abdominal pain. While the illness can be severe and debilitating, it is typically not life-threatening. However, those who experience Campylobacter illness are at increased risk of developing Guillain-Barre syndrome. This condition is characterized by rapid onset of muscle weakness in the extremities that progresses to the upper body. Difficulty in breathing can become life-threatening.

Clostridium botulinum

Botulinal toxins are neurotoxic proteins secreted by *C. botulinum* bacteria. These toxins are extremely lethal and require less than 1 µg to kill a 200 lb (90 kg) person. The toxin is produced during the germination of spores. Growth of *C. botulinum* bacteria requires an environment with little or no oxygen, a pH above 4.6, and temperature above 20 °C. Biologically, the toxin works by blocking release of the acetylcholine neurotransmitter and causes muscle paralysis. Eight types of botulinal toxin are known and denoted by the letters A through H. Specifically, types A, B, E, F, and H are associated with human illness. Symptoms of *Clostridium botulinal* poisoning are called botulism and vary with the type of toxin ingested. Effects of the toxin include muscle weakness, blurred vision, vomiting and diarrhea. The lethal effects of *C. botulinal* toxin result from muscle paralysis, causing cardiac and respiratory failure. The toxin can be destroyed by heating to 100 °C for 15 min. However, *C. botulinum* spores can survive for years in soils and other environments. Perhaps the oldest known example of botulism is associated with sausage. Fermented sausage has been employed as a means to preserve meat for centuries, but conditions within the product (i.e., absence of oxygen, favorable pH and temperature) can enable the outgrowth of *C. botulinum* spores and toxin production. Home canning of low acid foods (pH greater than 4.6) like green beans, potatoes, and other vegetables also represent a potential source of botulism poisoning. Canning of low acid foods requires a stronger thermal process (i.e., at 121 °C for 3 min) to insure safety of the food. Processing these foods in a pressure cooker or boiling the food just before serving also increases their safety. Spores of *C. botulinum* can be found in honey and corn syrup, but are unable to grow because of the low water activity in these foods.

Clostridium perfringens

Clostridium perfringens is a common cause of food poisoning in the United States and Europe. It is an intestinal inhabitant of most animals. Like *C. botulinum,* the *perfringens* species are spore-forming organisms. Spores are ubiquitous in soils and decaying vegetation. They are heat-resistant and remain viable for years. Spore germination in food occurs quickly and produces an enterotoxin responsible for gastrointestinal symptoms such as diarrhea and cramps. Fortunately, the enterotoxin of *Clostridium perfringens* is heat labile and destroyed by temperatures at or above 74 °C (165 °F). A common scenario for Clostridium perfringens-caused food poisoning is the big turkey dinner. Typically, a large bird is roasted and served. If the remaining portion is left for a few hours at room temperature, perfringens spores will germinate and produce numerous infective bacteria and their enterotoxin. Eating that leftover turkey without reheating is very likely to cause illness. More serious and life-threatening consequences can occur if the food contains large amounts of enterotoxin.

Escherichia coli O157:H7

E. *coli* bacteria inhabit your gut as part of its normal microbiome and play a beneficial role. Notably, E. *coli* prevent intestinal colonization by pathogenic microorganisms. However, some E. *coli* strains are themselves pathogens that may cause foodborne illness. Several forms of E. *coli* are recognized as pathogenic, based on virulence factors they produce. These include enterotoxigenic E. *coli* (**ETEC**), enteropathogenic E. *coli* (**EPEC**), enterohemorrhagic E. *coli* (*EHEC*), enteroinvasive E. *coli* (EIEC), and enteroaggregative E. *coli* (*EAEC*), and diffusely adherent E. *coli* (**DAEC**) (FDA Bad Bug Book, 2012). ETEC, EPEC, EHEC, and EIEC are responsible for milder forms of food borne illness that results in vomiting, diarrhea, and fever. A majority of illnesses (approximately 75%) are caused by E. coli belonging to the EHEC group. These produce Shiga toxin that causes infectious diarrhea (gastroenteritis), and inflammation of the digestive tract (enterocolitis). In severe cases, the infection can progress and cause loss of kidney function which is a life-threatening illness. The best known of this strain is E. *coli* O157:H7. It is responsible for thousands of hospitalizations and many deaths annually. A number of outbreaks due to E. *coli* O157:H7 have occurred in foods such as hamburgers from fast food restaurants, unpasteurized juices, and bagged salad greens. E. *coli* O157:H7 is spread by contaminated food, water, and person to person contact.

Listeria monocytogenes

L. *monocytogenes* is a pathogenic bacteria responsible for the foodborne illness known as listeriosis. The cause of listeriosis is a toxin called listeriolysin O produced by the bacteria. Listeriolysin O toxin is a virulence factor or a molecule that enhances the organism's ability to invade and colonize host tissues. Specifically, it enables the organism to enter the host's circulatory system and infect its cells and organs. *Listeria monocytogenes* is responsible for a variety of illness ranging from flu-like symptoms to more serious conditions. It causes sepsis, meningitis, spontaneous abortion, and encephalitis. Elderly, with a weakened immune system, and women who are pregnant are most at risk for developing listeria toxin diseases. Pregnant women are 20 times more likely than other healthy adults to contract listeriosis. Since the organism is able to cross the placental barrier, listeriosis can be deadly or cause life-long problem for the fetus. Listeria occurs in soil, water, and animals. It is often found in poultry, cattle, and unpasteurized milk. The organism can inhabit processing plant equipment and cause repeated contamination of processed food such as, meat, soft cheeses, smoked seafood, store-prepared salads, fruits, and vegetables. One of the most dangerous properties of the organism is its ability to grow at refrigerator temperature (2—4 °C).

Norovirus

Norovirus is found in food, water, and on the food preparation or serving surfaces. Eating utensils can carry the organism. Norovirus are very contagious and easily transmitted by person to person contact. In many instances norovirus outbreak are associated with restaurants, dormitories, cruise ships, nursing homes, and daycare centers. This supports the person to person contact cause. Norovirus is very hardy and resistant to most household disinfectants. The recommended disinfecting agent is a 5% solution of bleach. Symptoms of norovirus infection are projectile vomiting, diarrhea, and abdominal pain. An electrolyte imbalance can result from excessive fluid loss. Fever and headache may also result from this infection. Fortunately, the illness produced by norovirus is short lived. Most people recover in 1—3 days without long term health consequences. Prevention of norovirus illness requires several measures, including thorough hand washing, rinsing of fruits and vegetables with a dilute solution of dishwashing detergent, and cleaning and sanitizing utensils and food preparation surfaces.

Salmonella

Salmonella is responsible for two types of illness: Typhoid fever and non-typhoid salmonellosis. Typhoid disease is caused by fecal contamination of water used for drinking and food preparation. It is a very serious illness with high fever and spread of the infection to other organs. The mortality rate from typhoid fever can be 10%. Gastrointestinal illness caused by non-typhoid salmonellosis is a prominent cause of foodborne illness in the United States. Outbreaks of salmonellosis have been caused by contaminated eggs, meat, poultry, milk, cheese, or juice, alfalfa sprouts, melon, nuts and spices. Over 1 million people are affected by salmonellosis each year in the United States. Infection and illness are caused by eating food contaminated by animal feces. This is called the fecal-oral route. Contact with pet turtles and reptiles has also been associated with human infection. Cooking kills salmonella, but the degree of heat treatment is important. Salmonella may be viable in eggs with runny whites or yolks. The safe temperature for whole meat cuts (i.e., beef, lamb, and pork) is 63 °C (145 °F). Poultry, ground beef, egg dishes, and ground beef should be cooked to 74 °C (165 °F).

Shigella

Shigella bacteria are found in human feces. The organism spreads to foods through water or when handled by a person who does not wash their hands after a bowel movement. The illness caused by Shigella (shigellosis) causes diarrhea, fever, cramps, and typically subsides without treatment in a week. *Shigella* is the world-wide leader in causing diarrhea. Over 80 million cases of diarrhea and 74,000 deaths are attributed to infections caused by this organism annually. Children aged 2—4 years are most at risk. Shigella organisms invade the epithelial lining of the colon resulting in inflammation and dysentery. Once the gut lining has been perforated, *Shigella* enterotoxins can enter the blood and cause a potentially fatal condition known as hemolytic-uremic syndrome. This involves a combination of hemolytic anemia, acute kidney failure, and low platelet count.

Staphylococcus aureus

S. aureus produces a bacterial toxin while it grows on food. The toxin is a protein that causes gastrointestinal illness, vomiting and diarrhea that occur soon after contaminated food is eaten. The illness referred to as Staph poisoning is one of the most common types of food poisoning. Staph toxin is an enterotoxin with more than 20 identified isoforms. It is heat stable and resistant to digestive enzymes. Enterotoxins cause irritation and swelling of the small intestine in a condition known as enteritis. The onset of symptoms occurs within 6 h of eating toxin-containing food and often results in explosive vomiting with subsequent cramping and dehydration. An estimated at 25% of the population carry *S. aureus* bacteria on their skin and are without symptoms. Thus, food handlers can be a primary source of Staph poisoning unless appropriate sanitary precautions (e.g., handwashing, gloves) are taken. Keeping food in the safe zone above 60 °C (140 °F) or below 4 °C (40 °F) is critical. Staph poisoning is associated with a variety of foods, including meat, poultry, eggs, dairy products and some vegetables.

Vibrio vulnificus

V. vulnificus is one of several *Vibrio* pathogens genetically related to the cholera causing bacteria, *Vibrio cholerae*, but it does not cause cholera. *V. vulnificus* is found coastal marine environments where freshwater and salty sea water mix. More than 90% of the *Vibrio vulnificus* illnesses are associated with eating raw oysters from the Gulf coast. Illness can be caused by drinking the water or eating raw shellfish, such as oysters, shrimp, and clams contaminated with *V. vulnificus*. Initial symptoms of *V. vulnificus* illness are classified as acute gastroenteritis and include diarrhea, vomiting, and abdominal pain. Left untreated, the illness can progress to an immune response triggered by the infection as it enters the blood, this is called septicemia. *V. vulnificus* caused illness is fatal to about one-third of individuals in which the infection has spread to the blood. People with weakened immune systems, such as those with HIV/AIDS or those taking medicines to control arthritis, are at higher risk. Proper storage and handling of raw seafood is effective in reducing risk of infection. *V. vulnificus* grows rapidly on cooked food. For this reason, care should be taken to keep raw and cooked food from touching. *V. vulnificus* is destroyed by cooking to 63 °C (145 °F) for 15s. A quick blanching (1–2 min) in boiling water is a good precaution.

Mycotoxins

Mycotoxin is the general term referring to toxic substances produced by several types of fungi that infect commodities such as wheat, corn, sorghum, rice, cocoa beans, figs, spices (e.g., ginger and nutmeg) peanuts and tree nuts. Mycotoxins are ingested as a result of eating foods made from contaminated grains and cereals. Meat and milk products are also a human source of mycotoxin exposure that result from livestock fed contaminated silage and grain. Mycotoxins are produced by proliferating fungi. This condition is accelerated by high humidity. The major groups of mycotoxins include aflatoxin, ochratoxin, citrinin, ergot, patulin, and fumonisins.

Aflatoxins produced by *Aspergillus flavus* and *Aspergillus parasiticus* are potent carcinogens. They cause liver cancer in humans as a result of DNA mutation. Aspergillus species produce several subtypes of aflatoxin (B_1, B_2, G_1, G_2, M_1) and B_1 is the most potent carcinogen. Cow's milk is a source of aflatoxin resulting from aflatoxin-contaminated silage feed. In the cow's rumen, aflatoxin B_1 is converted to M_1 which is excreted into the milk. Crops with the greatest aflatoxin occurrence are peanuts, pistachios, and corn. Types A, B, and C ochratoxins are produced by *Penicillium* and *Aspergillus* species and occur as a contaminant of beer and wine. Type A ochratoxin is a carcinogen linked to tumors in the brain and urinary tract. Citrinin is a mycotoxin produced by *Penicillium* and *Aspergillus* species. Its mycotoxins are a contaminant in cheese (camembert) and fermented miso and soy products. Ergot are alkaloid type toxins produced by fungi of the *claviceps* genius. When ingested from sources such as bread made from contaminated flour, a disease known ergotism can result. Symptoms of the disease in the digestive system include spasms, nausea, vomiting, and diarrhea. Ergot alkaloid can also spread through the blood and result in swelling and death of peripheral tissues. Ergot poisoning can be transmitted from mother to child through breast milk. Patulin is another mycotoxin product of *Penicillium* and *Aspergillus* species. It is typically found in rotting fruit such as apples. It is often a contaminant in apple products such as juice, jam, and cider. The level of patulin in apple juice is used as an indicator of quality. A maximum concentration of 50 μg/L of juice is recommended as the upper limit of safety by the World Health Organization. Fumonisins, such as zearalenone, are toxins produced by *Fusarium* species most often found in corn. It is also found in wheat, sorghum, barley, rice and oats. Zearalenone-contaminated feed is of particular concern to livestock producers causing infertility and abortion in animals, with particular effect in swine (Table 7.4).

Mushrooms are prized for the flavors they contribute to foods, such as salads, soups, sauces, and cooked dishes. However, some poisonous varieties are difficult to distinguish from edible ones (Poisonous Mushrooms 2019). Incidents of mushroom poisoning most often involve unknowing individuals who gather wild mushrooms from the woods or the lawn. Visual identification of poisonous mushrooms is difficult. There have been cases when professionals have mistakenly picked a poisonous variety. The source of toxins is the fungi from which the mushrooms grow. Mushroom toxins are stable to typical processing treatments, such as cooking, freezing, or drying. They are grouped into four basic categories based on their physiological effects: protoplasmic poisons, neurotoxins, gastrointestinal irritants, and disulfiram-like toxins. Protoplasmic toxins represent the greatest risk because these can cause organ failure and death if treatment is not received soon after the mushroom is eaten. Illness from this type of toxin is characterized by a latent period without symptoms, followed by the sudden onset of severe symptoms from which the individual may not recover. Protoplasmic toxins include amanitin, hydrazine, and orellanine. Amatoxins, such as the octapeptide amanitin, are found in several mushroom species, including *Amanita phalloides* (Death Cap), *Amanita vberna* (Fools mushroom), and *Galerina autumnalis* (Autumn Skullcap). Poisoning is characterized by a 6—15 hr latent period after ingestion of the toxin–containing mushroom. The latent period is followed by a sudden onset of abdominal pain, vomiting, diarrhea, and loss of strength. The disease progression includes liver and kidney damage in its final phase. The mortality rate from amanitin type of toxin can be 90%. Hydrazine type toxins, such as gyromitrin, are found in *Gyromitra esculenta* (False Morel) mushrooms. The disease follows a similar pattern of an asymptomatic latent period followed by sudden onset of headache, vomiting and diarrhea, but is less severe. The mortality rate is 2%—4%. Orellanine type toxins are found in *Cortinarius orellanus* (Sorrel Webcap) mushrooms. The long latent period (3—15 days) is followed by multiple symptoms that include burning thirst, nausea, headache, chills, and loss of consciousness. Progression of this disease causes kidney failure and death in up to 15% of the cases. The remaining categories are lower in their life-endangering risk and most people recover from the toxin-caused illnesses. Those who are elderly or have existing health problems have a higher risk of death. Neurotoxins of note include muscarine, muscimol, and psilocybin. They are responsible for perfuse sweating, manic-depressive behaviors, and alcohol-like intoxications or hallucinations. Gastrointestinal irritants or toxins cause nausea, cramps, vomiting, and diarrhea. Disulfiram-like toxins result from production of an unusual amino acid (coprine) that infers with the metabolism of alcohol. The individual who has eaten mushrooms containing this toxin is typically asymptomatic unless they also drink alcohol within 72 h. The combination of coprine and alcohol results in nausea and vomiting (Bad Bug Book FDA, 2018) (Table 7.5).

Tetrodotoxin

Tetrodotoxin is a potent neurotoxin found in pufferfish, porcupinefish, and blue-ringed octopus. The toxin is produced by bacteria species such as *Vibrio alginolyticus* and *Pseudomonas tetraodonis*. Bacteria live symbiotically in these marine and freshwater animals. Pufferfish, also known as blowfish or fugu, are one of the most dangerous foods eaten. The poison is a very potent neurotoxin that blocks sodium ion channels and prevents brain signals from activating muscle contraction. The toxin is concentrated in the liver and sex organs of pufferfish. If the fish is improperly prepared, tetrodotoxin can contaminate the meat of the fish. Eating low doses of the toxin causes tingling sensations in the mouth fingers and toes. Higher doses produce nausea, vomiting, and respiratory failure. Lethality of this tetrodotoxin is similar to that of cyanide poisoning. Death can be caused by ingesting as little as 1 mg of the toxin. Tetrodotoxin also has an interesting pharmaceutical

Table 7.5 Mushroom toxins.

Toxin type	Toxins	Species	Effects
Protoplasmic Toxins	Amanitin	Death Cap (*Amanita phalloides*) Fools mushroom (*Amanita vberna*) Autumn Skullcap (*Galerina autumnalis*)	Latent period followed by vomiting and diarrhea, resulting in organ failure, mortality rate approx. 90%
	Hydrazine	False Morel (*Gyromitra esculenta*)	Latent period followed by vomiting and diarrhea, Mortality rate 2%—4%
	Orellanine	Sorrel Webcap (*Cortinarius orellanus*)	Long latent period (3—15 days), liver and kidney damage. Mortality rate approx. 15%
Neurotoxins	Muscarine Muscimol, and Psilocybin	*Clitocybe dealbata*	Perfuse sweating, manic-depressive behavior, hallucinations
Gastrointestinal irritants	Allenic Norleucine	Green Parasol (Chlorophyllum molybdites)	Nausea, cramps, vomiting, and diarrhea.
Disulfiram-like toxins	Coprine	Coprinopsis atramentaria	Similar to above

application as an agent to manage pain in cancer patients. Low doses of tetrodotoxin also block signals in nerve fibers responsible for pain.

Toxic metals in food

Metallic elements are minerals that originate in the earth's crust. In general, minerals leach into water and enter the food chain when incorporated into organic molecules by microorganisms. Plants take up minerals in various forms and subsequently pass them up the food chain to humans. Metallic elements, such as such as sodium, calcium, iron, potassium, zinc and others, are often referred to as trace elements because they are needed in small amounts for normal biological functions. Iron, for example, is essential to the oxygen carrying property of hemoglobin in blood. Calcium is required for bone strength and muscle contraction. However, elements such as lead, mercury, arsenic, and cadmium are classified as heavy metals and have no biological function. Aluminum, while technically not a heavy metal, has no biological function and is potentially toxic. The toxic effect is caused by binding to enzymes and other proteins and result in inactivation of biochemical pathways. Heavy metals are poorly eliminated from the body and this increases their toxicity. Lead, for example, has many toxic effects including inhibition of the synthesis of heme needed to make red blood cells. Toxicity of metals varies with the element and the amount ingested. Heavy metals are also suspect in causing cancer. They have genotoxic effects that damage DNA. It is unlikely that a single food represents a serious risk. Rather, the cumulative amount from all sources is more important. The following is a brief summary of the more significant toxic metals occurring in food. Additional information regarding metals in food and water can be found in the review by Tchounwou et al. (2012) and United States Food and Drug Administration summary on metals (Metals, 2018).

Aluminum (Al)

Aluminum is the most abundant metallic element in the earth's crust where it exists in the form of compounds such as aluminum sulfate or aluminum oxide. The metallic form of aluminum does not occur in nature, but is obtained from processing bauxite ore that is principally composed of aluminum oxide. Exposure to aluminum can occur by inhaling dust particles that contain the metal or from aluminum-contaminated drinking water and food. Small amounts are released from aluminum equipment used in food manufacture and preparation. If the material being cooked in an aluminum pot is acidic, higher levels of aluminum will be released. Fortunately, aluminum is poorly absorbed from the digestive track and most is eliminated in the feces. Investigations of aluminum's toxicity reveal it is a neurotoxin affecting cognitive and motor functions when fed at high levels to animals. This discovery led to a suggested link between aluminum and Alzheimer's disease, but analysis of data from several studies have not been conclusive. Aluminum is found in foods additives such as baking powder (sodium aluminum sulfate and sodium aluminum phosphate), anticaking agents (sodium aluminosilicate), and colorants (FD&C blue). Over the counter medications, such as buffered aspirin contain 10–20 mg/tablet and antacids contain 100–200 mg/dose, in the form of aluminum hydroxide. It is estimated that the average adult consumes about 7–9 mg/kg of body weight of aluminum per day. Most of this is from food additives and analgesics. This level of daily intake is equivalent to about 650 mg for a 175lb person. Although of neurotoxicity of aluminum has not been established for healthy individuals, a precautionary recommendation is to reduce the level of intake.

Arsenic (As)

Arsenic is a toxic heavy metal found in soil and water in organic and inorganic (salts) forms. Rice is a food with high potential for becoming contaminated with arsenic because the plants are very efficient in absorbing it from soil as they grow. The organic form of arsenic is methylarsonic acid. There are two inorganic arsenic forms: arsenite ($NaAsSO_2$) and arsenate (Na_3AsO_4). Inorganic arsenate (Na_3AsO_4) is the most toxic form. Its toxicity stems from its role as an easy substitute for phosphate which is a cofactor for enzymes involved in metabolism and other processes. Long term consumption of arsenate-containing food is associated with increased risk of cancers of the skin, bladder, and lungs. The risk of cancer is believed to be linked to arsenic inhibition of enzymes that repair DNA damage. The risk of cancer, while not dramatically high, increases with the frequency of consumption. Because rice is a food stable for much of the world's population, a limit of 100 ppb (parts per billion) has been set by FDA for arsenic in rice. The limit for arsenic in infant foods is much lower with 1 ppb in infant rice cereals and formula. Infant apple juice has a limit of 10 ppb arsenate.

Cadmium (Cd)

Cadmium is a toxic heavy metallic element widely found in the earth's crust. Mining, metal processing, and battery production are sources of cadmium exposure. Cigarette smoke also contains cadmium. However, food represents the major source of cadmium for most people. Cadmium is present at low levels in a variety of food, including green leafy vegetables, seaweed, potatoes, and grains. Animal sources include liver, kidney, and shellfish. Among these foods, rice, seafood, edible offal, and vegetables represent greatest source of cadmium. Ingestion of cadmium can cause kidney damage. Cadmium can also cause gastrointestinal symptoms ranging from nausea and vomiting to erosion of the gut epithelium. The later and more serious effect is thought to result from generation of reactive oxygen species (ROS) in reaction to cadmium. Adverse biological effects of cadmium, production of ROS, and inhibition of DNA repair mechanisms are sufficient to classify cadmium as a carcinogen. The maximum level of cadmium exposure for adults is 2.5 μg/kg of body weight. This amount is equivalent to approximately 205µg for a 175lb person. The maximum limit for cadmium in bottled water is 5 ppb.

Lead (Pb)

Lead is a toxic heavy metal that occurs ubiquitously in soil, water, food, and air as lead–containing dust particles derived from paint. Until 1996, lead was commonly added to gasoline as fuel for cars and trucks. Food crockery and crystal glasses commonly contained lead that could leach into food, especially if the food or beverage was acidic. Lead and other heavy metals have been shown to contaminate traditional Ayurvedic medicines used for centuries in India. Lead is responsible for a wide range of health problems including; anemia, kidney, and brain damage. Cardiovascular and reproductive system problems are also linked to lead in the environment. Among the more serious outcomes of lead ingestion results from its action as a neurotoxin. This is especially harmful to young children. Lead affects the developing brain with life-long consequences that include cognitive impairment and learning and behavioral problems, among other things. In view of lead's toxicity, FDA guidelines for the level of lead in food and water are very low, in the ppb (part per billion) range. The levels for lead in food and bottled water and drinking water are 5 ppb and 15 ppb, respectively.

A sad case of lead pipes and chemistry

A series of missteps and mistakes by government officials in Michigan precipitated a public health crisis resulting in the exposure of over 100,000 (40%) Flint Michigan residents to high levels of lead in their drinking water. The crisis began in 2014 when the city changed its source of municipal water to the Flint river. Compared to the original source of water, the Flint river contains high levels of the highly corrosive element chloride. The second mistake was not adding corrosion inhibitors, such as orthophosphate to the new water source. Orthophosphate forms a chemical bond with lead water pipes and prevents its oxidation and subsequent release of the toxic metal into drinking water. Without the protective coating, chloride in the water accelerated the corrosion process and caused even higher levels of lead to be released. The level of lead in Flint Michigan's water reached 100 ppb and persisted for almost two years. During this time, residents had to rely on bottled water for drinking and cooking. Outcomes of these mistakes have been costly. Millions of dollars have been spent to rebuild the water system and criminal charges have been filed against officials. Most costly of all is injury to the health of many Flint residents, especially children. Their blood lead levels remain high and the effects on their lives is potentially profound.

Mercury (Hg)

Mercury is a toxic metallic element found in soil, water, and air. The presence of mercury in the environment results from multiple sources, including leaching from the soil, industrial waste, volcanic eruptions, and emissions from coal burning energy plants. Mercury exists in inorganic or elemental mercury and organic or methylmercury forms. Elemental mercury is a liquid at room temperature and can volatilize. Exposure results from skin contact or inhalation. Organic mercury occurs as methylmercury which is a product of the action of microorganisms on the inorganic form. Major sources of mercury exposure result from eating fish and shellfish and from dental amalgams. Fish, especially long-lived species like albacore tuna, shark, and swordfish contain the highest level of mercury because they are bioaccumulators of methyl mercury. Shellfish and other aquatic animals feed on mercury-containing algae and plankton. These animals, in turn, are sources of food for larger fish. This effectively concentrates the level of mercury in their tissues. Once absorbed, mercury is poorly eliminated and is highest in fish at the top of the food chain. Mercury is a potent neurotoxin. Adverse effects of methylmercury include effects on cognitive functions of memory, speech, and fine motor skills. Children and fetuses are most susceptible to methylmercury during fetal development. It can pass through the placenta and enter the fetus. Methylmercury is a suspected carcinogen, linked to its ability to cause depletion of cellular antioxidants. The subsequent increase in reactive oxygen species (ROS) results in DNA damage that is a known initiator of cancerous processes. Exposure to inorganic mercury forms, such as elemental mercury or mercury compounds, can also cause adverse health effects. Symptoms of inorganic mercury exposure include headache, cognitive impairment, and muscle weakness. Exposure to high levels of mercury can result in respiratory failure and death. The FDA limit for mercury in bottled water is 2 ppb.

Process induced toxins

Acrylamide

Acrylamide is a neurotoxin and carcinogen formed in food materials as a result of cooking. Specifically, acrylamide is produced by heat-induced chemical reactions collectively known as the Maillard reaction. This reaction is widely recognized as responsible for the color and flavor of cooked foods. Discovery of acrylamide toxin-producing chemistry is rather recent (Tareke et al., 2002). The major reactants in food are the amino acid asparagine and reducing carbohydrates such as glucose and fructose. The amount of acrylamide produced in a food varies with temperature, highest at 247 °F (120 °C), type of sugar, fructose > glucose > sucrose, and moisture level, low moisture > high moisture. Foods such as French fries, potato chips, cookies, crackers, breakfast cereals and coffee have the highest levels of acrylamide. Acrylamide's neurotoxic effects include cognitive impairments such as memory loss, confusion, numbness in the limbs, and muscle atrophy. It has demonstrated ability to be genotoxic and cause cancerous tumors in mice. FDA and other regulatory agencies consider the present state of science regarding acrylamide's toxicity inadequate to make recommendations on the allowable limits in foods. However, recommendations regarding methods to reduce the amount of acrylamide formed in foods can be found in Guidance for Industry: Acrylamide in Foods, a 2018 publication from the Center for Food Safety and Applied Nutrition of the FDA.

Toxicants
Bisphenol A

Bisphenol A (BPA) is a component of materials used to make polycarbonate. These include hard, clear plastic, beverage containers and epoxy resins used to form protective coating inside metal cans. Concern has arisen over the unintended migration of bisphenol A from plastic bottles and can coatings into foods. Heating the food while in the container increases the level of BPA released. Concern over BPA ingestion is linked to reports of neurotoxicity and estrogen-like effects in animal studies. As a result of this concern about BPA in reusable plastic food containers, a number of states in the United States have banned BPA in bottles, mugs, and tableware used in infant feeding. The European Food Safety Authority has set a safe level termed the Tolerable Daily Intake of 0.05 mg per kg of body weight, equivalent to about 4 mg for a 175lb person. This level is about 100 times lower than the No Observable Adverse Effect Level (NOAEL) guideline used in toxicity assessment. It should be noted that BPA is unmodified and rapidly eliminated in urine. While safety and risk assessments are on-going, the regulatory agencies of the United States, Canada, Europe, and Australia have concluded that BPA is not expected to be a health risk. Precautions that individuals can take to avoid BPA exposure include buying and storing food in glass containers and refraining from use of plastic containers to microwave food.

Dioxin

Dioxins are a group organic compounds that are persistent environmental contaminants. They are man-made products typically resulting from incinerating chlorine compounds and hydrocarbon (petroleum) compounds. They are also formed as by-products of polyvinyl chloride (PVC) plastics. Dioxin is a potent carcinogen and has damaging effects on immune and reproductive systems system. Dioxins find their way into the food chain because of their hydrophobic nature. Dioxin in the soil is taken up by microorganisms and plants that are concentrated (bioaccumulated) in meat, milk, egg, and animal fats. Meat and milk can contain 10 to 20 times the dioxin level of soils. They may also persist in fat depots of the body. Because of dioxin prevalence in the environment, avoiding it is almost impossible. It is possible, however, to reduce dioxin intake by limiting consumption of animal-based foods and switching to a plant-based diet containing fruits and vegetables. A more comprehensive approach to reduction of dioxin exposure for everyone is to seek better waste incinerators with restrictions on the types of materials burned and limits on emitted dioxins. Additionally, animal feed and feedlot soils should be monitored for dioxin levels.

> ## Summary
>
> Food additives are used in processed foods to enhance safety, add color and flavor, modify texture, and preserve food. Food additives, including colorants, are a major source of confusion and controversy. For example, the terms natural and artificial (synthetic) have straight forward meanings to most, but are confusing when it comes to food labeling. Natural flavor, as defined by the US Code of Federal regulations is "a natural flavor is the essential oil, oleoresin, essence or extractive, protein hydrolysate, distillate, or any product of roasting, heating or enzymolysis, which contains the flavoring constituents derived from a spice, fruit or fruit juice, vegetable or vegetable juice, edible yeast, herb, bark, bud, root, leaf or similar plant material, meat, seafood, poultry, eggs, dairy products, or fermentation products thereof, whose significant function in food is flavoring rather than nutritional." Thus, any substance not included in this definition and added to another food is considered to be an artificial additive and the product must be labeled as such. For example, yogurt containing only

blueberries as flavoring can be labeled as "blueberry yogurt". However, if a blueberry extract is also added, the product must be labeled "naturally flavored blueberry yogurt with other natural flavors". It is no wonder that there is confusion. The controversy about the safety of food additives is made worse by claims of adverse effects in the popular press. The evidence for many of these claims is weak, re-evaluations of food additive safety with current science and knowledge, by food regulatory agencies, is warranted (Carocho et al., 2014).

Glossary

Adulteration: A food that does not meet a legal standard of identification or contains a poisonous substance

Amphiphilic: Substance possessing both hydrophilic and hydrophobic and is thus compatible with water and oil.

Artificial (synthetic) additive: Substance added to foods for any of the purposes listed above but is man-made.

Bacteriocin: Short peptides having bactericidal or bacteriostatic effects on other species.

Botulism: Disease resulting from ingestion of toxin secreted by *Clostridium botulinal*. The toxin blocks neurotransmitter release and prevents muscle contraction.

Certified color additive: Any dye, pigment, or other substance used impart color to a food, drug, or cosmetic.

Chelating agent: Substances that tightly bind metal ions such as copper, iron, calcium, and zinc through coordination of electrons.

Colloid: Homogeneous mixture of one substance dispersed in another, without separation.

Deliquescent: Property of a substance to absorb moisture from the atmosphere until it dissolves.

Direct additive: Substance added to a food for use in processing, packaging, or to provide a desired effect such as a; colorant, preservative, nutrient, antioxidant, or sweetener.

Enterohemorrhagic: Medical term meaning bloody diarrhea.

Fat mimetic: Biopolymer derived from protein, carbohydrate, or lipid modified by physical, chemical, or enzymatic means to imitate the eating properties of fat.

Food regulation: An enforceable law regulating the production, trade, and handling of food.

Free Radical Scavengers (FRS): Substances that stop free radical reactions by donating a hydrogen atom. Examples of FRS include ascorbic acid and alpha tocopherol, vitamin C and E respectively.

Free radical: An atom, ion, or molecule that has unpaired valence electrons. The high reactivity of a free radical comes from the need to gain or lose an electron and achieve a stable configuration.

Gastroenteritis: Illness caused by inflammation of the gastrointestinal tract. Vomiting and stomach cramps and diarrhea may also result.

Genotoxic: Property of chemical agents resulting in damage to DNA without causing a mutation. The damage may lead to cancer.

GRAS: Acronym standing for Generally Recognized As Safe and refers to a number of substances used as additives in food, drugs and cosmetics.

Heavy metals: Metallic elements characterized as having high density and poisonous properties.

Hydrophilic: Substance that is compatible with water, mixes in and or dissolves in water.

Hydrophobic: Substance that is not compatible with water and repels it. "water hating".

Hygroscopic: Property of a substance to absorb moisture for the atmosphere.

Indirect additive: Substance that migrates into food from the environment which can include; growing, processing, and packaging.

LD50 value: Amount of a toxic agent that is sufficient to kill 50 percent of a population of animals.

Lectin: Protein that binds to carbohydrate molecules on the surface of cells, a mechanism for the regulation biological processes such as immunity.

Mutation: Heritable alteration in nucleic acids of an organism.

Natural additive: Substance found naturally in some foods and extracted for use in others.

Nanoparticle: Very small particles ranging in size between 1 and 100 nanometers (10^{-9} meter).

Terpenoid: Class of organic compounds composed of a repeating, five carbon (isoprene) subunit. Their composition is limited to carbon and hydrogen atoms only and thus are not water soluble.

Toxicant: Person-made substances such as pesticides and herbicides which can be harmful to animals.

Toxin: Poisonous substance produced by organisms to provide biological advantage, such as protection against predation.

Vitamin Fortification: Addition of a vitamin to a food for the purpose of making it a superior source of that nutrient.

Vitamin Restoration: Addition of a vitamin to a level normally found in that food.

References

Al-Abri, S.A., Olson, K.R., 2013. Baking soda can settle the stomach but upset the heart: case files of the medical toxicology fellowship at the university of California, san francisco. J. Med. Toxicol. 9 (3), 255–258.

Bacciottini, L., Falchetti, A., Pampaloni, B., Bartolini, E., Carossino, A.M., Brandi, M.L., 2007. Phytoestrogens: food or drug? Clin. Cases .Min. Bone Metabol. 4 (2), 123–130.

Bad Bug Book, 2012. U.S. Food and Drug Administration, second ed. https://www.fda.gov/food/foodborneillnesscontaminants/causesofillnessbadbugbook/.

Banerjee, S., Bhattacharya, S., 2011. Food gels: gelling process and new applications. Crit. Rev. Food Sci. Nutr. 52 (4), 334–346.

Bishayee, A., Ahmed, S., Brankov, N., Perloff, M., 2011. Triterpenoids as potential agents for the chemoprevention and therapy of breast cancer. Front. Biosci. J. Vis. Lit. 16, 980–996.

Breiteneder, H., Mills, E.N., 2005. Molecular properties of food allergens. J. Allergy Clin. Immunol. 115, 14–23.

Carocho, M., Barreiro, M.F., Morales, P., Fereira, C.F.R., 2014. Adding molecules to food, Pros and Cons: a review on synthetic and natural food additives. Compr. Rev. Food Sci. Food Saf. 13, 377–399.

Dolan, L.C., Matulka, R.A., Burdock, G.A., 2010. Naturally occurring food toxins. Toxins 2 (9), 2289–2332.

Hung, W.L., Suh, J.H., Wang, Y., 2017. Chemistry and health effects of furanocoumarins in grapefruit. J. Food Drug Anal. 25, 71–83.

Ito, N., Fukushima, S., Tsuda, H., 2008. Carcinogenicity and modification of the carcinogenic response by BHA, BHT, and other antioxidants. CRC Crit. Rev. Toxicol. 15 (2), 109–150.

Jakszyn, P., González, C.A., 2006. Nitrosamine and related food intake and gastric and oesophageal cancer risk: a systematic review of the epidemiological evidence. World J. Gastroenterol.: WJG 12 (27), 4296–4303. http://doi.org/10.3748/wjg.v12.i27.4296.

McGee, H., 1984. On Food and Cooking: The Science and Lore of the Kitchen. Scribner, New York.

Park, J., Seo, J., Lee, J., Kwon, H., 2015. Distribution of seven N-nitrosamines in food. Toxicological Research 31 (3), 279–288. http://doi.org/10.5487/TR.2015.31.3.279.

Price, K.R., Johnson, I.T., Fenwick, G.R., Malinow, M.R., 2009. The chemistry and biological significance of saponins in foods and feeding stuffs. CRC Crit. Rev. Food Sci. Nutr. 26 (1), 27–135.

Remer, T., Manz, F., 1995. Potential renal acid load of foods and its influence on urine pH. J. Am. Diet. Assoc. 95, 791–797.

Ryssel, H., Kloeters, O., Germann, G., Schäfer, T., Wiedemann, G., Oehlbauer, M., 2009. The antimicrobial effect of acetic acid–an alternative to common local antiseptics? Burns 35 (5), 695–700.

Saha, D., Bhattacharya, S., 2010. Agents in food: a critical review. J. Food Sci. Technol. 47 (6), 587–597.

Shahidi, F., Janitha, P.K., Wanasundara, P.D., 2002. Phenolic antioxidants. Crit. Rev. Food Sci. Nutr. 32 (1), 67–103.

Suganthi, V., Selvarajan, E., Subathradevi, C., Mohanasrinivasan, V., 2012. Lantibiotic nisin: natural preservative from Lactococcus lactis. Int. J. Pharm. Pharm. Sci. 3 (1), 13–19.

Tareke, E., Rydberg, P., Karlsson, P., 2002. Analysis of acrylamide, a carcinogen formed in heated foodstuffs. J. Agric. Food Chem. 50, 4998–5006.

Tchounwou, P.B., Yedjou, C.G., Patlolla, A.K., Sutton, D.J., 2012. Heavy metals toxicity and the environment. EXS 101, 133–164.

USDA Food Safety and Inspection Service, 2017. Foodborne Illness: What Consumers Need to Know. https://www.fsis.usda.gov/wps/portal/fsis/topics/food-safety-education/get-answers/food-safety-fact-sheets/foodborne-illness-and-disease/foodborne-illness-what-consumers-need-to-know/CT_Index.

van Buul, V.J., Brouns Fred, J.P.H., 2014. Health effects of wheat lectins: a review. J.of Cereal Science 59, 112–117.

Wade, N., 1977. Laetrile at sloan-kettering: a question of ambiguity. Science 198 (4323), 1231–1234.

Younes, M., Aggett, P., Aguilar, F., Crebelli, R., Dusemund, B., Fillpic, M., 2017. Re-evaluation of mono- and di-glycerides of fatty acids (E471) as food additives. EFSA J 15, 1–43.

Internet resources

Acrylamide in Foods-Guidance for Industry, 2018. Food and Drug Administration (FDA). https://www.fda.gov/Food/GuidanceRegulation/GuidanceDocumentsRegulatory Information/ucm374524.htm.

European Food Additive Number Identifiers, 2018. https://en.wikipedia.org/wiki/E_number.

FDA Bad Bug Book, 2018. United States Food and Drug Administration (FDA), second ed. https://www.fda.gov/food/foodborneillnesscontaminants/causesofillnessbadbugbook/.

Food Additives (EFSA), 2018. European Food Safety Authority (EFSA). http://www.efsa.europa.eu/en/topics/topic/food-additives.

Food Additives (WHO), 2018. World Health Organization (WHO). http://www.who.int/mediacentre/factsheets/food-additives/en/.

Metals, 2018. United States Food and Drug Administration (FDA). https://www.fda.gov/food/foodborneillnesscontaminants/metals/default.htm.

Overview of Food Ingredients, Additives & Colors, 2018. United States Food and Drug Administration (FDA). https://www.fda.gov/Food/IngredientsPackagingLabeling/FoodAdditivesIngredients/ucm094211.htm.

Poisonous Mushrooms, 2019. https://www.namyco.org/mushroom_poisoning_syndromes.php#gastro.

Toxic Substances, 2018. CDC Agency for Toxic substances and Disease Registry) CDC, Center for Disease Control (ATSDR). https://www.atsdr.cdc.gov.

Further reading

Foodborne Illnesses, 2017. What You Need to Know. United States Food and Drug Administration (FDA).

Msagati, T.A.M., 2013. Chemistry of Food Additives and Preservatives, first ed. Wiley and Sons, Chichester UK.

USDA Food Safety and Inspection Service, 2018. Foodborne Illness: What consumers need to know. https://www.fsis.usda.gov/wps/portal/fsis/topics/food-safety-education/get-answers/food-safety-fact-sheets/foodborne-illness-and-disease/foodborne-illness-what-consumers-need-to-know/CT_Index.

Review questions

1. What are the important outcomes of the Pure Food and Drug Act of 1906?
2. Why is adding water to milk considered to be an act of adulteration?
3. Describe the term GRAS and its importance to food additives.
4. Give one example each of an acid and a base used as food additive.
5. What is "Chemical Leavening"? Give an example of how it works in a food system.
6. What are the major reasons for adding salt to food?
7. What is the function of acetic, propionic, and sorbic acids in food?
8. Give two reasons for adding nitrite to processed meats.
9. What is the health concern for nitrite in bacon?
10. Why is sulfite used in food systems? Give two examples.
11. What is a chelator and why are they added to foods?
12. Why is ascorbic acid added to food systems? Give two examples of it use.
13. Define the term hydrocolloid and give two examples of their use as food additives.
14. What is the importance of gels in processed foods? Give two examples of substances used to form gels in food systems.
15. What is rum caviar and how is it made?
16. Give two examples of emulsifiers and describe their use in food systems.
17. What is a fat replacer? Give an example.
18. Give an example of enzymes used as a food additives.
19. Define the terms toxin and toxicant. Give an example of each.
20. Give an example of a toxicant that results from the Maillard reaction.
21. What is trypsin inhibitor and where is it found?
22. Name 2 sources of trypsin inhibitor.
23. What is the difference between direct and indirect food additives?
24. What are mycotoxins? Give two examples.
25. Name three toxic elements and foods they might be found in.

CHAPTER EIGHT

Food colorants

Learning Objectives

This chapter will help you describe or explain:

- The difference between artificial and natural food colorants
- Rules and regulations concerning food colorants
- Health promoting properties of carotenoids, flavonoids, and chlorophyll
- The controversy about artificial colorants and behavior in children
- Examples of natural colorants as potential replacements for artificial colorants

Introduction

Color is the single most important factor influencing consumer food preference. An off color in a food is associated with spoilage and the item is likely to be rejected. Colorants are substances that absorb light in the visible part of the spectrum. Reflected wavelengths make up what is

Introduction to the Chemistry of Food
ISBN: 978-0-12-809434-1
https://doi.org/10.1016/B978-0-12-809434-1.00008-6

perceive as color. Color additives as defined by United States Food and Drug Administration (FDA) are any substance (natural or synthetic) that imparts color to a food or drink. Natural colorants are organic substances derived from a plant or animal sources. Natural colors can change in hue or fade with time. Many are also destroyed when food is processed. These limitations make it difficult to use natural extracts as a color additive. Examples of natural colorants include myoglobin and hemoglobin heme pigments, carotenoids, chlorophyll, anthocyanin, betalain, and some minerals. Paprika, saffron, and turmeric are examples of food ingredients containing natural colorants that have been used to color food for thousands of years. Artificial (synthetic) colorants are person-made substances used to make foods more appealing or to help identify flavors. Artificial colorants are preferentially used in processed foods and beverages because they can be blended to create a wide range of colors that are stable to processing and storage conditions. They also have the advantage of being less expensive. FDA requires that each batch of artificial colorants is tested for identity and purity before being used in food, drugs, or cosmetics. Plant and animal extracts that are considered to be nature-identical and some synthetic substances are classified by FDA as exempt from certification. Increasingly however, consumers are choosing foods without artificial colorants in preference for those with natural alternatives. This change is being driven by a desire for healthier diets and concern over the safety of synthetic substances.

This chapter includes questions that will help you explore and better understand more about colorants in food. For example:

- Why do people swirl their wine?
- What are the health promoting effects of carotenoids?
- Are chlorophyll and carotenoids possible alternatives to artificial food colorants?
- What is leghemoglobin and is it safe to eat?
- How are safe levels for artificial colorant use established?
- Do food artificial colorants contribute to hyperactivity in children?

Natural food colorants

Most natural colorants are derived from plant or animal sources. Examples of natural colorants include anthocyanins, betalains, carotenoids, chlorophylls, flavonoids, heme, and others (Table 8.1). Mineral compounds, such as titanium dioxide, are used to color food and considered to be natural colorants. In general, natural substances are less stable and more expensive than synthetics. Examples of natural colorant dyes include beta carotene, lycopene, and lutein (yellow-orange-red color from carotenoids), chlorophyllin (green

Table 8.1 Natural food colorants.

Compound	Sources	Color
Anthocyanin	Grapes, berries, cabbage	Red-blue-purple
Betalain	Beetroot, amaranth	Red-purple
Caramel	Caramelized sugar (glucose, fructose, sucrose)	Amber-Brown
Carminic acid	Cochineal insect	Carmine (Red)
Carotenoids	Annatto, beta carotene, lycopene, lutein	Yellow-orange-red
Chlorophyll	Na or Cu chlorophyll complex	Green
Curcumin	Turmeric plant rhizome	Yellow-orange
Heme	Myoglobin, Hemoglobin, Leghemoglobin	Red, purple, brown
Minerals	Titanium dioxide, Carbon Black Al, Ag, Au	White, black, varied colors
Phycocyanin	Spirulina, cyanobacteria/algae	Green

color from Na–chlorophyll), anthocyanins (flavonoids), betanin (red color from beetroot), turmeric, and carmine (red color from the cochineal insect). Additionally, spirulina, a green colorant and proposed health supplement, is derived from a microalgae cyanobacteria. The color of these substances, either within foods or as an additive, is affected by a range of biological and chemical factors summarized in the review by Sigurdson et al. (2017).

Anthocyanins

Anthocyanins belong to the flavonoid family of compounds that provides red, blue, purple, and violet colors to a large number of plants, foods, and flowers. They are the most extensively investigated members of the flavonoids group. Anthocyanins exist either as the aglycone or glycoside form. The term aglycone refers to the flavonoid molecule itself and the term glycoside refers to the flavonoid plus a covalently linked sugar (e.g., glucose). Galactose, rhamnose, and arabinose are also possible sugar substitutions in flavonoid glycosides. The chemical link between the flavonoid and sugar is termed a glycosidic bond as shown in Fig. 8.1A.

Flavonoids exist in both glycosylated and unglycosylated forms called an aglycone. The aglycone form of anthocyanin is called anthocyanidin (Fig. 8.1B). Additional modifications of anthocyanin include acylation with organic acids such as coumaric, caffeic, ferulic, synaptic, gallic, or hydroxybenzoic. Acylation changes its color and results in the large number (>600) of color variants typically seen in flowers. A desirable side effect of acylation is increased heat stability. The most commonly

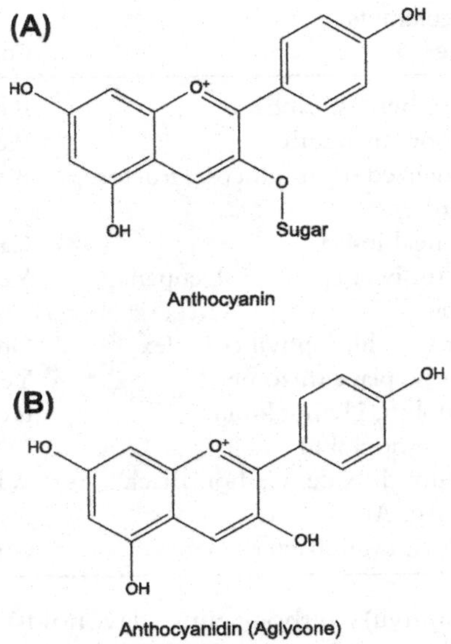

Fig. 8.1 Anthocyanin structure.

found types of anthocyanin in food are the aglycone forms that include pelargonidin, cyanidin, delphinidin, peonidin, petunidin, and malvidin. Anthocyanin colors are derived from excitation by visible light of double bonded electrons in their structure. While the structural complexities of anthocyanin molecules are significant, environmental factors such as pH, metal ions, and oxygen level have the greatest impact on color. Under acidic conditions (pH 1 to 3), anthocyanin has a positive charge localized on the oxygen atom. This form is known as the flavylium cation, is deeply red and has increased water solubility. When pH is increased to the 4 to 6 range, the molecule loses its positive charge and it changes to violet. In an alkaline environment, pH 8 to 10, the color changes to a deep blue with increased intensity (Fig. 8.2). Anthocyanins in wine, for example, are affected by several factors that can alter its color.

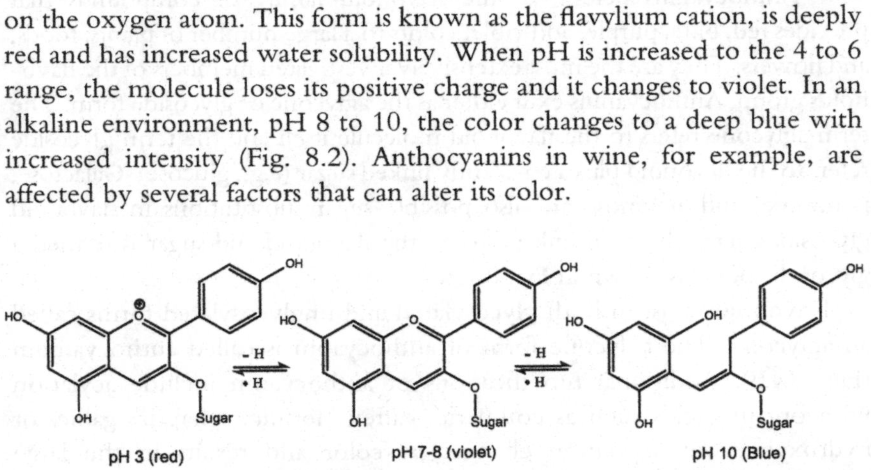

Fig. 8.2 Effect of pH on Anthocyanin Structure and Color.

Why do people swirl their wine?

An important step of wine tasting involves swirling the glass of newly opened wine before sampling its aroma and taste. It is said that swirling allows the wine to breathe, develop flavor, and darken in color. Swirling introduces oxygen and initiates a cascade of chemical reactions affecting wine's flavor and color. Oxygen causes oxidation of anthocyanin and other flavonoids, resulting in formation of polymers that create a desirable astringent mouthfeel. However, too much oxidation produces a very undesirable brown color.

Sulfite is another agent that has a strong effect on anthocyanin color. Sulfite reacts with and modifies anthocyanin's ring structure, resulting in a complete loss of color. Sulfite is used in making white wine. White grapes usually contain small amounts of anthocyanin that contribute a reddish hue. Low levels of sulfite are added to white wine to reduce redness and provide a uniform product color. Knowing this bit of chemistry can be handy when red wine is spilled on a white table cloth. A small amount of white wine added to the spot will usually cause it to disappear.

The reaction of anthocyanins with metal ions changes their color. Iron and tin ions solubilized from metal cans by acids are common culprits. Canned cherries, for example, are acidic and their tartaric acid is capable of solubilizing tin ions from the inside of cans. The tin-anthocyanin complex gives canned cherries a dark red hue. Heating also causes color change because it promotes the loss of hydrogen from hydroxyl groups and forms a negatively charged, more chemically reactive ion. Anthocyanin has better heat stability than anthocyanidin. Exposure to oxygen has a negative effect on anthocyanin color. For example, bottled grape juice can change from blue to brown when oxygen remains in the bottle head space. Replacing oxygen with nitrogen in packaging reduces formation of the brown color. Fruit juices naturally contain ascorbic acid (vitamin C) that can negatively affect anthocyanin color. The loss of anthocyanin color in the presence of ascorbic acid (an antioxidant) is unexpected. The explanation for the color change in fruit juice was found to results from a combination of ascorbic acid with trace amounts of copper. This produces hydrogen peroxide that destroys anthocyanin and causes color loss. Light is also detrimental to anthocyanin color. Exposure of grape juice or wine to the UV component of light causes red–violet colors to become brown. Anthocyanin (the glycosylated form) is more stable to UV light degradation compared to anthocyanidin (the unglycosylated form).

Health promoting properties of flavonoids

The purported health benefits of red wine consumption began in the early 1990s with the well-known French Paradox. Briefly stated, the paradox suggests that despite a diet rich in cholesterol and saturated fats, the morality rate for cardiovascular disease is lower in the French population compared to other Western countries. The proposed rationale for this observation is that wine is typically consumed as part of a meal and in greater quantity, rather than as a stand-alone drink (He and Giusti, 2010). Some suggest that the amount of wine needed to produce a significant resveratrol effect might be offset by the toxic effect of alcohol (Renaud and de Lorgeril, 1992). In vitro studies have shown that flavonoids bind to blood platelets and inhibit blood clot formation. They also have anti-inflammatory and antioxidant properties that are beneficial to prevention of cardiovascular disease (Khoo et al., 2017). However, in vivo studies of dietary anthocyanin and other flavonoids have not shown conclusive benefits.

Anthocyanin alternatives to artificial colorants

In response to increasing consumer preference for natural food colorants in place of artificial ones, several plant-derived colorants are being investigated as potential replacements for synthetic ones (Wrolstad and Culver, 2012). Anthocyanin-based colorants are now permitted for use in the EU and are classified as color additives that are exempt from certification in the United States. Grape skin extracts and grape color extracts are approved for wine beverage and non-beverage use respectively in the United States Anthocyanins in these extracts are polymerized and this results in darker color and somewhat better heat stability compared to the native compound. Similar anthocyanin extracts from red cabbage and purple sweet potato are being tested as artificial colorant replacements. Improved color stability provided by acylated forms of anthocyanin is a natural alternative to artificial colorants in processed foods (Zhao et al., 2017).

Curcumin

Curcumin is a flavonoid compound found in the spice derived from rhizomes of the turmeric plant (*Curcuma longa*), a member of the ginger family (Fig. 8.3). Dried and powered turmeric root (curry powder) is used in food both as a yellow colorant and for its flavor in Indian, Indonesian, and Thai foods. Turmeric is used for its flavor, color, and warm bitter taste

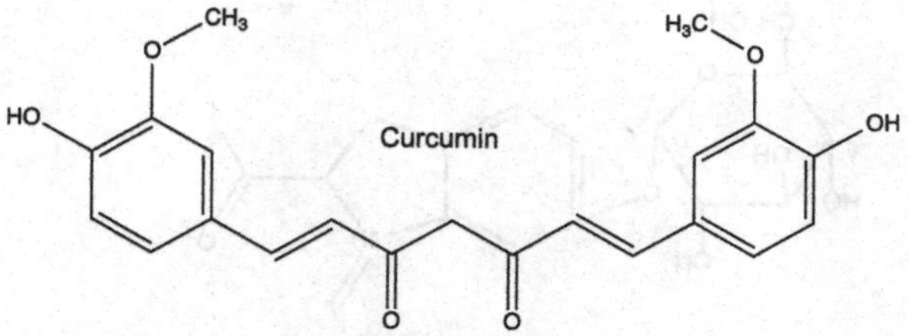

Fig. 8.3 Curcumin.

in curry, mustard, butter, and cheese. Curcumin constitutes about 2% to 9% of powdered turmeric root. Curcumin and ground turmeric are promoted as a health–promoting agent based on their antioxidant and anti–inflammatory properties. In vitro investigations have shown curcumin to be an effective scavenger of free radicals, ROS inhibitor, and anti–inflammatory agent (Curcumin, 2016). However, the poor solubility, uptake, and rapid excretion of curcumin suggest its effectiveness is likely limited to the gastrointestinal tract. Curcumin is proposed to be cancer preventative, based on animal studies using mice. Observational data from human studies do not support this claim. Curcumin supplied as an oil–in–water emulsion system could be a means to increase its bioavailability, but this form may also have increased loss through digestion (Kharat et al., 2017). Curcumin's stability is low at pH greater than 7.0. It is also degraded in the presence of light.

Betalain (Betacyanin)

Betalains are a group of water-soluble compounds contributing red, violet, orange, and yellow colors to beetroot (*Beta vulgaris*) and other plants. The predominant forms of betalain in beetroot are betacyanin and betaxanthin (Polturak and Aharoni, 2017). Betacyanins are intensely purple-red in color and represent the largest component of beetroot pigment. Betacyanins are glycosidic compounds. They contain a covalently attached glucose molecule (Fig. 8.4). Loss of the glucose group through hydrolysis converts it to betanidin, a molecule with lower water solubility. While betacyanin structure is similar to anthocyanin, they are substantially different compounds. Betacyanin contains a nitrogen atom and its color is stable over a wide range of pH (3–9). As a strong light absorber, betacyanin color is degraded by UV light. Its color is also sensitive to high temperatures. Enzymes, (i.e., polyphenol oxidase) found in beetroot and other foods cause oxidization of betacyanin's phenolic hydroxyl groups to quinones, changing its color to brown. Boiling betacyanin-containing food oxidizes the pigment, resulting in a darker color.

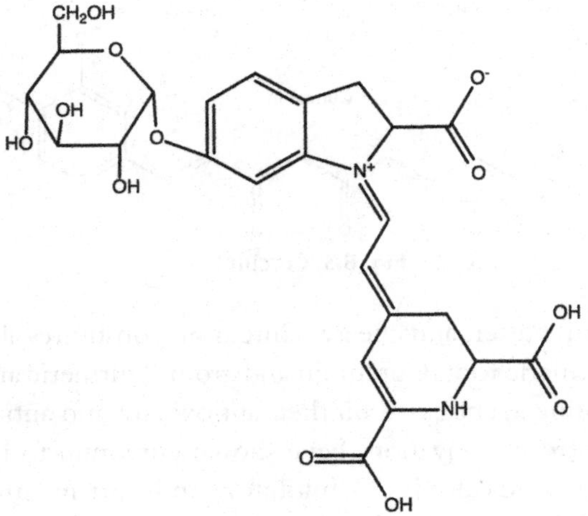

Fig. 8.4 Betacyanin.

Canned beets, for example, have a darker hue resulting from browns produced by the thermal process. Trace amounts of pro-oxidant metals, such as iron or copper, accelerate oxidative color changes. Despite these limitations, beetroot has become widely used as a natural food colorant. It is used in products such as ice cream and powdered drink mixes. Food applications can also be limited by its earthy flavor, so it must be used at a low level. Beetroot extract has health-promoting effects. In addition to the antioxidant activity provided by betacyanin, it contains vitamins, minerals, and blood pressure lowering nitrate. In-vitro studies show that betalain compounds have antioxidant and anti-inflammatory properties that reduce the level reactive oxygen species (ROS) and potentially decrease genotoxic damage (Clifford et al., 2015).

Caramel

Sugars, such as glucose, fructose, and sucrose heated under varying conditions, are used to make colorants called caramels, for use in a variety of food applications. The chemistry of this process is based on the Maillard reaction, initially described over 100 years ago (Louis Maillard). Colorants are produced by heating the sugar alone or in combination with other compounds including acid, base, and/or various salts. The color of caramel product can be varied from amber (reddish orange) to brown. Caramel colorants also contain compounds that contribute a range of flavors to food applications (Table 8.2). Soft drink beverages represent major applications for caramel colorants. It may surprise you that beer and other alcoholic beverages,

Table 8.2 Caramel colorants and food applications.

Class	E number	Description	Properties	Application
I	150a	Plain Caramel	Strong after taste	Whiskey, cookies
II	150b	Caustic Sulfite Caramel	Mild flavor	Tea, wine, some whiskeys
II	150c	Ammonium Caramel	Sweet aroma	Beer, BBQ sauce
IV	150d	Sulfite-Ammonium caramel	Mild flavor	Soft drinks, coffee, balsamic vinegar

Adapted from Caramel Color (https://en.wikipedia.org/wiki/Caramel_color#cite_note-10) and Fennema's Food Chemistry fourth ed CRC press 2007.

such as expensive single malt whiskey, also use caramel to create desired color in the product. Similarly, caramel colorants are used in bakery products, chocolate coatings, and syrups. Caramel colorants are large, soluble polymers that absorb a broad spectrum of visible light, resulting in brown color. These complexes are stable to heat and light and compatible with the wide range of pH in food. Essentially, there are four types of caramel products that are used as food colorants. Caramel colorants in the United States are classified as approved food additives by FDA (21 CFR 73.85) and in Europe with an E number that differ by the reactants used to make them (i.e., E150a, E150b, E150c, and E150d) and their applications.

Carmine/Carminic Acid

Carminic acid is a colorant extracted from the Cochineal insect, *Dactylopus coccus*. This insect inhabits the prickly pear cactus *(Opuntia cochenillifera)* native to Mexico, South America, and desert regions of the Southwest United States. The cochineal insect contains large amounts of carminic acid, as much as 20% of its body weight. Carminic acid is an anthraquinone compound containing a glucose molecule attached via a glycosidic bond. Unsaturated rings in the anthraquinone portion strongly absorb light and are responsible for its intense purple color (Fig. 8.5). Production of carmine colorant begins

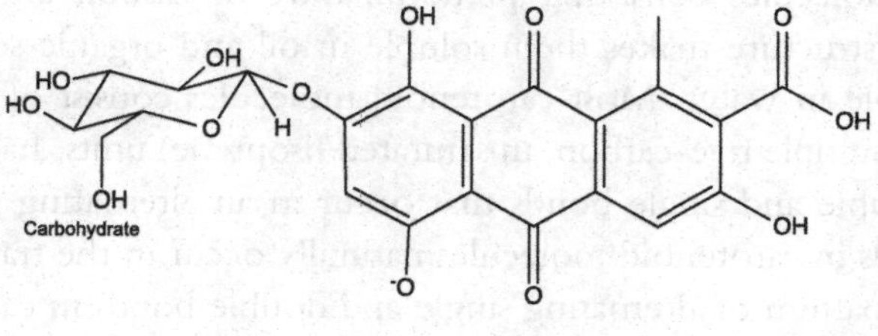

Carmine

Fig. 8.5 Carmine.

with a boiling water extraction of carminic acid from the insect. After filtration of the solution, an aluminum containing salt (alum) is added to precipitate the complex called carmine in the lake form. Aluminum forms a complex with carminic acid that changes the color to scarlet red. The precipitated complex, variably called carmine or cochineal, is dried, ground, and used as a food colorant. Carmine color is stable to a wide range of pH (3.5−8) and has excellent heat and light stability. In the United States, carmine is classified as a colorant exempt from certification by FDA. In Europe, carmine used in food is listed as cochineal, carminic acid, or Natural Red No. 4 and designated as E-120 in their list of approved additives. Major food uses for carmine include processed meat (sausages), artificial crab meat, cakes and pastries, yogurt, and liquors such as Campari™. Carmine is also used as a colorant in red-colored drinks and juices. While there may be a negative reaction to using compounds derived from insects in food, it has greater public acceptance than using synthetic colorants.

Carotenoids

Carotenoids are lipid type molecules synthesized by algae, fungi, plants, and photosynthetic bacteria. Over 700 forms of carotenoids have been identified and are responsible for red, orange, and yellow colors in many foods and flowers. Carotenoid and chlorophyll pigments occur together in the green tissues of plants, specifically in chloroplasts of plant leaves. Biologically, carotenoids protect chlorophyll and its function in photosynthesis by absorbing high energy light. Carotenoids retain their light absorbing properties long after chlorophyll has died away and are responsible for fall leaf color. Examples of foods whose color is derived from carotenoids include carrot, collard, kale, pumpkin, spinach, squash, sweet potato, tomato, and watermelon (Table 8.3). Animal foods, such as salmon and shellfish (lobster and shrimp), also contain carotenoids accumulated via the food chain from organisms that feed upon algae. Carotenoids are long chain, non-polar molecules consisting predominantly of carbon and hydrogen atoms. This structure makes them soluble in oil and organic solvents, but poorly soluble in water. Most carotenoid molecules consist of 40 carbons by joining multiple five-carbon, unsaturated (isoprene) units. Each isoprene contains double and single bonds that occur in an alternating pattern. All double bonds in carotenoid molecules naturally occur in the trans configuration. The pattern of alternating single and double bonds in carotenoids is responsible for the light absorbing properties seen as red, orange, and yellow

Table 8.3 Carotenoid Content of Foods, Expressed as mg per 100 g[a].

Food	α-Carotene	β-Carotene	β-Cryptoxanthin	Lycopene	Lutein + zeaxanthin
Carrots cooked/raw	2.8/3.8	5.3/5.1	0/0	0	0
Collards Cooked/raw	0/0	4.5/2.9	0	0	10.9/4.3
Kale, Cooked/raw	0	2.8	0.1	0/4.5	25.6
Pumpkin, canned	11.7	17	3.6	0	1.0
Spinach-raw	0	5.6	0	0	15.7
Squash, winter baked	0.9	3.5	0	2.9	4
Sweet potato	0	5.5	0	0	0
Tomato Canned/raw	0.1/0.1	0.3/0.1	0	28/2.5	0
Watermelon	0	0	0.2	4.5	0

[a]Adapted from US Department of Agriculture, Agricultural Research Service. USDA Nutrient Database for Standard Reference; Release 28. 2015.

colors (Melendez-Martinez et al., 2007). Carbon atoms at the ends of some carotenoid molecules alpha- and beta-carotene, lutein, and zeaxanthin are arranged in a six-membered structure known as an ionone ring (Fig. 8.6). The ionone rings of beta-cryptoxanthin, astaxanthin, lutein and zeaxanthin are further differentiated by addition of hydroxy (OH) groups. These oxygenated forms are referred to as xanthophylls. Lycopene is a linear carotenoid without ring structures at its ends. Carotenoids also exist in shorter chains forms called apocarotenoids, such as bixin. Bixin (not shown) contains just 25 carbons and has no ring structure.

Carotenoids in food

β-carotene, α-carotene, and β-cryptoxanthin are precursors of vitamin A and the most nutritionally important forms. These carotenoids are converted to vitamin A in the gut by enzymatic hydrolysis of a carbon–carbon bond at the molecule's midpoint, potentially yielding two vitamin A molecules for each β-carotene molecule. However, the efficiency of β-carotene conversion to the active retinol form is low and varies with the food matrix and individual differences in absorption. The best conversion of β-carotene to vitamin A occurs with dietary supplements in which the carotenoid is combined with oil. The oil containing form yields 1 μg of retinol for 2 μg

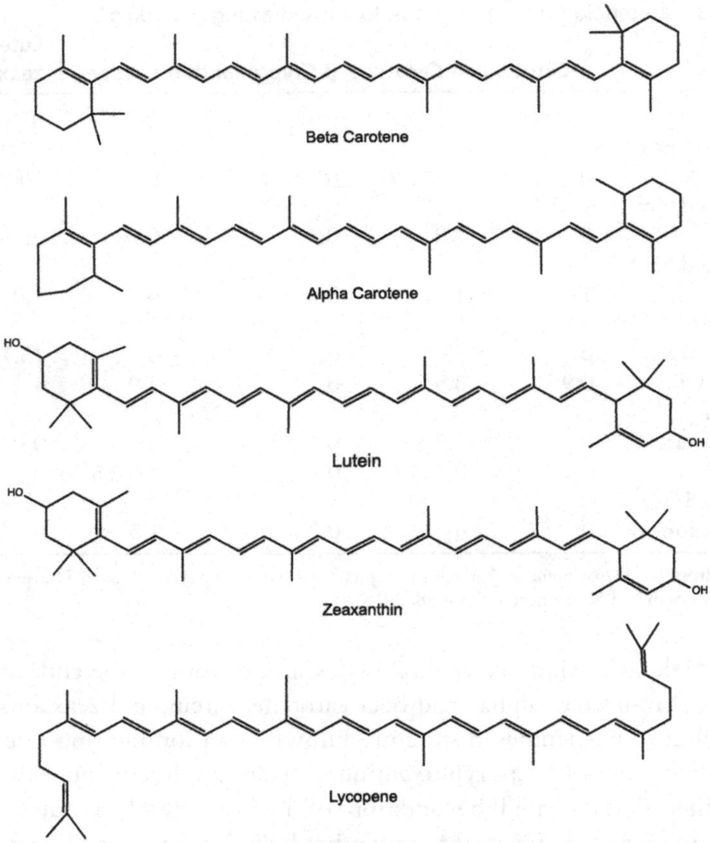

Fig. 8.6 Carotenoids.

of β-carotene. Conversion efficiency is considerably lower for β-carotene in a food matrix. Typically, it requires 12 μg of β-carotene to yield 1 μg of retinol. Conversion rates for α-carotene and β-cryptoxanthin carotenoids capable of generating retinol in food are even lower. The conversion ratio for these carotenoids is approximately 24 μg to 1 μg of retinol. Many red, orange, and yellow fruits and vegetables owe their color to carotenoid pigments. Examples of foods with high levels of carotenoids include tomatoes, sweet potatoes, and pumpkins. The level of carotenoids reaches a peak during ripening and maturation. Tomato, for example, transitions from green to bright red as production of the red lycopene carotenoid increases to its maximum. Green vegetables, such as spinach, peas, and kale, are also high in carotenoids, but their color is masked by high levels of chlorophyll. Salmon, lobster, and shrimp are among the few animal foods containing carotenoids, primarily as astaxanthin. As a result of eating aquatic plants, salmon accumulate carotenoids in their flesh, giving it a red color. Shellfish accumulate carotenoids in their exoskeleton, but are gray in the raw state because they are bound to proteins that block light absorption. The red to orange color of cooked shellfish becomes visible after cooking denatures the proteins and allows carotenoids to absorb light. While the color of carotenoids is reasonably stable in foods, the high content of double bonds makes them susceptible to chemical change. Isomerization of double bonds in the native carotenoid molecule from an all trans to a mixture of both cis and trans double bonds, results from high heat treatment such as frying or by exposure to intense light. Carotenoid double bond isomerization alters its light absorption characteristics and therefore its color. Double bond isomerization has a negative effect on β-carotene's potential for conversion to vitamin A and decreases its conversion by 50% or more. Conversely, heat treatment of carotenoid-containing foods can have a positive effect on the amount of carotenoid absorbed from the gut. For example, canning tomatoes used in making sauce, increases lycopene availability by two-fold or more (Table 8.3). Increased carotenoid availability after heat treatment is due to breakdown of cellular vacuoles stored within plant tissues. Oxidation is another cause of carotenoid destruction in food. It can occur either by exposure to oxidizing chemicals (e.g., oxygen and hydrogen peroxide) or as a result of free radical mediated lipid oxidation. The high double bond content of carotenoids makes them susceptible to free radical-caused lipid oxidation. Carotenoid peroxides formed from lipid oxidation undergo decomposition reactions resulting in loss of color and off flavors. Carotenoids are also substrates for oxidation caused by the enzyme lipoxygenase and have similar outcomes, such as loss of color, vitamin content, and generation of off flavors.

What are the health promoting effects of carotenoids?

Carotenoids are essential to several biological functions. Specifically, they are needed for normal growth and development, immune system function, and vision. Vitamin A derived from carotenoids is essential in maintaining normal vision. Evidence is also accumulating about the potential of carotenoids in health promotion and disease prevention. These attributes include lowering the risk of some cancers and preventing eye disease. There are some questions, however, regarding mechanisms by which these positive outcomes are achieved. Carotenoids are potent antioxidants as determined by in vitro tests, but the contribution of this chemical activity to disease prevention in vivo is unclear at present (Carotenoids, Linus Pauling Micronutrient Information Center, 2016).

Carotenoids and cancer

Prostate cancer is the most common form of cancer among men. Based on in vitro testing and observational studies, it has been suggested that dietary lycopene lowers risk for prostate cancer (Story et al., 2010, Viera et al., 2016). However, further examination of larger pooled data regarding the link between dietary lycopene and risk of prostate cancer was inconclusive. Similarly, taking supplemental forms of lycopene and other antioxidants was not effective in controlling disease progression. The association between dietary carotenoids and risk of developing lung cancer has also been studied. Risk of lung cancer was 20% and 15% lower for groups with highest consumption of β-Cryptoxanthin and lycopene, respectively (Gallicchio et al., 2008). In a rather startling contrast, dietary supplements of β-carotene given to ex-smokers and asbestoses workers was found to increase the risk of lung cancer by 16% to 28%, depending on the amount of β-carotene taken and length of the trial. As a result, high dose dietary supplementation of β-carotene is not recommended for ex-smokers and others with high cancer risk (Omenn et al., 1996).

Eye disease and macular degeneration

The incidence of blindness in elderly, called macular degeneration, is likely caused by long term exposure to light in the blue part of the visible spectrum. Blue light damages the macula, a part of the retina responsible for sharpness of vision in the central part of the field. The eye's retina contains protective lutein, zeaxanthin, and meso-zeaxanthin that are only derived from the diet. The anti-oxidant properties of these carotenoids is responsible for protection against light-induced damage of the retina. Analysis data from very large studies, such as the Nurse's Health Study (63,443 women) and Health Professional Health studies (38,603 men), found a 41% lower risk of Age-Related Macular Degeneration (AMD) for those receiving the highest level of lutein and zeaxanthin in their diet. Dietary supplements are also effective in reducing the risk of AMD. Supplements containing lutein, zeaxanthin in combination with antioxidant vitamins also substantially reduced the risk of developing AMD.

Carotenoids as alternatives to synthetic colorants

Carotenoids as alternatives to synthetic colorants have potential advantages, including thermal stability and antioxidant activity. β-Carotene, canthaxan-thin, and astaxanthin obtained by chemical synthesis methods are classified as Exempt from Certification and can be used at controlled levels in foods.

Additionally, the phrase Color Added must be included on the package. Astaxanthin, for example, is limited to animal feed and widely used for farmed salmon. Synthetic carotenoids are used in making margarines and oils. Extracts of paprika, red peppers, bell peppers, and carrots are approved for use in the United States Extracts of marigold flowers containing high levels of lutein are approved for use in the United States, but not the EU. Corn endosperm oil and dried algae meal are used in poultry and fish (salmon) feeds. Annatto is an orange–yellow colorant derived from the seeds of a tropical tree (*Bixa Orellana*). The compound responsible for its colorant properties is a shorter chain carotenoid (apocarotenoid) called bixin. Annatto is added to food as a colorant and for its sweet nutmeg-like aroma and peppery taste. It is used in processed foods such as butter, margarine, and cheese. Annatto is most widely used to provide the red-orange color to cheeses, such as Cheshire and Cheddar. Annatto colorant is stable to thermal processing and its use as a color additive is exempt from certification by FDA. Crocin is a carotenoid responsible for the golden yellow color in the saffron spice. Saffron is very expensive because it is only obtained by harvesting the stigmas of crocus plants (*Crocus sativus*). The plant flowers briefly once a year and the yield of saffron per plant is very small. It is estimated that stigmas from 70,000 flowers are required to obtain one pound of saffron. Given the extensive and costly inputs required to produce saffron, it is no wonder that its price per pound approaches that of gold. The high value of saffron makes adulteration attractive and lower cost examples are often diluted with paprika or turmeric. Saffron contains several carotenoids (e.g., zeaxanthin and lycopene) and numerous other compounds that contribute to its flavor, but its golden color results from the crocin content. There are two major forms of crocin in saffron, oil, soluble α-crocin and its glycosylated, water soluble form, α-crocetin. Crocin is stable to most heat treatments, but the color of saffron changes if anthocyanins and other pigments are destroyed by the treatment. Consuming saffron is proposed to have a number of health promoting benefits and is used in traditional medicine. Saffron carotenoids are effective antioxidants as determined by in vitro testing. However, there is little in vivo evidence at present to support these claims.

Chlorophyll

Chlorophyll is the component of plants and algae responsible for their green color. It is a porphyrin molecule that is similar in structure to heme found in the hemoglobin and myoglobin of animals. The chlorophyll

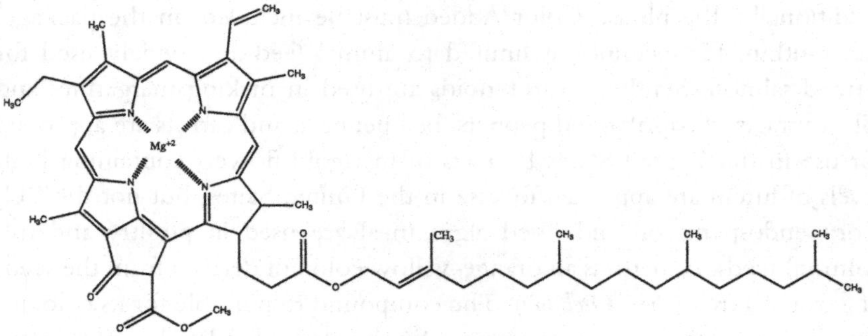

Fig. 8.7 Chlorophyll.

molecule is composed of four pyrrole rings, each containing a nitrogen atom oriented toward the center of the molecule (Fig. 8.7). Pyrrole rings are connected by methyl groups that bridge four pyrroles and make a tetrapyrrole structure called a porphyrin. Nitrogen atoms in the center of chlorophyll coordinate binding of a positively charged metal ion, typically magnesium. A 20 carbon long aliphatic chain called the phytol group is attached to a carboxylic acid group via an ester bond. The phytol group is very hydrophobic, making chlorophyll soluble in fat and poorly soluble in water. The ester link between chlorophyll and its phytol group can be chemically hydrolyzed by strong acid or base or by action of the enzyme, chlorophyllase. Chlorophyll minus the phytol group is water soluble. Chlorophyll exists naturally in two forms called chlorophyll a and b. These forms are differentiated by the presence of a methyl (CH_3) group in chlorophyll a and a formyl group (H–C=O) in chlorophyll b. Chlorophyll a occurs ubiquitously in plants, algae, and cyanobacteria, but the b form occurs primarily in plants. Chlorophyll a and b have slightly different light absorption properties and the ratio of a to b is about 3:1 in plants. Biologically, chlorophyll functions in photosynthesis by absorbing light energy and transferring it to the chloroplast. Photosynthesis traps energy from the sun and makes it available for use in the chemical conversion of carbon dioxide and water into sugars. Chlorophyll undergoes senescence-induced changes at the end of the plant's life cycle. These include opening of the porphyrin ring, loss of the ability to absorb light, and hydrolysis of the phytol group by the enzyme chlorophyllase. Plant and tree leaves lose their green color in the fall as chlorophyll is degraded, leaving carotenoids to display their red, orange, and yellow colors.

In food, chlorophyll color is sensitive to thermal processing and light. Green beans, for example, change from dull to bright green after a brief dip in boiling water followed by a dip in cold water. This process, known as blanching, causes denaturation of chloroplast-associated proteins, provides greater exposure to light, and a desirable bright green color. More intense heat treatments, such as canning, changes chlorophyll color to an undesirable olive brown. Olive brown color occurs when chlorophyll's magnesium ion is displaced by hydrogen ions from weak acids. The source of hydrogen ions responsible for this color change is not well established, but it likely comes from release of weak acids in plant tissues. For example, acetate released by heat-induced hydrolysis of its ester link to other molecules becomes acetic acid and a source of hydrogen ions and displaces the magnesium in chlorophyll. This chemistry can be demonstrated with peas or green beans by adding a pinch of baking soda (sodium bicarbonate) to the cooking water. Green color is preserved by sodium bicarbonate's ability to neutralize acid. The green color of vegetables such as spinach, broccoli, and chard will fade or become grayish-brown after prolonged storage. These color changes are accompanied by development of off flavors and loss of nutrient value. The chemistry responsible for this change is similar to that occurring as part of plant senescence, principally resulting from chlorophyll's action as a photosensitizer. Light energy absorbed by chlorophyll can produce the highly reactive form of oxygen known as singlet oxygen. Singlet oxygen, a significant Reactive Oxygen Species (ROS), damages food components in two ways. First, it is highly reactive toward molecules containing carbon-carbon double bonds. It destroys lipids, proteins, and essential micronutrients such as vitamins, essential fatty acids and amino acids. Second, singlet oxygen promotes formation of free radical ions that cause lipid oxidation and results in off flavors.

Can chlorophyll be used as an alternative to artificial food colorants?

Chlorophyllin is a semisynthetic chlorophyll created by addition of a mixture of sodium and copper salts. Chlorophyllins provide green colorant with good thermal and light stability. Their colors are also more resistant to change under acidic conditions and have good water solubility. Sodium-copper chlorophyllin is an approved food colorant in the United States as exempt from certification. Its use is limited to dry mix, citrus based beverages at a maximum level of 0.2%. Chlorophyllin is also used as a colorant for butter and other high fat foods (Wrolstad and Culver, 2012).

Chlorophyll and health

Native chlorophyll and chlorophyllin derivatives bind toxins, most notably aflatoxin, and block carcinogenic effects in experimental animal models (Egner et al., 2003). The anticancer effect of chlorophyllin results from formation of a tightly bound complex with toxins such as aflatoxin which is a known carcinogen and cause of liver cancer. The complex of chlorophyllin and aflatoxin is poorly absorbed from the gut and reduces the amount of toxin delivered to the liver. Similarly, both forms of chlorophyll provide protection from toxins in cigarette smoke and smoked meats by forming complexes that limited gut absorption (Nagini et al., 2015). It has been known since the mid 1950s that chlorophyllin has a therapeutic effect in accelerating wound healing when included as part of topical treatments. Chlorophyllin slows the growth of anaerobic bacteria and allows the healing process to proceed more effectively.

Phycocyanin

Phycocyanin is a pigment-protein complex that biologically functions cooperatively with chlorophyll in photosynthesis. Specifically, phycocyanin increases the efficiency of chlorophyll's oxygen production under low light conditions. Phycocyanins are blue in color which results from their absorption of red–orange wavelengths of light. They are found in cyanobacteria, a type of blue–green algae, that derives its energy from photosynthesis. Phycocyanobilin (Fig. 8.8) is responsible for blue-green color of the organism. Spirulina is the biomass of *Spirulina platensis* that is grown and harvested for its nutrient content. The dried and ground material contains 60% protein, 24% carbohydrate, 8% fat, and some vitamins and minerals. Spirulina is sold as a dietary supplement for humans and feed for some chickens and farmed fish.

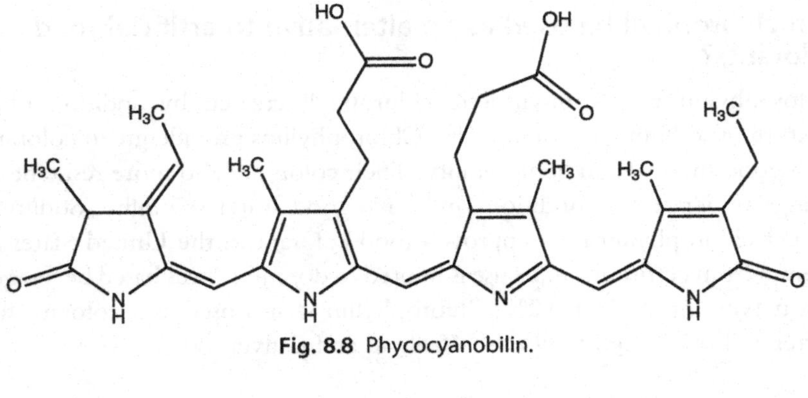

Fig. 8.8 Phycocyanobilin.

However, the safety of spirulina has come into question due to possible presence of microcystin, a toxin produced by the bacteria. Siprulina may also contain heavy metals such as lead, mercury, and arsenic. Microcystin toxins are commonly encountered in lakes and marine environments as a result of algal blooms. Microcystin toxins kill fish and cause health problems from those swimming in contaminated waters. In food, purified phycocyanin is one of few natural compounds that provides a blue color. It was approved by FDA as exempt from certification in 2013. The applications for phycocyanin as a colorant are limited because of its sensitivity to heat treatment unless sugars (preferably fructose) are present. Additionally, its blue color is degraded by pH environments outside of the range of 5.0—6.0. Food applications of phycocyanins include candies, ice cream, and other dairy products.

Colorants exempt from certification

Colorants classified as exempt from FDA certification pigments come from natural sources, including plants, animals and minerals. Natural substances, such as chlorophyll extract modified by adding sodium and copper salts to create Sodium-copper chlorophyllin, are also considered as exempt from certification. A list of colorants exempt from certification and CFR and E numbers are in included in Table 8.4. The term "exempt" means that each batch of these substances need not be tested by FDA. However, each substance must be approved by FDA before they can be used in food.

Heme

Myoglobin is a heme-containing protein responsible for carrying oxygen in muscle cells. It becomes a source of color in meat. Heme is a porphyrin molecule that is very similar in structure to chlorophyll. The heme porphyrin is composed of four pyrrole rings, one at each corner of the molecule (Fig. 8.9).

Each pyrrole consists of a 5 membered ring structure containing four carbons and a nitrogen. They are connected to each other through a methyl bridge and oriented with nitrogen facing the center. The function of nitrogen atoms is to hold a metal ion in the case of heme. Heme molecules are tightly bound to a protein molecule called the globin. A combination of heme molecule and a specific form of globin creates the functional proteins known as myoglobin, hemoglobin, and cytochrome. Biologically, hemoglobin carries oxygen from the lungs to myoglobin in muscle cells. It

Table 8.4 Colorants exempt from certification.

Colorant	Color	21 CFR #	E number	Use
Anthocyanins	Red-purple			
Grape skin extract		73.170	E163	Beverage
Grape color		73.169	E163	Non-beverage
Betalain (Beet powder)	Red	73.40	E162	General
Caramel	Brown	73.85	E150a-d	General
Canthaxanthin	Orange	73.75	E161g	General
Carmine	Red	73.100	E120	General
Cochineal Extract	Red	73.100	E120	General
Carotenoids	Yellow to			
Annatto extract	Orange to Red	73.30	E160b	General
Astaxanthin		73.35	- - -	Fish Feed
β-Apo-8′-carotenal		73.90	E160a	General
β-Carotene		73.95	E160a	General
Carrot oil		73.300	- - -	General
Paprika		73.40	E160c	General
Saffron		73.500	E164	General
Chlorophyll	Green			
Sodium-copper chlorophyllin		73.125	E141	Citrus-based dry beverages
Phycocyanin (Spirulina) cyanobacteria extract	Blue-green	73.530	- - -	Gum, candy, ice cream, frozen desserts
Tumeric Curcuminoid	Yellow	73.600	E100	General
Inorganic Colorants				
Ferrous Gluconate	Yellow-gray	73.160	E579	Ripe Olives
Ferrous Lactate	Green-white	73.165	E585	Ripe Olives
Iron Oxide	Yellow-orange, brown	73.200	E172	Hard & soft candy, Sausage casings
Titanium Dioxide	White	73.575	E171	General

Adapted from Sigurdson, GT.,. Tang, P., Giusti MM. 2017. Natural colorants: food Colorants from natural sources. Annu. Rev. Food Sci. Technol. 8:261–280.

also carries carbon dioxide back to the lungs for elimination. Myoglobin is responsible for holding onto oxygen until it is needed in respiration. This is the physiological process by which muscle cells breakdown glucose to make the energy required for contraction. Fresh meat color is strongly affected by the presence or absence of oxygen in myoglobin and the oxidation state of the iron atom in its heme group. A summary of these effects is shown in Fig. 8.10. Fresh beef, for example, has a desirable cherry red color

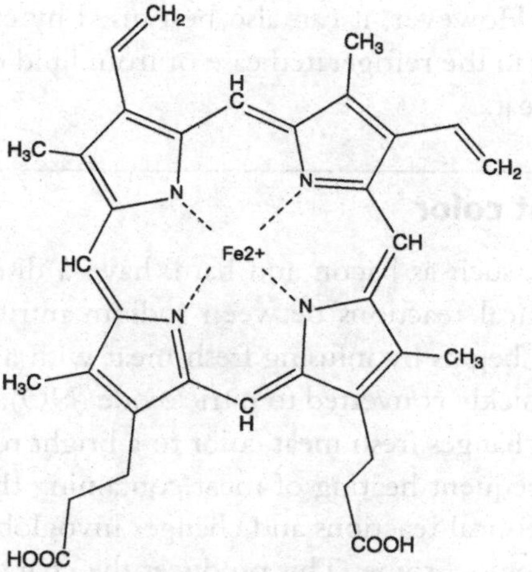

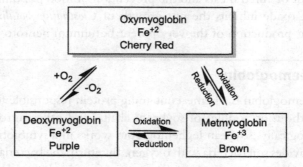

Fig. 8.9 Heme structure.

Oxymyoglobin
Fe^{+2}
Cherry Red

$+O_2$ / $-O_2$

Reduction / Oxidation

Deoxymyoglobin
Fe^{+2}
Purple

Oxidation
Reduction

Metmyoglobin
Fe^{+3}
Brown

Myoglobin Color in Fresh Meat
Adapted from Mancini and Hunt 2005

Fig. 8.10 Meat color.

called oxymyoglobin. This occurs when oxygen is bound to myoglobin and its iron atom is in the ferrous (Fe^{+2}) state. If oxygen is lost, the meat becomes a less desirable purple color called deoxymyoglobin as long as the iron atom remains in the ferrous state. The change from cherry red to purple is a reversible transition, depending on the amount of oxygen present. If the iron atom is oxidized to the ferric (Fe^{+3}) state, meat becomes an undesirable brown color known as metmyoglobin (Mancini and Hunt 2005, Suman and Poulson 2013). It is possible to reduce ferric iron back to the ferrous state and regain meat's red color by addition of a reducing agent such as ascorbic acid or nitric acid. Fresh beef that is brown in color may be a result of microbial spoilage. However, it can also be caused by exposure to intense fluorescent lighting in the refrigerated case or from lipid oxidation reactions occurring in the meat.

Cured meat color

Cured meats, such as bacon and ham, have a distinctive pink color produced by chemical reactions between sodium nitrite and myoglobin. The curing process begins by infusing fresh meat with a sodium nitrite solution where it is quickly converted to nitric oxide (NO). Nitric oxide binds to myoglobin and changes fresh meat color to a bright red known as nitroslymyoglobin. Subsequent heating of meat containing the NO-myoglobin complex causes chemical reactions and changes myoglobin to a compound known as nitrosylhemochrome. This produces the characteristic pink color of cured meats. Reactions involving nitric oxide are also responsible for the unique flavor of cured meats and the prevention of food poisoning. Specifically, nitric oxide inhibits the out-growth of *Clostridium botulinum* spores and prevents production of the very potent botulinum neurotoxin.

Leghemoglobin

Leghemoglobin is a heme-containing protein responsible for carrying oxygen in the root nodules of soybean, alfalfa, and other nitrogen-fixing plants. Biologically, soybean leghemoglobin works in a symbiotic relationship and provides soil bacteria with oxygen. In return, the bacteria synthesize heme which the soybean plant uses in nitrogen fixation. Recently (2016) the Impossible Foods™ company created a food product that looks and tastes like a hamburger, using soy leghemoglobin and other non-meat ingredients. Unlike veggie burgers of the past, leghemoglobin gives uncooked Impossible Burger™ the red color of fresh meat. When cooked, Impossible Burger™ turns brown and has the aroma, taste, and texture of a real beef burger. Impossible Burger™ superior meat-like flavor results from the chemical reactions involving heme supplied by leghemoglobin. Another interesting aspect of this product is that leghemoglobin protein does not come from soybean plants, it is made using recombinant protein technology. The gene encoding for soybean leghemoglobin is incorporated into yeast (*Pichia pastoris*) and produces large quantities of the protein via fermentation. A 3 oz serving of Impossible Burger™ contains 21 g of protein (leghemoglobin, wheat and potato), 13 g of fat (coconut), and 470 mg of sodium, and about 220 calories. Thus, the nutritional profile of Impossible Burger™ is similar to a beef burger, but contains no cholesterol. Its fat content is derived from coconut lipid that contains mostly saturated chain fatty acids.

Is leghemoglobin safe to eat?

The short answer is yes. Leghemoglobin is found in soybean root nodules. This part of the soy plant is not typically eaten and initially raised questions from the FDA regarding its safety. However, related heme proteins found in soybeans and bean sprouts have long been eaten without ill effect. Leghemoglobin produced in yeast was found to have little or no allergenic potential. Additionally, leghemoglobin is not classified as a GMO (Genetically Modified Organism) substance. Leghemoglobin is made by recombinant technology. This process is routinely used to make enzymes for food use (e.g., chymosin) and biologicals for use as medicine. Impossible Food™ requested and received approval of the leghemoglobin used in their products as GRAS (Generally Recognized As Safe) from FDA.

Heme and health

Iron is an essential mineral nutrient available in two forms from food: heme iron or non-heme iron. There are pros and cons concerning each type of iron source. Heme iron is complexed to myoglobin and hemoglobin and found in animal foods, such as meat, poultry, and fish. Iron in animal foods is more available than non-heme iron in plant foods. Meat myoglobin is broken down by the digestion process and approximately one-third of the iron is absorbed from the gut. In contrast, iron absorption from non-heme sources is about half that from myoglobin. The difference in iron absorption may not represent a risk for anemia unless the individual is on a strict vegetarian diet. Non-heme iron is the form found in plant foods, such as spinach, nuts, and beans. While the availability of iron from plant sources is lower, some say this may be a good thing. Western diets tend to have higher amounts of meat and thus higher iron intakes. Iron is a pro-oxidant mineral that potentially increases the level of reactive oxygen species (ROS), such as hydroxyl radical ($\cdot$OH), in blood and other tissues. Hydroxyl radical is a cause for concern because of its ability to react with and modify lipids, proteins, and most importantly, DNA. Hydroxy radical can modify DNA bases or cause breakage of the polymer. These modifications can be equivalent to a mutation having genotoxic effects. Based on correlational data, the risk of colorectal cancer is higher for those who consume more red meat (Aykan, 2015).

Minerals (inorganic food colorants)

Several mineral elements compounds are used as colorants in food. Specific examples include: titanium dioxide, iron oxide, aluminum, silver, and gold (Table 8.4). Titanium dioxide is a white powder used in foods, food ingredients, and pharmaceuticals through tablet coating. It has a high refractive index that contributes a bright white color and opacity. It is used with peroxides to brighten flour, powdered sugar, and dairy products. Its uses in food include hard candies, chewing gum, and mints. Calcium carbonate is an alternative to titanium dioxide for providing white color to many of the same applications. It provides firm texture to canned fruit and vegetables. Iron oxides are used to provide heat-stable red, yellow, black colors for candy, sausage casings, and pet food. Iron oxides are approved for use as alternatives to FD&C Red #40 and Yellow #5 synthetic colorants. Finely powdered metals like aluminum and gold are used to make colorant lakes. Metal powders are typically mixed with synthetic dye and other materials to form the lakes of various colors.

Artificial food colorants

Artificial colorants are synthetic substances used to provide uniform color and enhance the appeal of food. They have advantages over natural substances in stability and ease of blending to make a desired color. Foods containing artificial colorants include farmed salmon, yogurt, margarine, butter, cheese, ice cream, breakfast cereals, peanut butter, salad dressing, smoothies, and candy. Artificial colorants are also extensively used in cosmetics and pharmaceuticals. Unfortunately, the history in the late 19th and early twentieth century of artificial colorants is a sad tale of toxic substances used out of ignorance or deliberate deception and adulteration. During this time, there was little restriction or oversight of substances used to color food. Examples of toxic substances used to color food include textile dyes made from coal tar, lead and mercury compounds to color cheese, and copper, chromium, and arsenic compounds to color tea leaves. Processing practices were also lax in this era. A best-selling book in 1905 by Upton Sinclair entitled *The Jungle* exposed unsanitary and unethical practices of the meat packing industry. It was not uncommon for sick or dead animals to be processed into meat products. As a result of Upton's book and publicized stories regarding the harm caused by toxins and deceptive practices, public uproar prompted Congress to take action. Congress passed the first law (Pure Food and Drug Act of 1906) to prohibit buying and selling of adulterated or fraudulently labeled food and drugs. The legislation was titled "act for preventing the manufacture, sale, or transportation of adulterated or misbranded or poisonous or deleterious foods, drugs, medicines, and liquors, and for regulating the traffic therein, and for other purposes". Major accomplishments of the law included creating the Food and Drug Administration, defining misbranding and adulteration, and essentially requiring truth in labeling for food and drugs. A second major federal legislation, The Food Drug and Cosmetic act of 1938, gave FDA the authority to oversee the safety of food, drugs, and cosmetics. The law distinguishes between toxic substances that naturally occur in food and those that are intentionally added. It contains criminal and civil law that enabled the government and individuals to seek penalties for instances of misuse or adulteration. The 1938 Act established the first set of standards for artificial colorants with the effect of reducing the number of artificial colorants from 700 to seven. Currently, nine artificial colorants are approved for use in foods, cosmetics, and pharmaceuticals. FDA requires every batch of synthetic food colorant to be tested to insure it meets standards set by the agency. Approved substances are listed as certified color additives (Table 8.5).

How are safe levels for artificial colorant use established?

The safety of artificial colorants is determined through a process of assessing their risk to cause harm. Experiments are performed using animals to determine the maximum amount of substance consumed on a daily basis, without a causing an effect on their health. That amount of substance is defined as No Observed Adverse Effect Level (NOAEL). The dose corresponding to its NOAEL, is divided by 100 to establish the Acceptable Daily Intake (ADI) level for humans. Thus, ADI values are set using the assumption that we are 100 times more sensitive to the substance than test animals.

Artificial colorants come in two forms: dye or lake. A dye is simply the water-soluble substance used to color liquids. Approved artificial colorants in the United States are classified as certified color additives by FDA and assigned a specific Food Drug and Cosmetic (FD&C) number (Table 8.5). A lake is a dye that has been complexed with other substances, such as salts, to make them dispersible in oils and fats. The term lake is historically derived from the word "lac" which is a resinous secretion of the insect, *Laccifer lacca*. Lakes are formed by electrostatic attraction between the charged dye and an oppositely charged ion such as calcium (Ca^{+2}) or aluminum (Al^{+1}). Red

Table 8.5 Certified color additives FD&C and E numbers[a].

FD&C number	E number	Common name	Use
FD&C Blue No. 1	E133	Brilliant Blue	Beverages, frozen desserts, frostings
FD&C Blue No. 2	E132	Indigotine	Baked goods, snack foods, ice cream
FD&C Green No. 3	No E number	Fast Green	Cereal, baked goods, ice cream
Orange B	No E number		Only in hot dogs & sausage casings
Citrus Red No. 2	No E number		Only in orange peel
FD&C Red No. 3	E123	Erythrosine	Confections, beverages, frostings, ice cream
FD&C Red No. 40	E129	Allura Red	Cereal, beverages, gelatin, dairy products
FD&C Yellow No. 5	E102	Tartrazine	Confections, snack foods, baked goods
FD&C Yellow No. 6	E110	Sunset Yellow	Cereals, snack foods, beverages, gelatin

[a]Color additives certified as both dyes and lakes are indicated in bold.
Adapted from FDA Color Additives (2018).

number 40, for example, is a red dye known as Allura Red. The red number 40 compound contains a negatively charged sulfonate (SO_3^{-1}) group and its corresponding lake is made by adding sodium (Na^+) or potassium (K^+) ions. Lakes are used for colored coatings such as the outside of M&M™s and gum balls. Lakes also have the advantage of not bleeding color from one part to another. The essential structure of an azo molecule consists of two nitrogen atoms, each of which is bonded to a carbon atom. This structure is the basis for synthesizing numerous other colorants such as sunset yellow, tartrazine, brilliant black and others (Table 8.5). Azo compounds have advantages in regard to colorant synthesis. Their synthesis is easily varied to create a variety of colors. Azo compounds are generally water soluble and stable to the range of temperature and pH conditions occurring in food and its processing.

Certified color additives

Colorants are regulated separately by the FDA from other food additives in the United States. They are classified as certified or exempt from certification. Unlike other additives, there is no GRAS (generally recognized as safe) status for these colorants. Certified colorants are chemically synthesized compounds that are tested for identity and purity by FDA. Exempt colorants are those derived from plant, animal, or mineral sources. The European Union (EU) system for approving use of artificial colorants (and other additives) in food is regulated by the European Food Safety Authority (EFSA). Approved substances are given E numbers. Certified food colorants allowed for use in the United States and the EU are listed in Table 8.5 that lists their FD&C number, E number, and common name. Some colorants have use. For example, FD&C Red No. 3 is not permitted for use in food as a lake. Additionally, Orange B and Citrus Red No. 2 dyes are approved for use in limited applications. Orange B is a red colorant permitted for use only in hot dog and sausage casings. Citrus Red No. 2 is an orange colorant permitted for use only to color orange peelings. The artificial dyes providing blue color are FD&C Blue No. 1 and No. 2. Their common names are brilliant blue and indigotine, respectively. Indigotine is also called royal blue and is a substantially darker in color than brilliant blue. It is used to provide blue color to a large variety of foods including baked goods beverages, dairy products, jams, jellies, liqueurs and ice cream.

Do food artificial colorants contribute to hyperactivity in children?
The Feingold hypothesis
The proposed effects of eating food containing artificial colorants came to national attention in 1975 when physician and pediatric allergist Benjamin Feingold published a book entitled "Why Your Child Is Hyperactive" (Feingold, 1975). This widely read book asserts that hyperactivity in children is linked to food additives. A second book in 1979, A Cookbook for Children with Hyperactivity, proposed that eliminating all additives (colorants and flavorings, preservatives, and salicylates-aspirin) in a diet would alleviate hyperactive behaviors (Feingold, 1979). However, a review of several studies found Feingold's diet to be ineffective in alleviating hyperactive behavior (Kavale and Forness, 1983).

Update
The link between artificial food colorants and hyperactive behavior in children has been the subject of considerable debate and research. Results from studies regarding the potential link between artificial colorants and

hyperactive behavior in children have been reviewed by governmental agencies: FDA in the US and EFSA in the EU. A possible link between artificial colorants and hyperactivity was reported by McCann et al. (2007). They found increased hyperactivity for some children consuming foods containing a blend of artificial colorants, compared to the control group who consumed a placebo. The European Food Safety Authority responded by requiring a warning to be placed on food labels for the colorants identified as potentially associated with hyperactive behavior by the McCann study (EFSA, 2009). FDA's review of certified color additives using an advisory panel concluded that "a causal relationship between exposure to color additives and hyperactivity in the general population has not been established" (FDA, 2011). Members of the panel also stated that there may be a relationship between color additives and behavior in a small percent of the childhood population. A review by Arnold et al. (2012) concluded that artificial food colorants are not a major cause of attention deficit hyperactivity disorder (ADHD), but may have a small, but deleterious effect on children with additional susceptibility due to genetic and environmental factors (Millichap and Yee, 2012).

Summary

Humans are hard wired to make visual assessments in a split second. Thus, it is no surprise that color is the primary factor we use in selecting our food. The emerging field of neurogastronomy examines how color, taste, aroma, and texture are interpreted by the brain. This field is gaining prominence among psychologists, foodies, and food companies. Producers of food have long recognized the importance of color and relied upon synthetic substances to insure consistent and stable color in their products. Synthetics colorants have been the substances of choice because of their low cost and ease by which a wide variety of colors can be created. However, consumer preference has shifted away from synthetics to natural substances. The reasons for this change are complex, but include concerns over the potential adverse effects of synthetics and a desire for healthier diets. Natural colorants such as anthocyanin, chlorophyll (phycocyanin) and carotenoids represent promising sources for replacement of synthetic colorants in some food applications. For example, adding sodium to chlorophyll creates a more stable colorant. And, unlike synthetic colorants, flavonoids, chlorophyll, and carotenoids may contribute to health by providing antioxidant activity.

Glossary

Acceptable Daily Intake A measure of the amount of a specific substance in food or drinking water that can be ingested (orally) on a daily basis over a lifetime without an appreciable health risk. ADI is expressed as mg of substance per kg of body weight

Acylation The process of adding an acyl group (R-CO) to a compound. Acyl groups are carboxylic acids, minus the hydroxyl (OH) part.

Artificial (synthetic) colorant A substance that may or may not occur in nature but, is person-made.

Certified Colorants Synthetically produced compounds used to impart color to food and pharmaceuticals. Advantages include stability, intensity, and no change in flavor.

Colorants Exempt from Certification Pigments derived from natural sources including vegetables, animals, or minerals.

Colorant dye A soluble substance used to provide color in transparent liquids and other foods that are mostly water

Colorant lake An insoluble substance that provides color through dispersion of a dye and metallic salt complex. Lakes are more stable than dyes and typically used to color fats and oils

Genotoxicity A property of chemical agents that damages cellular DNA resulting in mutations that may result in cancer or other diseases.

In vitro and In vivo Latin terms meaning taking place; in glass (in vitro), and in a living organism (in vivo), respectively

Leghemoglobin The oxygen carrying protein of nitrogen-fixing plants. It is a functional analog of the animal protein, myoglobin in meat responsible for red meat color and flavor.

Natural colorant A substance derived from plant, animal, or mineral source

Nitroslymyoglobin A complex of myoglobin and nitric oxide that results in a red color to fresh meat

Nitrosylhemochrome A complex of myoglobin and nitric oxide that results in a pink (cured) meat color characteristic of ham and bacon

No Observed Adverse Effect Level (NOAEL) The highest amount of amount of a substance that shows no adverse effect in animal studies. Once a NOAEL is established, it is divided by 100 to give the ADI deemed safe for humans

Photosensitizer A substance that absorbs light and transfers that energy to other molecules.

Pigment A substance that absorbs a portion of the visible light spectrum and provides color as a result of reflected wavelengths

Synthetic colorant (See artificial colorant)

References

Arnold, L.E., Lofthouse, N., Hurt, E., 2012. Artificial food colors and attention-deficit/hyperactivity symptoms: conclusions to dye for. Neurotherapeutics 9 (3), 599—609.

Aykan, N.F., 2015. Red meat and colorectal cancer. Onco Rev. 9 (1). Oncol. 288.

Clifford, T., Howatson, G., West, D.J., Stevenson, E.J., 2015. The potential benefits of red beetroot supplementation in health and disease. Nutrients 7 (4), 2801—2822.

Curcumin, 2016. OSU Micronutrient Information Center. https://lpi.oregonstate.edu/mic/dietary-factors/phytochemicals/curcumin#biological-activities.

EFSA Panel on, 2009. Food additives and nutrient sources added to food (ANS) scientific opinion on the re-evaluation of sunset yellow FCF (E 110) as a food additive. EFSA Journal 7 (11), 1330.

Egner, P.A., Munoz, A., Kensler, T.W., 2003. Chemoprevention with chlorophyllin in individuals exposed to dietary aflatoxin. Mutat. Res. 523—524, 209—216.

FDA, March 30-31, 2011. Background Document for the Food Advisory Committee: Certified Color Additives in Food and Possible Association with Attention Deficit Hyperactivity Disorder in Children.

Feingold, B.F., 1975. Why Your Child Is Hyperactive. Random House. ISBN 0-394-73426-2.

Feingold, B.F., 1979. The Feingold Cookbook for Hyperactive Children. Random House. ISBN 0-394-73664-8.

Gallicchio, L., Boyd, K., Matanoski, G., 2008. Carotenoids and the risk of developing lung cancer: a systematic review. Am. J. Clin. Nutr. 88 (2), 372—383.

He, J., Giusti, M.M., 2010. Anthocyanins: natural colorants with health-promoting properties. Ann Rev. Food Sci. Techn. 1, 163—187.

Kavale, K.A., Forness, S.R., 1983. Hyperactivity and diet treatment: a meta-analysis of the Feingold hypothesis. J. Learn. Disabil. 16 (6), 324—330.

Kharat, M., Du, Z., Zhang, G., McClements, D.J., 2017. Physical and chemical stability of curcumin in aqueous solutions and emulsions: impact of pH, temperature, and molecular environment. J. Agric. Food Chem. 65 (8), 1525—1532.

Khoo, H.E., Azlan, A., Tang, S.T., Lim, S.M., 2017. Anthocyanidins and anthocyanins: colored pigments as food, pharmaceutical ingredients, and the potential health benefits. Food Nutr. Res. 61 (1), 1361779.

Mancini, R.A., Hunt, M.C., 2005. Current research in meat color. Meat Sci. 71, 100—121.

McCann, D., Barrett, A., Cooper, A., Crumpler, D., Dalen, L., Grimshaw, K., Kitchin, E.K., Porteous, L., Sonuga-Barke, E., et al., 2007. Food additives and hyperactive behaviour in 3-year-old and 8/9-year-old children in the community: a randomized, double-blinded, placebo-controlled trial. The Lancet 370 (9598), 1560—1567.

Meléndez-Martínez, A.J., Britton, G., Vicario, I.M., Heredia, F.J., 2007. Relationship between the color and the chemical structure of carotenoid pigments. Food Chem. 101, 1145—1150.

Millichap, J.G., Yee, M.M., 2012. The diet factor in attention-deficit/hyperactivity disorder. Pediatrics 129 (2), 330—337.

Nagini, S., Palitti, F., Natarajan, 2015. Chemopreventive potential of chlorophyllin: a review of the mechanisms of action and molecular targets. Nutr. Cancer 667, 203—211.

Omenn, G.S., Goodman, G.E., Thornquist, M.D., 1996. Risk factors for lung cancer and for intervention effects in CARET, the Beta-Carotene and Retinol Efficacy Trial. J. Natl. Cancer Inst. 88 (21), 1550—1559.

Polturak, G., Aharoni, A., 2017. "La Vie en Rose": biosynthesis, Sources, and Applications of Betalain Pigments. Mol. Plant 11, 7—22.

Renaud, S., de Lorgeril, M., 1992. Wine, alcohol, and the French paradox for coronary heart disease. The Lance 339, 1523—1526.

Sigurdson, G.T., Tang, P., Giusti, M.M., 2017. Natural colorants: food Colorants from natural sources. Annu. Rev. Food Sci. Technol. 8, 261—280.

Story, E.N., Kopec, R.E., Schwartz, S.J., Harris, G.K., 2010. An update on the health effects of tomato lycopene. Ann. Rev. of Food Sci. and Technol 1. https://doi.org/10.1146/annurev.food.102308.124120.

Suman, S.P., Poulson, J., 2013. Myoglobin chemistry and meat color. Ann. Rev. Food Sci. and Tech. 4, 72—99.

Vieira, A.R., et al., 2016. Fruits, vegetables and lung cancer risk: a systematic review and meta-analysis. Ann Oncol. Jan 27 (1), 81—96.

Wrolstad, R.E., Culver, C.A., 2012. Alternatives to those FD&C food colorants. Ann. Rev. Food Sci. Technol. 3, 59–77.

Zhao, C.L., YQ, Y., Chen, Z.J., Wen, G.S., Wei, F.G., Zheng, Q., Wang, C.D., Xiao, X.L., 2017. Stability-increasing effects of anthocyanin glycosyl acylation. Food Chem. 214, 119–128.

Further reading

Carotenoids, 2016. OSU Micronutrient Information Center. https://lpi.oregonstate.edu/mic/dietary-factors/phytochemicals/carotenoids.

Simpson, B.K., Benjakul, S., Klomklao, S., 2012. In: Simpson, B.K., Nollet, L.M.L., Toldra, F., Benjakul, S., Paliyath, G., Hui, Y.H. (Eds.), Natural Food Pigments. Food Biochemistry and Food Processing, second ed. John Wiley & Sons, Inc. Published 2012 by John Wiley & Sons, Inc.

Yellow, S., 2009. EFSA panel on food additives and nutrient sources added to food (ANS) scientific opinion on the re-evaluation of sunset yellow FCF (E 110) as a food additive. EFSA Journal 7 (11), 1330, 2009.

Review questions

1. Define the terms natural and artificial colorant. Give an example of each.
2. What are the advantages of artificial vs natural colorants?
3. Describe the importance of Upton Sinclair's book "The Jungle" to food additives.
4. How is a safe level for artificial colorant use established?
5. Define the term ADI and describe how it is established.
6. What is a "colorant exempt from certification"? Give an example.
7. Define the terms lake and dye.
8. What minerals are used as food colorants? Give two examples.
9. Which natural colorants also provide antioxidant activity?
10. Give 3 examples of foods in which flavonoid compounds are responsible for their color.
11. Name 3 factors that can cause flavonoid color to change.
12. What colorant is typically used in cola and beer? How is it made?
13. Is caramel a natural colorant? Explain your answer.
14. What is the source of the natural colorant carmine? What foods is carmine used in?
15. Name 3 foods that owe their color to carotenoids.
16. What part of the carotenoid structure is responsible for its color?
17. Which carotenoid has vitamin A activity? What foods are high in this carotenoid?
18. What health effects are provided by carotenoids?

19. Give two causes for the loss of carotenoid color in food.
20. Why does the color of raw shrimp change from gray to red-orange after cooking?
21. How does the magnesium ion in chlorophyll affect its color?
22. Explain why blanching green beans in baking soda-water make them bright green.
23. What are the health benefits of chlorophyll?
24. What is heme and how is it important to fresh meat color?
25. What mineral is essential to the color of red meat?
26. What is the source of leghemoglobin used in Impossible Burgers™?

CHAPTER NINE

Food systems and future directions

Learning objectives

This chapter will help you describe:

- Gut microbiome and its role in diet and health
- Influence of food components and additives on the microbiome
- Composition and properties of major food systems
- Properties of protein-rich plant foods
- Protein-rich microbial foods (fungi and algae)
- Novel plant-based animal foods (milk and meat) and edible insects

Introduction

The gut microbiome consists of a large and dynamic population of microorganisms (the microbiota), contributing much more than nutrient digestion to the host. It functions as part of our immune system, eliminates

Introduction to the Chemistry of Food
ISBN: 978-0-12-809434-1
https://doi.org/10.1016/B978-0-12-809434-1.00009-8

microbial pathogens, and produces substances that protect the intestinal lining. Diet can significantly affect microbiota populations and, consequently, the health of the host. High fiber diets increase the population of beneficial organisms, while high nutrient density/low fiber diets adversely affect the symbiotic relationship between microbiota and host. An imbalance of the gut microbiota (dysbiosis) is associated with irritable bowel disease, metabolic syndrome, and diabetes. Recent discoveries regarding the gut microbiome reinforces the connection between diet and health. The old adage "you are what you eat", was never more true.

This chapter reviews the composition of major animal and plant food systems. A detailed profile of macronutrients (i.e., protein, carbohydrate, and fat) and micronutrients (i.e., vitamins and minerals) is provided for meat, milk, egg, soybean, wheat, and rice systems. It also describes the functional properties provided by macronutrients. Proteins, for example, are functionally important to, fat emulsification in processed meats (myosin), egg white foams (ovalbumin), and the coagulum we know as cheese (casein).

The properties of protein-rich plant foods are also described because of their potential to meet the nutritional needs of a population predicted to increase by one-third in next few decades. Protein-rich plant foods offer a healthier food choice because most are a good source of dietary fiber. This group of plant foods includes, amaranth, canola, chia, flaxseed, lentils, peas, oats, peanuts, and quinoa. Additionally, protein-rich foods from novel sources, such as mycoprotein and algal protein, are described. These microbial protein sources are unique in their potential to produce large quantities of high-quality protein. The recent development of novel, plant-based foods as alternatives to traditional animal foods such as milk and meat, represents a major change. Plant-based products such as milk (soy and almond), non-dairy yogurt, and ice cream are increasingly popular in the marketplace. They appeal to consumers preferring vegan/vegetarian diets or needing to avoid lactose due to an intolerance. Novel foods, such as plant-based meat, clean-meat, and edible insects, are also growing alternatives to traditional foods. These alternatives to traditional animal foods have an inherent advantage in production sustainably.

The gut microbiome

The human digestive system (gut) is composed of two distinct regions, including the small and large intestine. The small intestine functions in the digestion and absorption of nutrients from the food we eat. The large intestine (colon) primarily functions to recover water and absorb minerals from indigestible materials before it are excreted. Although the gut has been primarily thought of as a digestive organ, it is populated by a very large number of microorganisms that contribute to health (Valdes et al., 2017, Davenport, et al., 2017). The number of micro-organisms inhabiting the gut (its microbiome) outnumbers the total cells in the human body by a factor of 10. Several trillion cells, referred to as the microbiota, composed primarily of bacteria and some fungi, viruses, and protozoa, inhabit our gut. In contrast to the 23,000 genes in the human genome, organisms of the gut microbiome have a collective genome numbering millions of genes (Ursell et al., 2012). Recently, it has been found that the gut microbiome contributes several beneficial functions to the host. Bacteria inhabiting the colon ferment oligosaccharides and resistant starches producing substances that stimulate immune function and suppress appetite. Beneficial products of microbial fermentation include nutrients such as B vitamins, biotin, cobalamin, folate, nicotinic acid, pantothenic acid, pyridoxine, riboflavin, and thiamin. Amino acids are also produced. Gut bacteria provide a further benefit by degrading toxic and potentially carcinogenic substances, such as N-nitroso compounds (Chassaing et al., 2015). Each individual's microbiome is initially inherited from the mother at birth and from breast milk. The microbial population of vaginally delivered babies resembles that of the mother's vagina within an hour of birth. However, the population of micro-organisms undergoes substantial change in parallel with the child's development and diet (Ursell et al., 2012). In recognition of the microbiome's contribution to nutrition and health, the National Institutes of Health (NIH) launched a multidisciplinary project, "Human Microbiome Project" (NIH HMP, 2015) in 2008 to facilitate research and communicate its findings. The HMP builds upon NIH's human genome project that provided a complete sequence of all genes of the human genome. Its web site represents a way to keep informed of progress in this rapidly developing field.

Prebiotics, fiber, and probiotics

Prebiotics are carbohydrates that are poorly broken down by human digestive enzymes, but serve as fuel for gut microbiota. Prebiotics include, carbohydrates, such as resistant starches, gums, pectins, and oligosaccharides. For example, the fructose-oligosaccharide (FOS) and galactose-oligosaccharide (GOS) in inulin are good substrates for fermentation by gut microbiota and thus termed prebiotic fiber (Slavin 2013). The oligosaccharide found in oats (beta-glucan) is composed of glucose units arranged in

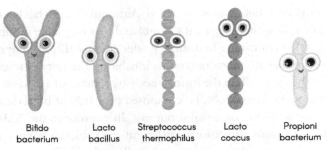

| Bifido | Lacto | Streptococcus | Lacto | Propioni |
| bacterium | bacillus | thermophilus | coccus | bacterium |

Fig. 9.1 Probiotics.

various linkages that resist breakdown by gut digestive enzymes. It is also a good source of prebiotic fiber. Fiber is defined as edible carbohydrate polymers, consisting of 3 or more monosaccharide units, that are resistant to digestion by the host's digestive system. The older definition of fiber referred to carbohydrates known as roughage that cannot be digested or absorbed. It is now known that many types of carbohydrate fiber are good substrates for fermentation by gut microbes and are included in the definition of prebiotics. Foods naturally high in fiber include garlic, onions, leek, chicory, asparagus, artichoke, bananas, sea weed, beans, and whole grains. Health benefits of prebiotic fiber include support of the growth of beneficial *Bifidobacteria* and *Lactobacilli* populations, increased calcium absorption, and promotion of the gut's innate immune system (Slavin, 2013; Carlson et, al., 2018). Probiotics (Fig. 9.1) contain living bacteria similar to the beneficial microorganisms found in the gut. They are taken as oral supplements to improve the population of bacteria in the gut. Probiotic supplements are most often composed of *Bifidobacteria* species that naturally occur in yogurts and other fermented foods.

Importance of the microbiome to health

Carbohydrate fiber and resistant starches are the fermentable substrates that gut bacteria utilize to produce beneficial substances. Among the many products of their fermentation are short chain fatty acids (SCFA), such as acetate, propionate, lactate, and butyrate. SCFA are acidic substances that lower the gut pH and create an environment favorable for the growth of desirable bacteria. Specifically, organisms of the *lactobacilli* and *Bifidobacteria* genera flourish, while the growth of the pathogens such as *Clostridium difficile* is inhibited (Salvin, 2013). SCFA have been shown to ameliorate bowel disorders (irritable bowel disease and diarrhea). However, the mechanism of this beneficial effect has not been established. The gut microbiome provides protection to epithelial cells of the colon by increasing the thickness of its mucus layer and stimulating immune responses that protect against pathogens. Butyrate and propionate fatty acids function by activating receptors in intestinal cells to secrete peptides that suppress appetite and improve glucose metabolism (lower blood glucose level). SCFA in the gut lumen serve as energy sources, contribute to a healthy profile of gut microbes, and increase the absorption efficiency of health-promoting flavonoids and carotenoids (Ríos-Covián et al., 2016; Ercolini and Fogliano, 2018).

Immunity and inflammation

SCFA produced through fermentation are not only a source of energy for the host, but also affect inflammation response. SCFA activate receptors in gut epithelial cells and increase the mucous layer thickness (a protective effect on the gut lining). The net effect of SCFA is a decrease in production of inflammatory factors and increase in production anti-inflammatory cytokines. Conversely, diets low in fermentable substrates result in a thinner mucus layer lining the gut lumen, increasing the susceptibility to infection of intestinal epithelial cells (Cani, 2018). SCFA also circulate to the lungs where they result in reduced inflammation and allergy symptoms (Ríos-Covián et al., 2016; Rooks and Garrett, 2016). Irritable bowel syndrome (IBS) is the one of the most commonly reported gastrointestinal disorders today. IBS affects the large intestine and causes gas, pain, bloating, and diarrhea. While some individuals are severely affected and seek medical attention, most are able to cope through dietary intervention (avoiding the offending food). IBS symptoms are common for those with an intolerance to lactose found cow's milk and other dairy products. Lactose intolerance results from a deficiency in the enzyme beta galactosidase (lactase). Research suggests that a major cause of IBS results from a dysbiosis (alteration of the gut microbiota) that may be mediated by altering its population (Kennedy et al., 2014). SCFA produced by bacterial fermentation may protect against development of colon cancer. Butyrate, in particular, has been found to induce apoptosis (cell death) of cancerous colon cells (Canani et al., 2011; Keku et al., 2015).

Gut-brain axis

The gut microbiome and the human brain are bidirectionally linked and thus termed the gut-brain axis. For example, substances produced by gut microorganisms enter the blood stream and provide signals to the brain that can activate or suppress appetite. Similarly, signals from the brain can simulate the gut to secrete digestive enzymes and activate contraction (peristalsis) in a Pavlovian response to the sight and smell of food. While studies of the microbiome gut-brain axis have been almost entirely conducted in animals, their results suggest the possibility of mediating diet-related diseases, such as obesity and diabetes, through manipulation of the gut microbiome (Niccolai et al., 2019). The gut microbiota has its own circadian rhythm and metabolites produced by gut micro-organisms can alter the host's circadian rhythm. High fat diets have the largest affect on circadian rhythm of the microbiome and the host.

Diet and health

The population of micro-organisms in gut microbiome varies substantially between individuals and changes with age and external factors (e.g., diet, exercise, stress, and medication). Among these external factors, diet is considered to be the most significant in altering an individual's microbiome (Cénita et al., 2014). Notably, the Western Diet is known to change the population of the gut microbiome in negative ways. Foods typical of the Western Diet are highly processed and high in nutrient density. Serving sizes tend to be large and it is not uncommon for a single meal to supply a whole day's calories. Carbohydrates in these foods have greater potential to increase glycemic load. It is therefore not surprising that the Western Diet can lead to diet-related diseases. For example, Western diet is linked to metabolic syndrome, a condition that significantly increases the risk of cardiovascular disease, stroke, obesity and type 2 diabetes (Palmnäs et al., 2014; Ríos-Covián et al., 2016; Zinöcker and Lindseth, 2018). Metabolic syndrome is very common in the United States and affects about one in every six people. The chance of developing one or more conditions related to metabolic syndrome increases with age. Additionally, alterations of the gut microbiome are suggested to be linked to non-alcoholic fatty acid liver disease and psychiatric disorders.

Effects of food additives on gut microbiota

Food additives have been extensively investigated by FDA and regulatory agencies in other countries. While the safety and toxicological properties of additives approved for use in food are well known, relatively little is known about their effects on the gut microbiome. Most information about the effects of food additives on the gut microbiota is limited to animal studies. Food additives, including artificial sweeteners, emulsifiers, antibiotics, and pesticides, have negative affects on the gut microbiome (Roca-Saavedra et al., 2017, Culley et al., 2019). Artificial sweeteners (aspartame, acesulfame K, saccharin, and Splenda) have become very popular and are widely used in carbonated beverages, diet drinks, and other foods to avoid excess calories and weight gain. However, artificial sweeteners at the level equivalent to two to three cans of soft drink, was found to alter the gut microbiome in mice. Specifically, feeding artificial sweeteners to mice and rats decreased the population of beneficial bacteria (*Lactobacilli* and *Bifidobacteria*), reduced insulin tolerance, increased blood glucose level, and resulted in weight gain. Emulsifiers such as carboxymethyl cellulose and polysorbate 80 are widely used as additives in foods such as ice cream, margarines, bread, and other baked products. These emulsifiers are two examples shown to have negative effects on the gut microbiome. Specific effects on mice included a decrease in protective mucus layer of the gut epithelium and increased inflammation. It is proposed that emulsifiers weaken cell membranes of the intestinal wall, allowing micro-organisms to cause infection. Orally administered antibiotics are widely prescribed medications that are targeted to kill organisms causing illness. However, they also kill beneficial organisms, significantly altering and reducing microbial diversity in the gut microbiota. It is of interest to note that antibiotics have been added to livestock feed to improve their weight gain since the 1970s. Antibiotics act as a growth promotor and increase the rate of weight gain over controls without antibiotics. Mice fed low levels of antibiotics showed alterations in the population of their microbiota. These animals had lower levels of beneficial organisms (e.g., *Lactobacilli* and *Bifidobacteria*) and increased levels of pathogens and tended to be obese (Roca-Saavedra et al., 2017). Additionally, residual levels of pesticides (chlorpyrifos) and herbicides (glyphosphate) in foods are suspected to be the cause of similar negative effects on the human gut microbiota (Roca-Saavedra et al., 2017).

Microbiota-directed food (MDF)/Personalized nutrition

Greater knowledge of how the microbiota transforms food components into health-promoting components has presented the possibility of developing foods that promote a healthy microbiome. Microbiota directed foods (MDF) are defined as foods designed to promote a beneficial profile of microorganisms in the gut (Green et al., 2017). For example, the oligosaccharides found in human milk might be added to infant formula as an MDF food. MDF foods represents a challenge regarding how they should be classified by the FDA. However, the concept of adding prebiotic fiber and other types of carbohydrate that promote a heathy microbiome is a potentially beneficial step in the goal of personalized nutrition.

Animal food systems and their composition

Meat, milk, and egg represent the dominant foods in many cultures. A summary of their nutritional content (e.g., protein, carbohydrate, fat, vitamins and minerals) is included in this chapter. Examples of functional properties provided primarily by their proteins and fat is described.

Meat

Eating meat provided an evolutionary advantage to our ancestors through high nutrient content and availability. The word meat commonly means red meats, such as beef, lamb, and pork, but also may include poultry and fish (Fig. 9.2). Meat products available today are vastly different from those a century ago. Most muscle foods were only available as fresh product because of the limited availability of refrigeration. Salting, drying, and fermentation were the predominant means to prevent this highly perishable commodity from spoiling. Meat animal production practices have changed dramatically since the middle of the last century. Selective breeding of meat animals and more efficient production practices have increased quantity and improved quality. During this time period, changes in processing technology led to development of the large variety of ready to eat meat products available today. Meat is an energy-dense food containing 200—250 calories

Fig. 9.2 Meat.

per 100 g, depending on the species. It typically contains 70%–75% water, 20% protein, and 5%–6% fat. The nutritional quality of meat protein is high because of its essential and branched chain amino acid content. Meat is a good source of micronutrients, such as the B vitamins (B_6, B_{12} and niacin), and minerals (iron, zinc, potassium, and phosphorous). Meat is a superior source of iron because it is more efficiently absorbed when complexed with the heme group of myoglobin. Once in the gut, enzymes breakdown myoglobin and release the iron bound to its heme group. The lipid content of meat is the most variable component of its composition. The amount of lipid in muscle foods varies from approximately 2% in fish to 15% or more in beef. Meat lipids are predominantly composed of acylglycerides with very small percent as free fatty acids. Most fatty acids in meat lipids are saturated, i.e., molecules without double bonds between carbon atoms. The level of saturated fatty acid is highest (38.2%) in beef versus 26.6% for chicken and 19.5% for fish. Palmitic (C16:0) and stearic (C18:0) are the predominant saturated fatty acids in all three species. The predominant unsaturated fatty acid is oleic acid (C18:1) for beef, chicken, and fish. Two classes of polyun-saturated fatty acids (PUFAs), known as omega 6 and omega 3 fatty acids, are associated with the risk of cardiovascular disease. Omega 6 fatty acids have a double bond located 6 carbons from the methyl end and include linoleic (C18:2), gamma linolenic (C18:3). Omega 3 fatty acids have a double bond located 3 carbons from the methyl end and include alpha linolenic (ALA), eicosapentaenoic (EPA), and docosahexaenoic (DHA). arachidonic acid (C20:4). The ratio of omega 6 to omega 3 fatty acids in the diet is corre-lated with the risk of cardiovascular disease (CVD) obesity, and diabetes. Specifically a low omega 6 to 3 ratio is associated with a reduced risk of these diseases. Red meats, such as beef, have a high omega 6 to 3 ratio. Fish, like salmon, are one of the few muscle foods with a significant content of omega 3 fatty acids and a low omega 6 to 3 ratio.

Muscle is composed of three distinct classes of protein that are differen-tiated by their function: sarcoplasmic, myofibrillar, and stromal. The sarco-plasmic fraction is analogous to the cytoplasm of other cells, but its composition is specialized to support muscle contraction. Sarcoplasm contains myoglobin and enzymes that biologically function as oxygen carriers and energy providers. Myoglobin is a heme containing protein responsible for fresh and cooked meat color. Myoglobin content varies with animal species and the type of muscle fiber. The color of fresh beef, lamb, pork, poultry, and fish decreases in that order, as a result of their respective myoglobin content. Cooked meat color and flavor results from myoglobin. For example, the distinctive color and flavor of ham and bacon results from heat-induced chemical reactions involving heme-bound additives (i.e., nitrite). Sarcoplasmic enzymes are essential to metabolizing glycogen for energy, and synthesizing new protein, turning over old proteins. Nucleotides, such as ATP (adenosine triphosphate) and GTP (guanosine triphosphate), are molecules that serve as energy-rich stores needed to support contraction. In cooked meat, nucleotides and amino acids serve as major sources of umami flavor. Nutritionally, the sarcoplasm is a source of proteins, vitamins, and minerals. In total, the sarcoplasmic fraction constitutes about 30% of muscle protein. The myofibrillar fraction is composed of proteins that enable muscle contraction.

The principle contractile proteins myosin and actin are responsible for converting chemical energy into movement. In meat, myosin is noted for its exceptional ability to act as an emulsifying agent. This is a functional property essential to making processed meat products. Myosin is a salt-soluble protein. Large amounts of salt (about 5% by weight) is added to make processed meat products. The addition of salt and vigorous mixing maximizes myosin extraction and creates a paste-like product that gels upon heating with little loss of fat or moisture in the cooked product. The myofibrillar fraction constitutes about 50 % of the total protein in muscle. The stromal fraction of muscle is predominantly composed of collagen and elastin proteins. Biologically, collagen is the major component of connective tissue that surrounds muscle fibers and bundles. This tissue provides the connection between bones and is responsible for transmitting the force of contraction that enables movement. Collagen is an unusual protein. Its amino acid composition contains about 33% glycine and 60% proline plus hydroxyproline. Collagen can have negative effects in the eating qualities of whole meat cuts. Specifically, connective tissues in the meat of older beef and lamb animals cause toughness. Collagen-toughness is caused by highly cross-linked collagen molecules that are resistant to breakdown by fast cooking methods such as grilling or frying. High moisture cooking methods, such as boiling or stewing, are required to loosen connective tissue networks and soften meat texture. Collagen extracted from skin, bone, and cartilage has numerous food uses. Purified collagen, also known as gelatin, has several applications in food. Gelatin is used to form reversible gels, emulsify fats, thicken beverages, and form protective coatings. Supplements prepared from purified collagen and its hydrolyzed peptides are suggested to reduce joint pain and strengthen bones for individuals with osteoarthritis. As a whole, the stromal fraction constitutes about 15—20 % of the total muscle protein.

Effects of meat on microbiome

A high level of trimethylamine oxide (TMAO) in the blood is a strong indicator of increased risk for heart attack. A recent clinical study reported high levels of TMAO in trial subjects eating a diet high in red meat (Koeth et al., 2018). TMAO is produced from trimethylamine when gut bacteria digest red meat components of choline, lecithin, and carnitine. Elevated levels of TMAO in blood result from gut bacteria production and decreased elimination in the urine caused by TMAO's inhibition of kidney function. Regular consumption of red meats is associated with an increase the risk of heart attack mediated by the gut microbiome. Evidence suggests that the gut microbiome is involved in the development of cardiovascular disease (CVD) resulting from dysbiosis of the gut microbiota (Fu et al., 2015).

Milk

Milk is legally defined in the United States as the lacteal secretion, free of colostrum, resulting from the milking cows (Fig. 9.3). Whole milk contains about 61 calories per 100 g. Milk from domesticated animals (i.e., cows, goats, and other mammals) has been a source of human food for thousands of years. Early humans learned how to separate milk fat and make butter. It is only since the mid 19th century that commercial production of milk and other

Fig. 9.3 Milk.

dairy products became available. Milk is an energy dense food designed to sustain and support the development of the specie's young. Cow's milk is composed of 4.6% to 5 % lactose, 3% to 4.5 % fat, 3% to 4 % protein, and 1.7 % minerals (principally calcium and phosphorus). The water content of milk is about 85 % (± 2 %), depending on the variation in composition of other components. Milk's largest component, lactose, is a disaccharide composed of galactose and glucose. While lactose is an energy source for most infants and children, it can be a problem. The enzyme lactase (beta galactosidase) is needed to breakdown lactose into constituent monosaccharides. It declines with age and some people carry a genetic mutation that makes them lactase deficient. Without the necessary enzyme, lactose transits to the large intestine where it is fermented by gut bacteria and produces gas, discomfort, and diarrhea. This condition is an example of a food intolerance.

Milk fat is composed of a large number (400+) of fatty acids originating from synthesis by microorganisms in the rumen and secretory cells of the mammillary gland. Seasonal variation of a cow's diet also influences the profile of fatty acids. The fatty acids of milk fat include large proportions of short chain (less than 6 carbons) and long chain (greater than 12 carbons) molecules. Milk fatty acids are predominantly saturated (no double bonds). Myristic and palmitic acids represent the greatest percent of the unsaturated fatty acid group. Some milk fatty acids contain an odd number of carbon atoms and are thus unusual because they are synthesized from 2 carbon units. The carbon-carbon double bonds of unsaturated fatty acids are mostly in the cis configuration, but some occur in the trans configuration, as a result of bacterial action. Oleic acid (C18:1) is an unsaturated fatty acid with a double bond in cis configuration. It makes the single largest contribution to the profile of milk fatty acids. A notable fatty acid found in milk of ruminant animals (cows particularly) is rumenic acid, also known as conjugated linolenic acid (C18:2) or (CLA). The unusual part of CLA is the location and configuration of its two double bonds. The double bonds are located adjacent to one another and thus the name "conjugated". Additionally, one of its double bonds is in the cis configuration and the other is in trans configuration. CLA is notable for its anti-cancer activity in animal studies. CLA inhibits the growth of cancerous cells in the colon (Rodríguez-Alcalá et al., 2017). Cow's milk is the predominant food source of CLA, but also exists at a lower level in beef. The cancer-preventing activity of CLA is somewhat doubtful because the level in cow's milk is only 10% of that required to provide anti-tumor activity in animal studies. Almost all milk fatty acids exist in the form of acyl-glycerols, about 1% as free fatty acids in milk. The lipid component of cow's milk is contained within fat globules that are surrounded by the milk fat globule membrane (MFGM).

Cooling freshly harvested milk is an essential step in milk production because it slows the growth of spoilage microorganisms. However, low temperature can rupture the MFGM and cause release of its lipid and rise to the surface. This phenomenon, known as creaming, was a substantial problem for fresh milk producers until the process of homogenization was introduced. In modern milk production, homogenization forces milk under pressure through small orifices. The shearing action caused by homogenization causes released lipid to form a stable complex with milk proteins and prevents separation. Milk fat is an important component in dairy products such as butter and cheese. Butter is made from the high fat (cream) fraction of milk. A centrifugation process developed over 100 years ago separates the lighter fat component from the aqueous milk fraction. Milk fat is separated into cream with varying levels of fat. For example, light cream, whipping cream, and heavy cream contain approximately 20%, 35%, and 38% fat, respectively. Butter is made by agitating the heavy cream milk fraction. The physical process of churning disrupts the milk globular membranes and releases its lipid content into the milk. The hydrophobic nature of lipid promotes its coalescence into droplets called seeds. Continued churning results in coalescing fat seeds into larger solid masses and expels most of aqueous milk. At this point the solid fat (butter) contains about 10%—15% water and milk protein in the form of a water-in-oil emulsion. Cultured butter is made by adding bacteria (e.g., *Streptococcus* and *Lactococcus*) and allowing fermentation to take place. Cultured butter has an acidic and buttery flavor due to lactic acid and diacetyl produced by fermentation. Sweet butter is made by pressing and packaging the solid fat. Salt can be mixed in (about 1%) prior to pressing to make salted butter. Clarified butter or ghee is made by heating until melted and separating the lipid fraction from the watery and salt-containing material on the bottom of the vessel. The heating process give clarified butter a nutty, richer taste compared to butter.

Cow's milk proteins are grouped into major fractions, caseins and whey proteins that compose about 80% and 20% of total milk protein, respectively. Caseins are composed of several similar, but distinct gene products, alpha, beta, kappa, and gamma. In solution, they spontaneously form a spherical cluster called the casein micelle. Casein proteins have little or no three-dimensional structure, but exist as highly flexible random coils. Caseins are phosphorylated, meaning that phosphate groups are covalently attached to their polypeptide chains, contributing a substantial negative charge. Micelles formed from casein proteins are stabilized by electrostatic interactions between negative charges on the proteins and positively charged calcium ions. Alpha and beta caseins form the core of the micelle while gamma and kappa are distributed on or near the surface. Casein micelles are able to maintain a stable suspension in milk's aqueous environment because hydrophobic caseins (alpha and beta) are buried in the interior away from water. Hydrophilic caseins (i.e., kappa) are on the surface. The stability of the micellar suspension is lost when the environment becomes acidic. Specifically, caseins precipitate when milk pH reaches about 4.6. Acid conditions disrupt ionic links between proteins and calcium phosphate, resulting in protein aggregation and precipitation. This change occurs naturally when milk sours. Bacteria can produce sufficient acid from metabolizing milk lactose to cause precipitation of its caseins. Conversely, the acid precipitation of milk caseins is desirable in making cheese and yogurt. Caseins are also precipitated by the action of protease enzymes on the micelle-stabilizing protein, kappa casein. Chymosin is a proteolytic enzyme that cleaves a single bond in kappa casein. Its action releases the C-terminal (hydrophilic) portion of kappa casein, causing casein proteins to aggregate and precipitate. In case of cheese making, acid is also added to achieve complete precipitation of the casein fraction.

Precipitating caseins with acid and/or enzyme treatment has been used for a thousand years in making cheese. In the past, the protease enzyme was extracted from the stomach of calves, but is now made through recombinant technology. Cheese making is a combination of technology and art. Hard chesses, like cheddar, are made by adding lactobacilli bacteria to produce lactic acid that lowers milk pH to about 5.5. Bacterial lipase enzymes from the starter culture hydrolyze milk acylglycerol substrates and release the fatty acids (especially short chain ones) responsible for cheese flavor. The subsequent addition of chymosin causes casein precipitation as the pH reaches 4.6. The precipitated curd is salted and pressed to remove excess whey liquid. Hard cheeses are often held for months or even years in a process known as aging to fully develop its flavor. Knowing what to add and when is essential to the art of making cheese. Whey proteins are defined as those remaining soluble after precipitation of caseins from milk. In addition to the proteins, the whey fraction contains lactose, minerals, and riboflavin (vitamin B_2). The major protein constituents of whey are beta lactoglobulin, alpha lactalbumin, and bovine serum albumin. In addition to protein, the whey fraction from cheese making contains substantial amounts of lactose. Beta lactoglobulin composes about 50% of the protein in whey and 9% of the total milk protein. In the past, the whey fraction from cheese making was either fed to animals or simply used as a source of fertilizer. However, beta lactoglobulin has good nutritional qualities based on amino acid content. It is now extracted and purified for use as a major protein ingredient in processed foods. Beta lactoglobulin and whole whey are also purified and marketed as nutritional supplements. Yogurt is a popular food made from bacterial fermentation (*Lactobacillus and Streptococcus*) of milk. These organisms produce lactic acid, causing the pH to decrease and milk proteins to precipitate. Yogurt's soft texture is achieved by adding additional milk protein and controlling the fermentation process to produce a coagulum. Enzymes derived from fermentation bacteria produce enzymes that are responsible for yogurt's unique flavors.

Egg

Eggs have been a food staple for millennia. Eggs were used as food and as food ingredients in Roman times (Fig. 9.4). Chickens are easily domesticated as a source of eggs and meat. Commercialization of egg foods did not begin until the mid-nineteenth century. Eggs are high in nutritional value, containing the all the components (i.e., proteins, lipids, vitamins, and minerals) necessary for development of the chick. Whole egg contains about 16% protein, 16% fat, 1.5% carbohydrate, and 60% water. The egg is divided into two very distinct parts: the white and yolk. Egg white, also referred to as egg albumin, represents about two-thirds of an egg by weight and is composed of thick, highly viscous and thin albumin layers. Egg white contains 88% water, 10% protein, 1% carbohydrate (glucose), and a trace of fat. The predominant proteins of egg white are ovalbumin, conalbumin,

Fig. 9.4 Egg.

and ovomucoid. Ovalbumin, a phosphoprotein, contains attached carbohydrates and constitutes the largest percent of egg white protein. It is readily denatured by heat and responsible for the observed transition from clear to translucent when frying or boiling. Conalbumin is a glycoprotein. This means that is has carbohydrates (mannose) attached to its structure. Conalbumin binds metal (i.e., iron and copper) ions and is similar in biological function to lactoferrin in cow's milk. Both proteins have anti-microbial activity and transport iron to the developing embryo. Conalbumin's copper binding ability can affect the foaming properties of egg white. Copper bowls are superior for making egg white foams. This is suggested to result from the complex of copper ions with conalbumin. Copper has oxidizing properties that promote the formation of disulfide cross-links between conalbumin's cysteine amino acids. Cross-linked proteins add stability to the foam. Lysozyme is a minor constituent of egg white, but is important for its ability to lyse bacterial cell walls and control microbial growth. Egg white protein is noted for its complete compliment of essential amino acids and is easily digested. It is often used as a comparative standard in the nutritional evaluation of other protein sources. Egg yolk contains 53% fat, 16% protein, 1% carbohydrate, and about 30% water. Egg lipids are entirely found in the yolk, composing about 70% of its weight. Egg yolk lipids are composed of acylglycerides (47%), sterols (37%), and phospholipids, also known as lecithin (20%). The color of egg yolk results from its carotenoid content typically derived from corn and other grains in the diet. Yolk is a good source of fat-soluble vitamins A, D, E, K, and the B vitamin group, especially B_{12}. Egg yolk is composed of several classes of protein, including lipoproteins, livetins, phosphoproteins, and enzymes. The lipoproteins lipovitellin and lipovitellenin are phosphoproteins complexed with phospholipids and acylglycerides. The amount of lipid associated with this group of proteins is high (17%−30%). Livetin proteins are composed of albumins and globulins and the most important are immunoglobulins and transferrin. Transferrin binds iron and functions in the transport of this essential mineral. The phosphoprotein class is principally composed of phosvitin, a heavily phosphorylated (10% by weight) protein. Lipase, an enzyme that releases fatty acids from acylglycerols, is a major component of the enzyme class of yolk proteins. Egg yolk lipids are composed of triacylglycerides (46%), phospholipids (20%), and sterols (37%). The profile of fatty acids in egg yolk triglycerides parallels that of the adult bird. The fatty acid composition of egg yolk lipid is about one-third saturated and two-thirds unsaturated. About 73% of the phospholipid in egg yolk is in the form of phosphatidyl choline or lecithin. Lecithin's combination of polar and nonpolar properties makes it an excellent emulsifying agent. Egg yolk lecithin is used to make mayonnaise and salad dressings. It is essential in making cookies, moist cakes, pancakes, and other baked goods.

Egg white proteins are noted for their food applications. Egg white albumin, for example, is responsible for making high volume foams and providing a light and airy texture to meringues, soufflés, and angel food cake. The physical action of whipping or beating egg white causes ovalbumin to unfold or denature at the air-water interface. Bubbles formed in this process are surrounded the by a thin membrane containing water and ovalbumin. Unfolded ovalbumin forms a network of protein molecules that prevents collapse of the foam and provides elasticity to the membrane surrounding the bubble when it expands with heat. Sugar added to egg white during foaming increases heat stability as noted in meringue type products. Liquid egg protein denatures and forms soft gels when heated, as demonstrated when a fresh egg is dropped into boiling water. Heat also causes the carbonate dissolved in egg white to become CO_2 gas and substantially expands the product's volume. Anyone who has tried to cook an egg in a microwave oven has observed this chemical reaction. Egg white protein also has a substantial level of sulfur-containing cysteine and methionine amino acids that chemically decompose on heating to hydrogen sulfide (H_2S). Boiling eggs for example, produces an odor some identify as rotten egg from the H_2S produced. Hydrogen sulfide is also responsible for the discoloration that occurs in refrigerated boiled eggs. The black discoloration on the yolk is due to iron sulfides formed by a reaction between H_2S and iron.

Plant food systems and their composition
Soybean

Soybean is a member of the legume family and has been a traditional food in Asian diets for many centuries (Fig. 9.5). Upon germination, soybeans are seeds intended to supply the developing plant with essential nutrients for growth. Soybeans are composed of protein (38%), carbohydrates (30%), oil (18%), with moisture and minerals making up the remainder of the bean. The protein of soybean is concentrated within the seed's protein bodies. Soybean protein primarily consists of globulins (glycinin and beta-conglycinin) and albumins (water soluble proteins). The principle products obtained from soybean are protein and oil. Soy protein is harvested from crushed beans following extraction of the oil content with solvents such as hexane. The dry powdered defatted soy flour resulting from this process

Fig. 9.5 Soybean.

has a light cream color. It contains about 55% protein, 18% carbohydrate, and represents the basis from which other concentrate and isolate protein fractions are made. Soy concentrate is made by removal of soluble carbohydrates, such as stachyose and raffinose, from soy flour. Removing these oligosaccharides is beneficial because they cause gas (flatulence) when fermented in the gut. Soy concentrate contains about 70% protein and 25% carbohydrate, most as insoluble fiber. Soy isolates are made by extracting soy flour with a high pH (9-10) water solution. Alkaline pH disrupts the seed's protein bodies and releases its proteins. Materials that are not solubilized are removed and discarded by centrifugation. Proteins in the soluble fraction are precipitated by lowering the pH to 4.5 with acid. A second centrifugation separates and collects precipitated protein that is subsequently mixed with a small amount of water and neutralized with sodium hydroxide. The final soy protein isolate product contains 90% protein and is the form of soy protein most often used as a food ingredient.

Major proteins of the soybean include glycinin, beta-conglycinin, and vicilin. Soy protein is considered to be nearly complete in essential amino acids, but low in methionine and lysine. Antinutritional components of soy include phytate and lectins. Protein inhibitors of digestive enzymes are potential limitations to soy's use in foods. Phytate is an inositol polyphosphate organic molecule containing multiple phosphate groups that tightly bind minerals such as iron, calcium, and zinc. Its biological function in the seed is to serve as a source of phosphorus. Ingested phytate is considered an antinutritional component of legumes because it limits the availability of iron and zinc and could affect growth. Phytate can be eliminated by the process of fermentation. Unfortunately, the phytase enzyme responsible for degrading phytate is only present in micro-organisms, such as yeasts and some bacteria. Lectins are proteins that bind carbohydrates attached to other proteins and serve as signals for biological processes. Soy lectins, for example, bind proteins in the membranes of intestinal cells, causing inflammation, cell death, and lesions. Lectins are stable to dry heat treatments, but wet methods, such as pressure cooking, can inactivate them.

Fermentation is another way to eliminate soy lectins. Soy protein contains Bowman–Burk and Kunitz trypsin inhibitor types. Trypsin, the major digestive enzyme of the small intestine, is responsible for breaking down dietary protein in the small intestine. However, its activity is greatly reduced in the presence of either of these inhibitors. Consequently, the pancreas must make more enzyme to maintain an effective level. Diets high in trypsin inhibitor can result in hyperplasia of the pancreas. Heat treatment is only partially effective in denaturing or inactivating trypsin inhibitors. Additional constituents of note in soybean include the genistein, daidzein, and glycitein isoflavones. Isoflavones are flavonoid type compounds with structural similarity to the estrogen hormone. They are often referred to as a phytoestrogens due to their plant origin. Isoflavones have been a potential concern for those eating large amounts of soy food in their diet because of their potential hormone activity. The isoflavone intake varies between Asian and non-Asian countries. For example, an average intake for Japanese adults is about 30 mg/per day, while that of adults in Europe and the United States is about 3 mg per day. However, isoflavones exhibit only weak biological activity and should not be equated with the human hormone (Messina, 2016).

Food products made from soybeans include tofu, miso, soy sauce, soy milk, tofu, textured products, and gluten-free food. Miso is a traditional Japanese seasoning made by fermenting soybeans with *Aspergillus oryzae* fungus. It is described as having a salty/savory flavor and used in soups and meat dishes. Making soymilk involves soaking soybeans overnight in water and grinding. The mixture is then boiled and insolubles are filtered out to produce the milk-like product. Various additions are used to give the milk a sweet, savory, or salty taste. Sugar and vanilla flavoring are the most preferred additions in the United States. The nutritional composition of soy milk depends on preparation, but is approximately 2%–4% protein, 2% fat, and 1% carbohydrate. Tofu is made from soy milk by adding calcium sulfate to create a coagulum. The positive charges of calcium ions create electrostatic links between proteins and transform the liquid into a soft gel that is pressed to make the firm product known as tofu. Tofu is cream colored and bland in taste. It is typically composed of 10% fat and 13%

protein. Soy protein isolate is used in making textured, meat-like products using extrusion cooking. Soy protein, starch, and small amounts of water are compressed at a temperature of 140 °C and pushed through a narrow die. The pressure of extrusion keeps water in the liquid state until it exits holes in the die, causing a transition from liquid to steam. This physical change expands the product and provides it with the texture of ground meat. Beef or chicken flavorings are added to the hydrating solution, creating the desired taste. Soy flour is used in making gluten-free foods. It is also used to fortify the protein content of wheat, rice, corn-based breakfast cereals, and pasta. Foods such as tortillas nutritionally benefit from soy addition by supplying sulfur-containing amino acids missing in corn. Soy flour added to baked goods increases their protein content. Soybean hulls can be added to bread to increase its fiber content. Soy protein isolates are used as ingredients in protein drinks, soup bases, and gravies. Soy protein isolate (90% protein) is an excellent emulsifying agent and widely used in making processed meat products.

Soybean is a major source of food oil used in applications ranging from baking and frying to products such as mayonnaise, margarine, and salad dressings. Soybean oil is made from soybeans that have been cracked and heated to about 90 °C. A crude oil fraction is extracted from the dried material using solvents such as hexane. The solvent is removed from the oil using vacuum distillation. Polar impurities are removed from the oil by adding a small amount of water. Added water does not mix with the oil, but attracts its lecithin polar phospholipids and makes them easier to remove. The oil is further purified by washing with an alkali solution to remove free fatty acids, gums, and colorants. The last filtration over activated carbon step removes remaining off flavor or odor components. Soybean oil is composed of 16% saturated, 23% monounsaturated, and 58% polyunsaturated fatty acids. Linoleic acid (C18:2) is the predominant fatty acid in soybean oil. The ratio of omega 6 to omega 3 fatty acids in soybean oil is about 7. Soybean oil is therefore high in omega 6 fatty acids associated with increased inflammation response, atherosclerosis, and arthritis (Patterson et al., 2012).

Hydrogenation is a chemical process developed early in the twentieth century that uses hydrogen to convert vegetable oils to shortening or fat. Double bonds between carbon atoms in fatty acids are converted single bonds in this reduction reaction. The reaction can be controlled to limit the number of double bonds reduced to single bonds, thus creating fats with a soft texture. Fats made by this process are termed partially hydrogenated. The application of this chemistry to vegetable oils was enormously successful in the last century and one of its major products is margarine. Margarine with added flavorings is made to taste like butter and have a butter-like yellow color due to added beta carotene or FD&C yellow no. 5 colorants. Margarine is a softer in texture compared to butter because it has less saturated fat. It is also lower in cost. The profile of fatty acids in margarine contains more polyunsaturated compared to butter's higher proportion of saturated fatty acids. Switching from butter to margarine was thought by many to be nutritionally beneficial and healthy. However, partial hydrogenation of fatty acids also causes isomerization in the configuration of double bonds from cis to trans. For example, elaidic acid, the trans isomer of oleic acid, is created by partial hydrogenation of linoleic acid (C18:2). Dietary trans-fats pose a greater risk for developing cardiovascular disease than saturated fats (Dhaka et al., 2011). The Food and Drug Administration took action in 2015 declaring that trans-fats are not safe and required food manufactures to remove them from their products by 2018 (FDA, 2015). It is important to note that shortening, previously made from soybean and other vegetable oils by hydrogenation, can made without trans-fat. Shortening is widely used and the preferred fat for baked foods such as cookies, cakes, biscuits, and pies. The process of interesterification makes shortening and margarine without the risk of creating trans-fat. The method involves heating two or more acylglycerols with differing saturated and unsaturated fatty acid content to make a hybrid. For example, tristearin, composed entirely of saturated fatty acids and triolein, composed entirely of the unsaturated fatty acids, are mixed to create a new population of acylglycerols. This process is useful for making acylglycerols with desired functional properties (i.e., melting point), and without trans fatty acids.

Wheat

Wheat is a cereal grain that archeologists suggest has been a food source for thousands of years. Some suggest it is one of the first crops grown when agriculture began 10,000 years ago. Today, wheat is a major source of carbohydrate and protein in the diet of many cultures (Fig. 9.6). Wheat is the most widely grown and traded food crop in the world and the major ingredient of bread, pasta, cake, cookies, and numerous baked goods. Its composition is approximately 70% carbohydrate, 13% protein, and 1%—2% water. Wheat also contains vitamins and minerals. A wheat kernel contains substantial amounts of vitamins: B_1 (Thiamine), B_3 (Niacin), B_6 (Pyridoxine), and

Fig. 9.6 Wheat.

B₅ (Pantothenic acid). Its mineral content includes potassium, phosphorous, magnesium, and calcium. Wheat is a good source of dietary fiber, containing 12 g per 100 g. The kernel is divided into three parts: bran, endosperm, and germ. Bran constitutes the outer layer protecting the endosperm and germ. It represents about 13% of the wheat kernel's weight and contains its cellulose and hemicellulose carbohydrate fiber and minerals. The endosperm contains protein and starch carbohydrate that serve as the energy source for the developing plant. Protein bodies and starch granules are distributed throughout the endosperm. Nutritionally, wheat protein is low. The essential amino acid lysine and its Protein Digestibility Amino Acid Score (PDCAAS) 0.41, is about half that of milk casein. The germ represents the embryo that will develop into the mature plant. It contains wheat's vitamins, enzymes, and most of its lipid. Milling is the process by which wheat is separated into its major fractions of flour, bran and germ. Dry milling removes the bran and germ from the endosperm which is further processed into refined flour fractions. Cleaning is the first step in milling and is designed to remove stones, metal fragments, and dirt. Wheat kernels are next tempered to increase moisture content, making it easy to fracture them into particles in the grinding operation. Roller mills used to grind wheat consist of closely spaced steel cylinders rotating in opposite directions. They break kernels into fragments as they pass between them. Fragmented kernels of bran, endosperm and germ are passed through a series of sifters to separate particles on the basis of size. The process separates wheat's bran and germ from the endosperm fraction that contains most of protein and almost all of the starch. Endosperm flour particles are repeatedly repassed through rollers and sieved. The end result is a flour with very fine particle size. It is possible to separate starch granules from proteins in flour with a technique known as air classification. This process uses a stream of air to separate particles on the basis of shape and density, producing flours differing in their ratio of starch to protein. Eighty percent of the proteins in wheat are represented by the glutenin and gliadin. Together, these proteins are responsible for much of wheat's quality in making bread and pasta. There are two types of glutenin proteins that differ principally in their physical size: high- and low-molecular weight glutenins. The gliadins are relatively smaller globular proteins characterized by intermolecular disulfide links or covalent bonds between cysteine amino acids. Adding water and mixing forms a complex between glutenins and gliadins called gluten and the flour to become a dough. In this complex, glutenin is responsible for the strength and cohesion of wheat dough and gliadins are responsible for its elasticity. Approximately 20% of wheat proteins are water soluble globulins and albumins. This group of proteins contains enzymes (i.e., proteases, lipases, and amylases) that function during seed germination and thus have little effect on the bread making properties of wheat. The enzyme lipoxygenase, however, may contribute off flavors through catalyzing the oxidation of unsaturated fatty acids.

Starch is the major component of wheat and exists as granules located in the endosperm. The size and shape of starch granules varies between wheat varieties. Starch is composed of amylose and amylopectin types of polysaccharide. Amylose is a linear glucose polymer connected by alpha 1—4 links. These links cause the polymer to become a coiled structure. Amylopectin is also a glucose polymer, but its structure is branched. Amylopectin molecules have numerous branches (linked alpha 1—6) that result in a shrub-like appearance to the overall structure. Amylose and amylopectin make up the semi-crystalline structure of wheat starch granules. The amylose content of starch granules is fairly constant at about 23 to 27% for most wheat varieties. Bread making begins when water is added to flour, followed by mixing. Added water is quickly taken up by starch granules. They can adsorb up to 50% of their weight. Gluten interacts with hydrated starch to make the dough. When dough is heated, hydrated starch granules swell and gelatinize. In this way the intermolecular bonds between starch molecules are broken. Amylose molecules diffuse from the granule and adsorb lipids from the wheat flour or that added lipid in the recipe. Starch granules near the bread's surface are less hydrated and experience higher temperature that contribute to formation of the crust. High temperatures at the surface cause flavors and brown color resulting from Maillard chemistry. In contrast, the interior of bread has a soft texture attributed to the hydrated state of its starch, especially the amylose molecules. A few days after baking, however, bread texture becomes stiff or even tough due to reformation of molecular hydrogen bonds interactions between starch molecules in a process called retrogradation. This change is commonly known as staling. Pasta and pie crust are made in a similar way, except that lipids (oil in the case of pasta and fat in the case of pie crust) are added first. Mixing lipid and flour before water causes the oil or fat to coat flour particles and inhibits protein interactions. The result is a pasta with softer texture when cooked. Wheat is a good source of carbohydrate fiber and constitutes about 12% of the kernel. Wheat dietary fiber originates in the bran and is classified as insoluble or soluble. Wheat grains contain about 10% and 2%–4% of these classes respectively. Water-insoluble wheat fiber is composed of hemicelluloses, lignin, and beta glucans. Water-soluble wheat fiber is composed of pentosans that are polymers of five-carbon sugars such as xylose and arabinose.

Rice

Rice (*Oryza sativa L.*) is a staple in the diet of at least three billion people worldwide (Fig. 9.7). The consumption by China and India account for at least half of the total. Rice grain is processed following harvesting to remove the inedible hull and reveal the rice kernel containing bran, germ and endosperm fractions. Rice at this stage of processing is termed brown rice and is composed of 6.8% protein, 2.7% fat and 74% carbohydrate.

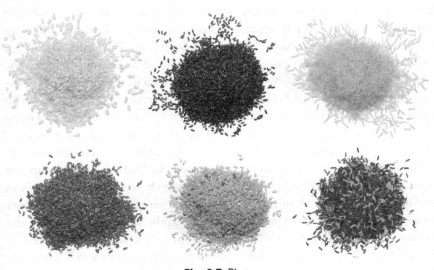

Fig. 9.7 Rice.

Further processing to remove the bran layer and germ results in the product known as white rice that is composed of 6% protein, less than 1% fat, and 78% carbohydrate. Rice is most often consumed directly as a food. However, brown rice can be processed into additional products such as oil, protein, and fiber. The bran portion of the kernel contains approximately 25% fat, 13% protein, 6% fiber, 8% crude fiber, and 22% insoluble dietary fiber. Bran also contains mostly B group vitamins and minerals (i.e., potassium, magnesium, calcium, and iron). Brown rice has a superior profile of nutrients, tan color, and nutty flavor compared to white rice. The fiber content of brown rice is about 3 times that of white rice. Rice proteins, as classified by the Osborn system, consist predominantly of 60% to 80 % glutelins, followed by 4% to 22% albumins, 5% to 13% globulins, and 1% to 5% prolamin (Hoogenkamp et al., 2017).

The protein fraction of rice, especially its bran, can be separated and used in various food applications (Hoogenkamp et al., 2017). Purified rice bran protein isolates are noted for hypoallergenicity and digestibility. Rice protein contains a full complement of essential amino acids, but is limiting in lysine. Rice bran protein is noted for its high level of arginine and branched chain amino acids. It also contains high levels of glutamic acid that contribute savory umami flavor to foods. Rice flavor is derived from its lipid fraction. Specifically, the oxidation products of unsaturated fatty acids are responsible for the volatile flavor notes. Thus, brown rice is stronger in flavor than white rice. Rice protein added to processed meats improves the gelling and water holding capacity of these products. These proteins are also used in a variety of health drinks, sauces, and gravies. Rice protein digestibility and hypoallergenic properties make it a good choice for use in infant foods. Treating rice bran protein with protease enzymes coverts most of the protein into shorter peptide fragments. Enzymatically hydrolyzed rice bran is valued for its use as a nutritional protein supplement and flavoring. While rice has numerous benefits related to diet and health, it also contains antinutritional and toxic substances. Specifically, alpha amylase inhibitor is a heat stable protein that can limit the digestion of starch in the gut. More importantly, amylase inhibitors can be allergenic. Additionally, the rice plant grows submerged in water for a period of time during its development. A serious concern is that rice plants are effective at accumulating and translating minerals (i.e., arsenic) from the soil into rice kernels. Arsenic is a very toxic and carcinogenic element.

Protein-rich plant foods and their composition

The world-wide demand for plant-based protein-rich foods is increasing in response to several factors. Briefly, factors include feeding the expected two to three billion increase in worldwide population, lowering the environmental impact of food production, and providing healthier alternatives for the human diet. The sources described in this section represent some of the protein-rich plants with potential to meet those needs.

Amaranth

Amaranth is a grain used as a food source for thousands of years, primarily in South America and Mesoamerica (Fig. 9.8). The Aztec culture that flourished in central Mexico used amaranth as food stable and as part of religious ceremonies for hundreds of years. Dry amaranth seeds contain approximately 14% protein, 65% carbohydrate, 7% fat and about 2% dietary fiber (Orona-Tamayo and Paredes-Lopez, 2017). Uncooked amaranth is indigestible. Boiling, however, hydrates the grain and substantially improves its digestibility. Amaranth is higher in dietary protein than wheat and other cereals. Its profile of essential amino acids is lysine rich, but low in leucine and threonine. Pairing amaranth with low in lysine wheat represents a good option for increasing the nutritional quality of cereal-based food. Amaranth does not contain gluten proteins and works well in gluten-free food applications. The carbohydrate component of amaranth is primarily in the form of waxy starches. Waxy starch creates highly viscous solutions when heated. This is a desirable property for soups made with amaranth.

Fig. 9.8 Amaranth.

The lipid fraction of amaranth is high in linoleic acid, an essential (omega 6) fatty acid. The fatty acid composition of amaranth lipid is approximately two-thirds unsaturated and one-third saturated. Amaranth is a good source of vitamin B6, pantothenic acid, folate and minerals. It is high in manganese, magnesium, phosphorous, potassium, and iron. Amaranth contains substantial levels of phenolic antioxidants (e.g., anthocyanin and others) that are of potential health benefit. Less desirable components of amaranth include phytate and tannin. Phytate is composed of inositol to which 6 phosphate groups are bound. Each phosphate carries a strong negative charge making the molecule a strong binder of dietary minerals such as calcium, iron, and zinc. Phytate is stable to most processing treatments and thus there is risk for mineral deficiencies when consumed at a high levels. Tannins are polymers of phenolic acids noted for their ability to bind proteins. High tannin foods are considered be antinutritional because they can bind digestive enzymes and inhibit their activity. Amaranth grains are similar in size to wheat, but are superior in nutrient content. Food uses of amaranth include addition of amaranth flour into cookies and bread baked goods or in pasta. Adding amaranth to baked goods increases their protein content. Amaranth is widely used in creating gluten-free versions of pasta and bread products. Its high starch and protein content are an advantageous combination in soups and meatless dishes. Adding amaranth thickens and increases the nutritional value of the food at the same time. Amaranth seeds can be popped like popcorn and eaten directly. Popped amaranth has a nutty flavor, soft texture, and is nutritionally superior to corn. Amaranth production is more sustainable than other cereal grains. Most amaranth varieties are well-adapted to unfavorable conditions. Amaranth plants are able to tolerate unfavorable high salt and/or alkalinity soils. Once the plants are established, it grows well under low moisture conditions and requires less fertilizer. Compared to corn, amaranth requires less than half the amount of water. It is well suited to planting in semi-arid (less than 20 inches of rain per year) regions of the world, such as the high plains of the United States, Australian outback, and Africa. Amaranth crops fared well during the historic drought of 2012 in the upper Midwest of the United States (Santra and Schoenlechner, 2017).

Canola

Canola protein is derived from the rapeseed plant which is primarily cultivated for its oil. The plant is a member of the *Brassicaceae* (Cruciferae) family (Fig. 9.9). Archeological evidence suggests that the plant has been

Fig. 9.9 Canola/rapeseed.

cultivated for medicinal use since prehistoric times. The common name used for this oil-seed crop, canola is a contraction of the phrase Canada oil low acid. The term low acid refers to erucic acid which is a fatty acid that can cause negative health consequences when consumed at high levels. Canola oil is therefore required to contain less than 2% erucic acid. This is the level deemed safe by Canadian, United States, and other regulatory agencies. Rapeseed is primarily grown for its oil content and the protein fraction remaining after oil extraction has traditionally been used as animal feed. The composition of rapeseed is approximately 40% oil, 20%—25% protein, 10% carbohydrate, and 7%—8% ash. The canola oil fraction is considered to be a nutritionally healthy oil. It has a low level of saturated fatty acids. The ratio of mono-to poly-unsaturated fatty acids is about 2:1. Its ratio of omega-6 to omega-3 fatty acids is also desirable, about 2:1. The carbohydrate component of rapeseed meal (the fraction remaining after oil extraction) is its major component. Rapeseed canola meal is high in dietary fiber (32%) and non-starch polysaccharides (15%). Dietary fiber in rapeseed meal contains beneficial carbohydrates such as pectin, hemicellulose, and beta glucan. The ash component of rapeseed meal is high in potassium and phosphorous in complex with phytate. Rapeseed meal also contains the antinutritional factors glucosinolate, phytate, and phenolics. Glucosinolate is a substrate for the enzyme myrosinase that causes production of thiocyanates and a sharp stinging sensation when the meal is eaten. Glucosinolate in animal feed can trigger avoidance and may affect thyroid function. Phytate is a complex of inositol and phosphate that binds essential minerals such as calcium, iron, and zinc. Phenolic tannin compounds have a strong affinity for binding proteins and can decrease the level of digestive enzymes in the gut.

The protein fraction of rapeseed is composed of a cupin protein, cruciferin, and a napin protein that is a member of the prolamin family. Cruciferin is a seed storage protein and represents the largest percent (90 plus) of the protein in rapeseed. It is a complex protein, consisting of several subunits. Napins are much smaller proteins compared to cruciferin and composed of two subunits linked by disulfide bonds. Napins are also thermally stable, requiring temperatures above 75 °C to denature them. Other minor rapeseed proteins include oleosins and rapeseed digestive enzyme trypsin and chymotrypsin inhibitors. Oleosins are termed oil-body proteins because their function is to surround oil droplets to prevent the oil from coalescing in the seed's structure. Rapeseed meal contains several types of trypsin and chymotrypsin inhibitors. Some of these are heat stable and have the potential to limit dietary protein utilization. Food uses of rapeseed protein are limited by treatments used in the oil extraction process. The combination of organic solvent and heat needed to efficiently extract oil results in protein denaturation and loss of functional properties important for its use in food products. Techniques have been developed to process rapeseed protein into fractions composed predominantly of cruciferin or napin proteins. These protein products are commercially available under the names of Puritein™ and Supratein™, respectively. Both of these rapeseed protein sources have been granted GRAS designations by the US Food and Drug Administration. The essential amino acid profile of rapeseed protein shows tryptophan and methionine as limiting amino acids. Methods used to extract and purify these proteins may reduce their lysine level resulting from Maillard reactions. Isolated cruciferin protein forms gels when heated and has potential use in processed meat and cheese type foods. Unfortunately, those applications are limited by the relatively high temperature (75 °C) required for gelation in these food matrices. Purified napin proteins foam well and show promise for use in beverages and baked products. At present, the largest potential use for rapeseed protein is in protein supplements. Limitations to the use of rapeseed protein in food applications include reduction in functionality, such as gel formation, emulsification, and foaming, as a result of the oil extraction process. Additionally, the napin fraction of rapeseed protein contains allergens. Specifically, Bra n 1 and Bra r 1, have been identified as rapeseed allergens (Mills et al., 2003). Rapeseed production is one of the more sustainable crops. It grows well in a wide variety of climates found in the Northern hemisphere. Water irrigation is required for adequate yields in climates without sufficient rain or with elevated temperatures.

> ## Chia

Chia protein is derived from the plant *Salvia hispanica,* which provides omega fatty acids, antioxidant, and fiber, and protein (Fig. 9.10). Salvia is perhaps best known as an annual plant with white or purple flowers. Salvia is also used as a herb to season foods and in folk medicine. However, its seeds are high in protein. Chia production is highest in Mexico, Central America, and the Southwester US where salvia is naturally found. Chia seeds are small (about 2 mm) and contain approximately 17% protein, 41% carbohydrate and 25%–40% fat. Chia protein is composed predominantly of globulin-type proteins and lesser amounts of albumin, prolamin, and glutelin. The essential amino acid content of chia flour protein is deficient in tryptophan, methionine, and cysteine. The carbohydrate fraction of chia is rich in dietary fiber, constituting about three-fourths of its carbohydrate fraction. The lipid fraction of chia is high in polyunsaturated fatty acids. Notably, two-thirds of the PUFS are omega 3 fatty acid (i.e., alpha linolenic acid). The vitamin and mineral content of chia seeds is high in thiamin and niacin and contains moderate amounts of riboflavin and folate. Chia is high in calcium, manganese and phosphorus minerals. Chia seeds also contain phenolic antioxidants such as rosmarinic acid.

Chia was used as a food source by the Aztecs and other Mesoamerican cultures. Chia seeds can be sprinkled on other foods dishes or added to smoothies and other beverages in the ground form. The carbohydrates found in chia contains mucilage polysaccharides that function well as hydrocolloids in food applications. Chia mucilage can be used to increase moisture content and improve texture in foods such as processed meat and cheese. Isolated fractions of chai protein (flour or concentrates) have promise for future food applications.

Fig. 9.10 Chia.

Flaxseed

Flaxseed (*Linum Usitatissimum L*) is composed of about 45% oil, 20% protein, 28% dietary fiber, and 3%–4% ash (Fig. 9.11). The oil fraction contains 53% alpha linolenic acid (ALA), 17% linoleic acid, 19% oleic acid, 3% stearic acid, and 5% palmitic acid. About 70 % of the fatty acids are polyunsaturated and less than 10% are saturated. Flaxseed fatty acids have a beneficial ratio of omega 6 to omega 3 fatty acids of about 1:4. Flaxseed protein has an amino acid profile similar to that of soybean. It is limiting in the essential amino acids, lysine, threonine, tyrosine, and tryptophan. Flaxseed carbohydrate is rich in dietary fiber composed of cellulose, hemicellulose, lignin, and gums. The carbohydrate fraction also contains arabinoxylan and acidic rhamnogalacturan polysaccharides. Flaxseed fiber absorbs substantial amounts of water, providing a feeling of satiety and facilitating passage of the solids through the gut. It is a substantial source of vitamins (E and niacin) and minerals (calcium, magnesium, and phosphorus). Flaxseed lignans secoisolariciresinol (SECO) and its secoisolariciresinol diglucoside (SDG) have antioxidant activity that are linked to the health benefits of flaxseed. The antinutritional components of flaxseed include trypsin inhibitors, cyanogenic glycosides, and phytoestrogens. Trypsin inhibitors slow the action of this important digestive enzyme in the gut and may not be inactivated by heat processing. Cyanogenic glycosides are substrates for the enzyme myrosinase producing thiocyanate, a chemical that can inhibit thyroid function. Phytoestrogen are flavonoid compounds with structural similarity to estrogen and

Fig. 9.11 Flaxseed.

have a weak hormone effect. Flaxseed protein is composed of 65% albumins (water soluble), 22% glutelins (dilute acid soluble), and 18% globulins (salt soluble). Several types of flaxseed protein preparations are used in food applications, such as protein drinks and baked goods, such as bread, cookies, and muffins. Depending on the composition of the flaxseed ingredient, the addition may provide enhanced level of protein nutrition, fiber, and/or beneficial fatty acids. Flaxseed oil is highly unsaturated and is readily oxidized, resulting in off or rancid flavors. Additionally, a unique cyclolinopeptide peptide composed of 8 amino acids can be found in flaxseed oil. This bitter peptide has a cyclic structure, meaning that its N and C terminals are joined in a peptide bond. One of its constituent amino acids, methionine, is oxidized to methionine sulfoxide, resulting in a very bitter flavor.

Lentils

Lentils are small disc-shaped seeds derived from the legume plant (*Lens Culinaris*) (Fig. 9.12). This ancient plant has been grown in most parts of the world. Australia, Canada, China, Ethiopia, India, Nepal, Turkey, and the United States are all major producers of lentils. Canada is the largest producer. Lentils are higher in protein than other legumes, such as soybean. Red or green lentils are the most common. Lentil color is due the presence of lutein and zeaxanthin carotenoids. The composition of lentils is approximately 28% protein, 63% total carbohydrate (47% of which is starch and 12% dietary fiber), and only about 1% fat. Lentils contain vitamins (i.e.,

Fig. 9.12 Lentils.

thiamin, riboflavin, niacin, pantothenic acid, vitamin B_6, and folate) and minerals (i.e., predominantly iron, phosphorus, zinc, and calcium). Minor components of lentils include phytochemicals, such as flavonoid antioxidants. Lentil starch composes about two-thirds of its carbohydrate content and has a low glycemic index classification (approximately 35). One-third of lentil carbohydrate is in the form of dietary fiber that is beneficial to the gut microbiota and its host. Principal proteins in lentils are characterized as albumins and globulins that make up most of its extractable protein. Lentil protein has a high nutritional quality. While it contains all the essential amino acids, it is limiting in methionine and tryptophan. The protein digestibility-corrected amino acid score (PDCAAS) ranges from 0.52 to 0.71, but is lower than that of soy protein (0.90).

Food applications for lentils include its traditional combination with rice. This combination provides a complete profile of essential amino acids. Lentils are commonly used to make soups. Lentil starch thickens the soup as it is boiled and its protein provides good nutrition. Lentil flour added to bread at about 3% increases the protein content without compromising loaf volume. Lentil flour, in combination with soy protein and other ingredients, are extrusion processed into snack foods and textured vegetable protein bars. Flavor is a major limitation to the use of lentils in food. Lentil hull is indigestible and bitter tasting. Like most legumes, lentil flavor suffers from the oxidation of unsaturated fatty acid by the enzyme lipoxygenase. This enzyme causes production of volatile aldehydes and ketones, contributing to beany and other off flavors in lentil products. Additionally, some lentil proteins have been identified as allergens, specifically Len c 1, and Len C 2. Boiling does not eliminate the potential for allergic reaction in those who are sensitive to its proteins. There is the added possibility of an allergic reaction to lentil-containing food by individuals with allergy to other legumes (e.g., peas or peanut) because of structural similarity in the proteins.

Peas

Peas (*Pisum sativum L*) are legumes grown for use as food and animal feed in many parts of the world (Fig. 9.13). Some pea varieties are grown for specific food applications, such as canned or frozen foods. Others are intended for fresh garden use. Peas grow well in fertile, well-drained soil and are strongly dependent on adequate moisture during germination. Pea plants fix nitrogen in the soil and can be used as fertilizer for subsequent crops. The composition of nutrients in peas is approximately 60%

Fig. 9.13 Peas.

carbohydrate, 23% protein, 1%—2% fat. Pea carbohydrates are composed of 49% starch, 21% dietary fiber, and 6% sugars. Pea starch is about two-thirds amylopectin and one-third amylose. They are a good source of dietary fiber (hemicellulose, cellulose, pectin, and gums) that provides health benefits. Dietary fiber promotes cholesterol elimination and serves as an energy source for gut microbiota. Other sucrose and raffinose carbohydrates found in peas compose about 2%—3% of this fraction. Pea proteins are classified under the Osborne scheme into five classes: albumins, globulins, prolamines, glutelins, and insoluble residue. Globulins are the most abundant and consist of vicilin and legumin type proteins. Albumin and globulin proteins are most important to the food applications of pea protein. Peas are a good source of vitamins and minerals. Vitamins in peas include folate, niacin, thiamin, ribo-flavin, and several forms of vitamin B_6. Peas are also a source of phosphorous, magnesium, and calcium. Pea lipids are composed mostly of linoleic acid and oleic acid. Peas contain several antinutritional components of phytate, lectins, tannins, saponins, and trypsin inhibitors.

Food applications of peas typically utilize pea flour (25% protein), pea concentrates, or isolates. Pea flour contains substantial amounts of starch (50%) and dietary fiber (10%—20%). Its applications include baked goods, bread, pasta, and extruded or canned pet foods. Pea protein provides func-tional properties, including water-binding, emulsification, and gelation. Pea isolate is almost entirely (90%) protein and used to improve the nutritional value or functional properties, such as water-binding, emulsification, gela-tion, and foaming. Foams created with pea isolate are often superior to soy isolates. Enzymatic treatment of pea isolate can make its functional prop-erties equivalent to egg white. Pea protein addition in baked goods and bakery foods increases their moisture content and is beneficial to the texture of these products. This protein has long been used in traditional Chinese noodles (e.g., vermicelli). Pea isolate works well in this application and also in wheat-based pastas. Pea protein combined with starch and other ingredients can be processed by extrusion into snack foods and breakfast cereals. It is also used to create pea plant-based meat products. A mixture of pea protein, starch, and other ingredients results in a product with the texture of real meat when processed by extrusion.

Oats

Oat (*Avena sativa*) is a member of the grass family (*Gramineae*) with two major species used in foods (*A. sativa L* and *A. sativa* var. *nuda*). Oats are among the top ten cereal crops produced in the world. They are high in protein and do not contain gluten (Fig. 9.14). Oat-containing foods are well tolerated by individuals with celiac disease or gluten sensitivity. Health promoting components of oats include dietary fiber (beta glucan) and phytochemicals with antioxidant activity. Whole oats are composed of 66% carbohydrate, 17% protein, and 7% fat. Oats are also a source of vitamins (i.e., thiamin, riboflavin, pantothenic acid, and folate) and minerals (manganese, phosphorus, magnesium, iron, and zinc). Oat carbohydrates are composed of starch and fiber. The type of starch found in oats is slowly digested and therefore has minimal effects on blood sugar level or glycemic index. About one-third of oat carbohydrate is in the form of dietary fiber which is an energy source for beneficial gut microbiota. Oat protein is composed mostly of the salt-soluble globulins (70%–80 %) with prolamins

Fig. 9.14 Oats.

and albumins making up the remainder. Globulins are the major storage proteins of oats and are used by the plant to support growth during germination. The major globulin protein of this group has a complex structure similar to other plant storage proteins, such as soybean glycinin. Oat globulins are notably stable to heat denaturation, but are sensitive to extremes in pH. Oat prolamins (avenins) are smaller in size relative to the globulins and constitute about 14% or less, of total oat protein. Nutritionally, oat protein is high in the essential amino acid lysine and contains substantial amounts of cysteine and methionine. This composition of essential amino acids makes it a complimentary source for blending with legumes or other cereals. Oat protein digestibility is good in comparison to other cereals, but its digestibility corrected amino acid score (PDCAAS) 0.59 is substantially lower when comparison to soy and milk protein (Peterson, 2011). Oats contain higher amounts of lipid compared to other cereals, except corn. The fatty acid profile of oat lipid is dominated by omega 6 fatty acids linoleic (35%) and oleic (31%). The palmitic saturated fatty acid in oat lipid is present at about 21%. Avenanthramides (i.e., anthranllic acid amides) are an unusual and potentially beneficial compound found in oats. These phenolic compounds have antioxidant and anti-inflammatory activities. Avenanthramide-containing oat extract is effective when applied topically in reducing skin irritation and itch (Meydani, 2009). Extracted oat avenanthramide is used in skin, hair, and sun-care products. Major food applications for oats include baked products and breakfast cereals. In bread, modest amounts (3%—6 %) of added oat protein, increase bread nutritional quality (protein content) and its loaf volume while not affecting its crumb texture. Oats are widely used in other baked goods (e.g., cookies and muffins). Smoothies and other beverages make use of oat powders for their protein and fiber content.

Peanut

Peanuts (*Arachis hypogaea*) are legumes grown in many parts of the world (Fig. 9.15) China, India and Nigeria are world leaders in peanut production. Peanuts contain 48% fat, 25% protein, and 21% carbohydrate. The fatty acids composition of peanut fat is composed of 17 % saturated fatty acid, 46% monounsaturated fatty acid, and 32% polyunsaturated fatty acid. Unsaturated fatty acids of peanut lipid are all omega 6. Peanut proteins are composed partially of globulins, of which arachin and conarachin fractions can be isolated. Adding salt to an aqueous extract of peanut flour is used to obtain a partially selective precipitation of arachin from conarachin.

Fig. 9.15 Peanuts.

Peanut protein contains all 20 of the common amino acids. Arginine content is higher than all other amino acids, and the sulfur containing amino acids, methionine and cysteine, represent the limiting essential amino acids. Its protein digestibility corrected amino acid score (PDCAAS) is greater than that of soy. Carbohydrates in peanut contain about 12% dietary fiber and 8% sugars. Peanuts contain very little starch and a low glycemic index. Micronutrients in peanuts include vitamins (i.e., thiamine, riboflavin, niacin, pantothenic acid, folate, vitamins B_6 and E) and minerals (i.e., calcium, magnesium, phosphorus, and potassium). Peanut phytonutrients include antioxidant flavonoid resveratrol and phytosterols.

Food applications of peanuts are almost entirely as whole nuts and the ground product known as peanut butter. Peanuts are added to cookies and snack foods. Peanut butter is made by grinding roasted peanuts. The grinding operation blends protein and fat into a stable emulsion.

Peanut flour made from oil-extracted peanuts has good emulsifying properties, water retention, and ability to foam. Peanut flour can be incorporated into noodles. The flavor of roasted peanuts is a desirable attribute and considered to be the driving force in peanut consumption (Arya et,al., 2016). The major limitation to the use of peanut in food applications is allergenicity. Peanut allergy affects 1%–2% of the population in the United States. Often allergic reaction to peanuts is severe and can be life-threatening. The major allergens of peanut are Ara h 1, Ara h 2, and Ara h 3. These protein allergens make up a large portion of the total peanut protein.

Quinoa

Quinoa (*Chenopodium quinoa*) is an annual seed-producing plant grown as a grain crop (Fig. 9.16). It was originally grown in the Andean region of South America, but its cultivation has spread to North America,

Fig. 9.16 Quinoa.

Europe, Asia, Africa, and Australia. Quinoa seeds contain 14% protein, 64% carbohydrate, and 6% fat. Quinoa protein is composed of albumins and globulins (seed storage type proteins). Albumin and globulin fractions each contain a dominant protein. Together, these two proteins constitute 70% of the total quinoa protein. The nutritional quality of quinoa protein is high. It has a balanced profile of essential amino acids for adults. However, quinoa is limiting in leucine, lysine, threonine and valine for young children. Quinoa carbohydrate is primary composed of starch having a high amylopectin to amylose ratio. Quinoa starch provides high viscosity when heated and has good resistance to retrogradation. The mono-and di-saccharides glucose, fructose, ribose and maltose components of quinoa carbohydrate are present at low levels. Dietary fiber in quinoa composes about 10% of the carbohydrate fraction and is made up of galacturans, xyloglucans, and cellulose. The carbohydrate composition in quinoa makes it a low glycemic food. The fatty acid profile of quinoa lipid is composed of 60% polyunsaturated fatty acid, linoleic (omega 6) and alpha linolenic (omega 3), 25%–30% monounsaturated (oleic), and 10% saturated (palmitic). Quinoa is higher in vitamins and dietary minerals, compared to many grains. Quinoa vitamins include thiamin, riboflavin, niacin, ascorbic acid, folate, and alpha tocopherol. Its mineral content is high relative to cereal grains and includes phosphorous, potassium, magnesium, calcium, iron, zinc, and copper. Quinoa also contains antinutritional substances: 1% phytate, 0.5% tannin, and a high content (5%–7%) of saponins. Phytate and tannin are important because they limit absorption of minerals and form inhibitory complexes with digestive enzymes, respectively. Saponins can be problematic when consumed at high levels because of their ability to increase permeability of the intestinal wall, a condition called leaky gut. Saponins are also very bitter tasting and this limits the acceptability of quinoa as a food and in its applications.

Food applications of quinoa are limited to saponin-reduced forms (i.e., protein isolates and concentrates). A majority of quinoa saponin is located in the seed coat, thus processing technologies that remove it are advantageous to its use as an ingredient in food dishes. Purified forms of quinoa protein (i.e., concentrate, 70% protein) have promising functional properties, such as foaming, gelation, and emulsification, and a much lower level of saponin. High protein saponin preparations have potential applications in baked goods, protein beverages, and foods of the modernist cuisine genre.

Microbial Protein-rich foods (fungi and algae)

Microbial sources described here (mycoprotein and algal protein) are of particular interest because they can be produced in large quantity with lower inputs and cost. Their protein is complete, containing all the essential amino acids, but limiting in cysteine and methionine. The mycoprotein known as Quorn™ has been successfully produced by industrial scale fermentation since the 1970s. Several applications of mycoprotein as food products (meat-meatballs and patties) have been available since 1985. Microalgae have been dried and used as a food source for hundreds of years. Microalgae accumulate large amounts of protein or oil (approximately 60% by weight) and have recently been the focus of efforts to develop them as a source of edible protein and oil on a commercial scale. Much of the oil production seems to be directed toward biofuels. Algal protein has been granted GRAS status by FDA in the United States. Its food applications however, are limited at present by its intensely green color.

Mycoprotein

Mycoprotein is the basis of the widely known product known as Quorn™). Mycoprotein used in this product is derived from fermentation carried out by the *fusarium venenatum* fungus. Unlike many fungi, this organism does not produce toxins. The development of mycoprotein was initially an off shoot of efforts begun in the 1960s to create single cell protein in response to the perceived need of future populations. It was then predicted that there would be a world-wide shortage of protein-rich foods by the end of the century. The filamentous fungus *fusarium venenatum*, discovered in 1967, was subsequently used to develop a fermentation process for production of mycoprotein on a commercial scale. Today, mycoprotein of quorn is made using continuous fermentation. Glucose from roasted barley malt, plus nitrogen from ammonia, provides the nutrients needed for growth of the organism and mycoprotein production. The fungal mycelium becomes branched during its growth. Branched morphology is desirable because it is responsible for the meat-like texture in the final product. Upon harvest, the fungal mass is heat shocked to induce the production nuclease enzymes that in turn break down its ribonucleic acid content. Calcium and a binding protein are added to increase firmness of the mass. Egg white albumin was the protein of choice for this effect, but has since been replaced by potato protein to make the mycoprotein product vegan. Flavorings can also be added at this stage, but may not be necessary because the residual nucleic acid content provides a pleasant umami flavor to the product.

Mycoprotein has a good nutritional profile and is high in protein, dietary fiber, and low in fat. Specifically, it contains 11% protein, 3% carbohydrate and 3% fat. Mycoprotein has a full complement of essential amino acids. While mycoprotein is limiting in methionine and cysteine content, its PDCAAS score is 0.86. The carbohydrate component contains about 25% fiber consisting of two-thirds chitin and one-third beta glucan. Low levels (less than 1%) of glucose, mannose, galactose, and arabinose are also present. The fat of mycoprotein is predominantly unsaturated consisting of oleic, linoleic, and linolenic (omega 3). Stearic and palmitic saturated fatty acids are about 1.5% of the total. Mycoprotein vitamins consist of thiamin, riboflavin, niacin Vitamin B_6, folate, and biotin. Minerals contained in mycoprotein are calcium, phosphorous, magnesium, iron, and zinc. Quorn is a meat substitute that can be made into products resembling whole pieces of beef or chicken. It can also be made into a ground beef-like product and used in numerous vegan applications, such as meatless lasagna, tacos, curries, and burgers.

Algal protein

Algae have existed for a hundred millions years or more. They are photosynthetic organisms making their own metabolic energy from sunlight. Algal organisms occur in two major forms: macroalgae (seaweed) and microalgae (unicellular, microscopic organisms) (Bleakley and Hayes, 2017). Microalgae are common inhabitants of ponds, lakes, and other open waters. Until recently, their primary use has been in making hydrocolloids, animal feed, and for wastewater treatment. Algae were used as a food source by ancient Mesoamericans cultures, microalgae have gained attention for their potential use in foods. Specifically, two microalgae species,

Spirulina (*Arthrospira platensis*) and Chlorella (*Chlorella vulgaris*), are being used to make edible protein and oil using commercial scale processes. Spirulina and Chlorella microalgae species can accumulate high levels of protein (60%) or oil (60%–70%). They also contain substantial amounts (25%) of dietary fiber (beta glucan). Algal protein has a high nutritional value, containing a full complement of the essential amino acids. The essential amino acid profile is similar to eggs, but algal protein is limiting in cysteine and methionine. The lipid fraction of microalgae is high in polyunsaturates, most importantly the omega 3 fatty acids, alpha linolenic (ALA), eicosapentaenoic (EPA), and docosahexaenoic (DHA) acid. These long chain fatty acids are only produced in plants. The essential fatty acid content of algal oil is passed up the food chain from crustaceans to fish as they are consumed. Microalgae are also a good source of vitamins and minerals. Carotenoids, such as zeaxanthin, lutein, and astaxanthin, are essential to photosynthetic in microalgae and thus also end up in other food sources. The color of salmon, for example, is due to the astaxanthin content.

While limited at present, food applications of algal protein are beginning to increase. The strong green color and slightly fishy flavor notes in Spirulina are limiting to its use. However, it is possible to take advantage of microalgae's ability to grow using carbohydrates instead of sunlight as their energy source. Chlorella algae grow well using carbohydrate as the energy source and consequently produce very little chlorophyll (Fig. 9.17). The result is less green color in the protein product (Caporgno and Mathys, 2018). Protein derived from Chlorella species has been granted GRAS status by FDA in the US and is considered as a "Not Novel" protein source in the EU. Microalgal protein is suggested to have functional properties, such as

Fig. 9.17 Algal protein (Spirulina and Chlorella).

emulsification, foaming, and gelation in food (Chronakis and Madsen, 2011). The protein can be added to baked goods, such as cookies and bread, with good acceptance because the color and flavor are little noticed. A potential concern for use of algal protein in food is microcystin, a toxin produced by cyanobacteria (*Microcystis aeruginosa*). This organism is a blue–green algae that releases microcystin toxin into the water as they lyse. Microcystin is a small peptide composed of 7 amino acids. When consumed, microcystin is dangerous to the liver and also causes skin rash and eye irritation when people come in contact with contaminated water. Microcystin levels become high in lakes and ponds due to phosphate runoff and elevated temperatures that cause algal bloom. Major algal bloom incidents occurring in the recent past are thought to result from climate change. However, microcystin is not a problem for algal protein produced in confined environments like ponds. This is because the temperature and phosphate level triggers for microcystin production are controlled.

Novel foods, plant-based animal foods

Plant-based foods represents the largest segment of growth in food product sales for the US during 2018. Consumer acceptance of plant-based foods has increased substantially for several reasons. They are compatible with vegan and vegetarian diets, offer potential for a health–promoting composition, and are more sustainably produced compared to equivalent animal products. The market for all plant-based foods in the United States increased 17% in 2018, totaling $3.7 Billion. Retail sales of plant-based dairy products such as milk, butter, cheese, yogurt, and ice cream, represented the largest segment of plant-based food sales and amounted to $2.6 Billion in 2018. Plant-based milk (e.g., soy, almond, and oat milk) are milk substitutes made with protein from plant sources. The popularity of plant milks is growing because they represent alternative for those with lactose intolerance or milk allergy. Milk caseins are among the eight most common causes of food allergies. It is of interest to note that plant milks have gained sufficient market share to prompt some states to consider laws barring use of the word milk on their label. FDA is also reviewing the labeling issue in response to objections from dairy industry groups.

Plant-based meat

The market for plant-based meat is considerably less than that for plant-based dairy. Sales of plant-based meat ($684 million) in 2018 amount to 25% of those for plant-based dairy. The major companies offering plant-based meat products (Impossible Burger™, Beyond Meat™) have created products with flavor and texture very close to that of beef. Both Impossible Burger™ and Beyond Meat™ products are similar in their composition of protein and fat. Impossible Burger™ originated using a combination of soy, wheat, and potato proteins in the product. The composition was reformulated in January 2019 to replace wheat protein with soy protein, making it a gluten free product. Impossible Burger™ now also contains less salt. Its fat content is provided by coconut oil that adds to its flavor and juiciness. A key component of the product's success is the unique soy leghemoglobin protein. Leghemoglobin is a heme-containing protein originating in root nodules of the soybean plant. Biologically, soy leghemoglobin is essential to the nitrogen fixation process in soils. However, the leghemoglobin of Impossible Burger™ behaves like myoglobin in meat. It binds oxygen from the air, providing the red color of fresh beef to impossible burgers. Leghemoglobin is also responsible for desirable meaty flavors and aromas created when impossible burgers are cooked. Heating induces chemical reactions involving the globin's heme iron that generate flavors and aromas associated with grilled beef. Leghemoglobin protein is produced using a large-scale fermentation (*Pichia pastoris*) process. While leghemoglobin is not a common dietary protein, it has been granted GRAS status by FDA. Beyond Meat™ is a plant-based meat made from rice and pea protein, canola and or coconut oil, and seasonings from yeast extract. It is also a gluten-free food. The red color of Beyond Meat™ burgers comes from added beet extract. Beyond Meat™ is expanding their product applications to include sausage and meat crumble products for pizza topping. Both of these plant-based meat products are available in grocery stores and restaurants.

The spectrum of plant-based foods intended to compete with animal products, is expanding to include eggs. The JUST™ company has created "scramble", a plant-based egg substitute made from mung bean protein isolate, canola oil, and lecithin, a plant-based emulsifier. The enzyme transglutaminase is used to cross link proteins and provide a desirable texture. The yellow color of JUST™ "scramble" comes from added carrot and turmeric extracts. The liquid product coagulates when heated, creating a soft solid that closely resembles traditional scrambled eggs. JUST™ can also be used as an egg replacer in other applications (e.g., making French toast). The JUST™ company also produces an egg-less mayonnaise product made using canola oil, modified starch, vinegar, lemon juice, salt spice (garlic flavoring).

Clean meat

Clean meat is an animal-based protein food that uses tissue culture methods to make a meat-like product (Friedrich, 2016) (Fig. 9.18). Several synonyms are used as a label for this future muscle food, including cultured meat, in vitro meat, and test tube meat. The technology of clean meat is based on muscle cell culture methods developed in the 1970s as a tool to understand the biology of muscle growth. It is now being adapted to large scale production of meat tissue for development into muscle foods (Cassaiday, 2018). A major advantage of clean meat production is uncoupling from the conventional production and harvest of animals for their meat. The inputs required and environmental impact of conventional meat production (especially beef) is large, compared to that of an equivalent amount of proteins and fats nutrients from plants. Additionally, clean meat is safer from microbial pathogens, such E coli and salmonella, and does not contain antibiotics.

Making clean meat is a complex and challenging process that begins with harvesting progenitor cells and growth in monolayer culture. As they grow, cells differentiate into fibroblast and myoblast cells, making collagen and muscle proteins, respectively. Fibroblast collagen serves as a scaffold for attachment of fused myoblasts that contain the contractile proteins of muscle. Myoblasts subsequently differentiate into elongated tubular structures with the ability to shorten and generate force. Repeated cycles of myotube contraction increases the level of protein synthesis and net gain of protein mass in the culture. While clean meat technology offers promise for making muscle foods in a more sustainable way, there are several hurdles

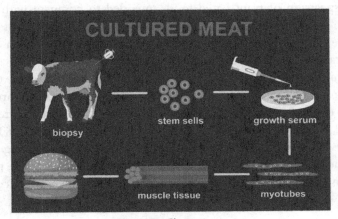

Fig. 9.18 Clean meat.

e.g., flavor, color, and cost, that must be overcome for it to become a successful and economically viable food source. Muscle tissue made in cell culture has little fat requiring addition of lipids to achieve the taste and juiciness comparable to beef or chicken. Muscle cells grown in culture media are pale in comparison because of low myoglobin content. The addition of red colorant, such as carmine, may be required to achieve the red color of fresh beef. The largest hurdle is cost. The first product demonstration of a clean meat muscle food (hamburger) was provided by Dr. Mark Post of Maastricht University in 2013. (Lab-grown burger, BBC news 2013). It is estimated that the cost of this single lab burger was about $300,000. However, dramatic reductions in the cost of production has been achieved through scaling up of the process technology and refinement of ingredients used in clean meat production. Companies such as Mosa Meat^{TM}, Memphis Meat^{TM}, and American Food Company^{TM} are actively developing clean meat muscle foods at lower cost. For example, Memphis Meats produced meat ball and chicken nugget products in 2016, costing about $6,000 per pound. More recently, American Food's clean chicken product is expected to hit the market at a cost that is just 30 % higher than conventional chicken. Clean meat products have a promising future. They have the potential to be less a less costly and more sustainable alternative to conventional muscle food (Plant-Based Market Overview, 2019).

Novel foods, edible insects

Insects are an animal-based protein source that are rarely part of Western diets, but represent important sources of nutrition in many parts of the world (Fig. 9.19). Entomophagy, the practice of using insects as food, has been practiced for a long time. Evidence from prehistoric cave art and surviving documents from ancient Egypt depicting honey gathering suggest that insects were part of their food supply. Today, the use of insects as food is common practice in over 100 countries. It is estimated that insects are important nutrients in the diets of 2 billion people worldwide. More than 2000 insect species belonging to the orders of Coleoptera, Lepidoptera, Hymenoptera, and Orthoptera are common food sources (Costa-Neto and Dunkel, 2016). Insects are a good source of dietary protein because it makes up the largest percent of the insect body. The level of protein in insects ranges from 20% to 76% and much of it found in the exoskeleton (Williams et al., 2016). Insect protein digestibility is only slightly lower than egg and meat (Churchward-Venne et al., 2017). When examined by rat feeding studies,

Fig. 9.19 Edible insects.

insect protein contains all essential amino acids, but is limiting in methionine. Both methionine and arginine were found to be limiting in growing chicks. The difference in these findings may due to the fact that rats can synthesize some arginine, while chickens cannot. The fat content of insects varies widely (from 10% to 50%). The profile of fatty acids in insect acylglycerols contains a relatively high content of C18 fatty acids. Notably, the content of linoleic acid (C18:2) and linolenic acid (C18:3) is relatively high in lepidoptera insects. Insects can also be a source of vitamins (i.e., riboflavin, biotin, pantothenic acid, folic acid) and minerals (i.e., iron and zinc) albeit at low levels. The carbohydrate content of insects is predominantly in the form of chitin from the exoskeleton. Chitin is a long chain polysaccharide composed of N-acetyl-glucosamine monomers. Chitin prevalence is second only to cellulose as the most abundant carbohydrate in the world. Chitin and its derivative chitosan can be a source of dietary fiber and provide cholesterol absorbing properties in the gut. Chitin is fermentable in the gut by constituent microbiota, producing beneficial short chain fatty acids (Borrelli et al., 2017).

The options for insect food applications are presently limited to eating them whole or as flour or protein isolate preparations added to other foods. The psychologically averse reaction to eating whole insects represents a significant limitation to their use as food. Insect flavor varies greatly with the species, developmental stage, and preparation. The most commonly consumed whole insects are grasshoppers, grubs, meal worms, and crickets. They are typically seasoned to increase their flavor and desirability. Insect developmental stage is important because transient compounds, such as pheromones, can be a source of objectionable flavor. Similarly, the crunchy texture of insect exoskeleton can also be objectionable, if unexpected. Insect flour and protein isolates (especially from cricket) are the most widely used form in food applications. Foods made with these preparations include smoothies, protein bars, cookies, and snack foods. A major obstacle to the wider use of insects as dietary protein is their potential risk of food allergy (Downs et al., 2016). Individuals with shellfish or dust mite allergy may experience an allergic reaction to insects upon eating them for the first time. This unexpected reaction is due to the structural similarity of proteins between phylogenetically related organisms and is termed cross-reactivity (Ribeiro et al., 2018). For example, the proteins tropomyosin and arginine kinase in house dust mites, shrimp, and insects are very similar in structure. It is therefore possible for someone allergic to dust mites to experience an allergic reaction after eating insects for the first time.

Summary

The function of the gut was thought to be limited to digestion and absorption of nutrients. However, it is now known that gut microorganisms are of importance to the host in several ways. The microbiome is an essential part of our immune system, controls microbial pathogens, and produces substances that are linked to appetite, mood, and protection of the intestinal lining. It follows that the food we eat can alter the microbiome and therefore affect the well-being of its host. Discoveries being made in this rapidly advancing field will better define the link between diet and health.

Major food systems (i.e., meat, milk, egg, soybean, wheat, and rice) represent the traditional food sources of the Western diet. Knowledge of their composition is important to assessing nutritional quality and the potential for transforming these commodities into food products. However, change is rapidly occurring in the world of food. Present consumer demand and the future need for additional resources have spawned the examination and development of plants, microbes, and insects to fill the need for protein-rich foods. These sources have the inherent advantage of being more sustainable in their production, compared to most animal-based foods. Plant-based alternatives to animal foods, represent the largest area of growth in food product sales for the US during 2018. Plant milk food products are reported to be about 10% of the total market for fluid milk. More recently, plant-based meat products (Impossible Burger™ and Beyond Meat™) are rapidly gaining acceptance. They are presently available in restaurants and grocery stores. Clean meat is an alternative muscle food made with cell culture technology. It offers the possibility of "real" meat grown without the environmental impacts associated with animal production and processing. Clean meat has substantial hurdles to over-come before it can be commercially viable, but progress is being made. Edible insects represent a nutritious food source. About one-fourth of the world's population currently consumes insects in their diet. Insects (e.g., crickets, mealy bug larva, and soldier flies), produced by modern agricultural methods, are commercially available as whole products or purified protein extracts.

Glossary

CLA: Acronym for conjugated linolenic acid, a fatty acid originated in the foregut of ruminants. CLA may protect against certain forms of cancer.

Cytokines: Small molecules (often peptides) that signal a variety of biological processes, such as regulating immunity, inflammation, and cell growth.

Dysbiosis: A microbiota population associated with a disease state

Entomophagy: Insects as a source of Food

Eruicic acid: A toxic fatty acid found in rapeseed (the source off canola oil).

Extrusion: A food processing technology using compression and heat to cook and transform a mixture of food ingredients into food products.

GRAS: Generally Recognized As Safe is a status granted to food substances after thorough review by the Food and Drug Administration of the US.

Gut-brain axis: The bi-directional connection between gut and brain. Each can influence the other.

Hydrocolloid: A mixture of substances in which one substance (usually polysaccharide) is uniformly dispersed in another.

Leghemoglobin: A heme containing protein responsible for soybean plant's ability to fix nitrogen from the atmosphere into soil. It is also provides meat-like taste and color in the Impossible burger.

Microbiome: The collective community of micro-organisms inhabiting a particular region of the body (e.g., mouth, gut, skin, vagina, etc)

Microbiota: Population of micro-organisms that comprises a specific microbiome

Microcystin: A small toxic peptide produced by some algae during growth. It is a problem during periods of algal blooms in lake and ponds.

Mycoprotein: A type of protein derived from fungal fermentation.

PDCAAS: Protein digestibility corrected amino acid score is a method to measure the nutritional quality of a protein. The ranking of a specific food protein is determined in comparison to the amino acid profile a standard amino acid profile The highest possible score is 1.0

Phytate: A salt or ester of phytic acid containing important minerals (calcium, iron, zinc).

Prebiotic: Non-digestible ingredient that promotes growth of beneficial bacteria

Probiotic: Organisms such as bacteria and yeasts that support the digestive system. Lactobacillus and bifidobacteria are examples of bacteria that are beneficial to the gut and the health of the host

Pulses: Legume-based foods such as beans, peas, lentils, beans, and chickpeas

Saponin: Plant substance characterized by the ability to form foams and emulsions. They are toxic to fish.

SCF: Acronym for Short Chain Fatty acids

TAMO: Acronym for the compound trimethylamine oxide. The level of TAMO in the blood is a marker for cardiovascular disease.

Tannins: Polymers of phenolic compounds with yellow to brown color. Bind enzymes in the gut and are also noted for antioxidant activity.

References

Arya, S.S., Salve, A.R., Chauhan, S., 2016. Peanuts as functional food: a review. J. Food Sci. Technol. 53 (1), 31—41. https://doi.org/10.1007/s13197-015-2007-9.

Bleakley, S., Hayes, M., 2017. Algal proteins: extraction, application, and challenges concerning production. Foods 6 (5), 33. https://doi.org/10.3390/foods6050033.

Borrelli, L., Coretti, L., Dipineto, L., Bovera, F., Menna, F., Chiariotti, L., Nizza, A., Lembo, F., Fioretti, A., 2017. Insect-based diet, a promising nutritional source, modulates gut microbiota composition and SCFAs production in laying hens. Sci. Rep. 7 (1), 16269. https://doi.org/10.1038/s41598-017-16560-6.

Canani, R.B., Costanzo, M.D., Leone, L., Pedata, M., Meli, R., Calignano, A., 2011. World J. Gastroenterol. 17 (12), 1519—1528.

Cani, P.D., 2018. Human Gut Microbiome: hopes, threats and promises. Gut 67 (9), 1716—1725, 2018.

Caporgno, M.P., Mathys, A., 2018. Trends in microalgae incorporation into innovative food products with potential health benefits. Fron. Nutr. 5, 58. https://doi.org/10.3389/fnut.2018.00058.

Carlson, J.L., Erickson, J.M., Lloyd, B.B., Slavin, J.L., 2018. Health effects and sources of prebiotic dietary fiber. Cur. Dev. Nutr. 2 (3) nzy005.

Cénita, M.C., Matzarakia, V., Tigchelaarab, E.F., Zhernakovaab, A., 2014. Rapidly expanding knowledge on the role of the gut microbiome in health and disease. Biochem. Biophys. Acta. 1482 (10), 1981—1992.

Chassaing, B., Koren, O., Goodrich, J.K., Poole, A.C., Srinivasan, S., et al., 2015. Dietary emulsifiers impact the mouse gut microbiota promoting colitis and metabolic syndrome. Nature 519, 92—96.

Chronakis, I.S., Madsen, M., 2011. Algal proteins.2011. In: Phillips, G.O., Williams, P.A. (Eds.), Handbook of Food Proteins. Woodhead Publishing Series in Food Sciences, Technology and Nutrition, pp. 353—394.

Churchward-Venne, T.A., Pinckaers, P.J.M., van Loon, J.J.A., van Loon, L.J.C., 2017. Consideration of insects as a source of dietary protein for human consumption. Nutr. Rev. 75 (12), 1035—1045, 2017 Dec 1.

Costa-Neto, E.M., Dunkel, F.V., 2016. Insects as food: history, culture, and modern use around the world. In: Dossey, A.T., Morales-Ramos, J.A., Guadalupe Rojas, M. (Eds.), Insects as Sustainable Food Ingredients. Academic Press, p29—60.

Culley, M.K., Li, X.S., Fu, X., Wu, Y., Li, L., DiDonato, J.A., Wilson-Tang, W.H., Garcia-Garcia, J.S., Hazen, S.L., 2019. l-Carnitine in omnivorous diets induces an atherogenic gut microbial pathway in humans. J. Clin. Investig. 129 (1), 373—387.

Davenport, E.R., Sanders, J.G., Song, S.J., Amato, K.R., Clark, A.G., Knight, R., 2017. The human microbiome in evolution. BMC Biology 15 (1), 127.

Dhaka, V., Gulia, N., Ahlawat, K.S., Khatkar, B.S., 2011. Trans fats-sources, health risks and alternative approach - a review. J. Food Sci. Technol. 48 (5), 534—541.

Downs, M., Johnson, P., Zeece, M., 2016. Insects and their connection to food allergy. In: Dossey, A.T., Morales-Ramos, J.A., Guadalupe Rojas, M. (Eds.), Insects as Sustainable Food Ingredients. Academic Press, pp. 255—268.

Ercolini, D., Fogliano, V., 2018. Food design to feed the human gut microbiota. J. Agric. Food Chem. 66, 3754—3758.

FDA, 2015. Final Determination Regarding Partially Hydrogenated Oils (Removal of Trans Fat). https://www.fda.gov/food/food-additives-petitions/final-determination-regarding-partially-hydrogenated-oils-removing-trans-fat.

Fu, J., Bonder, M.J., Cenit, M.C., Tigchelaar, E.F., Maatman, A., Dekens, J.A., Brandsma, E., Marczynska, J., Imhann, F., Weersma, R.K., Franke, L., Poon, T.W., Xavier, R.J., Gevers, D., Hofker, M.H., Wijmenga, C., Zhernakova, A., 2015. The gut microbiome contributes to a substantial proportion of the variation in blood lipids. Circ. Res. 117 (9), 817—824.

Green, J.M., Barratt, M.J., Kinch, M., Gordon, J.I., 2017. Food and microbiota in the FDA regulatory framework. Science 357 (6346), 39—40.

Hoogenkamp, H., Kumagai, H., Wanasundara, J.P.D., 2017. Rice Protein and Rice Protein Products. Sustainable Protein Sources Nadathur, S. Sustainable Protein Sources [Vital-Source Bookshelf]. Retrieved from. https://bookshelf.vitalsource.com/#/books/9780128027769/.

Human Microbiome Project, 2019. https://www.genome.gov/27549400/the-human-microbiome-project-extending-the-definition-of-what-constitutes-a-human/.

Keku, T.O., Dulal, S., Deveaux, A., Jovov, B., Han, X., 2015. The Gastrointestinal microbiota and colorectal cancer. Am. J. Physiol. Gastrointest. Liver Physiol. 308 (5), G351—G363.

Kennedy, P.J., Cryan, J.F., Dinan, T.G., Clarke, G., 2014. Irritable bowel syndrome: a microbiome-gut-brain axis disorder? World J. Gastroenterol. 20 (39), 14105—14125.

Koeth, R.A., Lam-Galvez, R.B., Kirsop, J., Wang, Z., Levison, B.S., Gu, X., Copeland, M.F., Bartlett, D., Cody, D.B., Dai, H.J., König, D., Oesser, S., Scharla, S., Zdzieblik, D., Gollhofer, A., 2018. Specific collagen peptides improve bone mineral density and bone markers in postmenopausal women-A randomized controlled study. Nutrients 10 (1), 97.

Messina, M., 2016. Soy and health update: evaluation of the clinical and epidemiologic literature. Nutrients 8 (12), 754.

Meydani, M., 2009. Potential health benefits of avenanthramides of oats. Nutr. Rev. 67, 731—735.

Mills, E.N.C., Madsen, C., Shewry, P.R., Wichers, H.J., 2003. Food allergies of plant origin-their molecular and evolutionary relationship. Trends Food Sci. Technol. 14, 145—156.

Niccolai, E., Boem, F., Russo, E., Amedei, A., 2019. The Gut⁻Brain Axis in the neuropsy-chological disease model of obesity: a classical movie revised by the emerging director "microbiome. Nutrients 11 (1), 156.

Orona-Tamayo, D., Paredes-López, O., 2017. Amaranth Part 1—sustainable crop for the 21st century: food properties and nutraceuticals for improving human health. In: Nadathur, S.R., Wanasundara, J.P.D. (Eds.), Laurie Scanlin, Sustainable Protein Sources. Academic Press, pp. 239—256.

Palmnäs, M.S., Cowan, T.E., Bomhof, M.R., Su, J., Reimer, R.A., Vogel, H.J., Hittel, D.S., Shearer, J., 2014. Low-dose aspartame consumption differentially affects gut microbiota-host Metabolic Interactions in the diet-induced obese rat. PLoS One 9 (10), e109841. https://doi.org/10.1371/journal.pone.0109841.

Patterson, E., Wall, R., Fitzgerald, G.F., Ross, R.P., Stanton, C., 2012. Health implications of high dietary omega-6 polyunsaturated Fatty acids. J. Nutr. Metab. 2012, 539426.

Peterson, D.M., 2011. Storage proteins. In: Webster, F.H., Wood, P.J. (Eds.), Oats: Chemistry and Technology. American Association of Cereal Chemists, St. Paul, MN (123—142).

Ribeiro, J.C., Cunha, L.M., Sousa-Pinto, B., Fonseca, J., 2018. Allergic risk of consuming edible insects: a systematic review. Mol. Nutr. Food Res. 62, 1700030. https://doi.org/10.1002/mnfr.201700030.

Ríos-Covián, D., Ruas-Madiedo, P., Margolles, A., Gueimonde, M., de Los Reyes-Gavilán, C.G., Salazar, N., 2016. Intestinal short chain fatty acids and their link with diet and human health. Front. Microbiol. 7, 185.

Roca-Saavedra, P., Mendez-Vilabrille, V., Miranda, J.V., Nebot, C., Cardelle-Cobas, A., Franco, C.M., Cepeda, A., 2017. Food additives, contaminants, and other minor components: effects on human gut microbiota-A review. J. Physiol. Biochem. 74, 69—83.

Rodríguez-Alcalá, L.M., Castro-Gómez, M.P., Pimentel, L.L., Fontecha, J., 2017. Milk fat components with potential anticancer activity: a review. Biosci. Rep. 37. https://doi.org/10.1042/BSR20170705. BSR20170705.

Rooks, M.G., Garrett, W.S., 2016. Gut microbiota, metabolites, and host immunity. Nat. Rev. Immunol. 16 (6), 341—352.

Santra, D.K., Schoenlechner, R., 2017. In: Nadathur, S. (Ed.), Amaranth Part 2—Sustainability, Processing, and Applications of Amaranth.

Slavin, J., 2013. Fiber and prebiotics: mechanisms and health benefits. Nutrients 5 (4), 1417—1435, 2013 Apr 22.

Ursell, L.K., Metcalf, J.L., Parfrey, L.W., Knight, R., 2012. Defining the human microbiome. Nutr. Rev. 70 (Suppl. 1), S38—S44.

Valdes, A.M., Walter, J., Segal, E., Spector, T.D., 2017. Role of the gut microbiota in nutrition and health. Br. Med. J. 361. https://doi.org/10.1136/bmj.k2179.

Williams, J.P., Williams, J.R., Kirabo, A., Chester, D., Petersen, M., 2016. Nutrient content and health benefits of Insects. In: Dossey, A.T., Morales-Ramos, J.A., Guadalupe Rojas, M. (Eds.), Insects as Sustainable Food Ingredients. Academic Press, pp. 61—84.

Zinöcker, M.K., Lindseth, I.A., 2018. The western diet-microbiome-host interaction and its role in metabolic disease. Nutrients 10 (3), 365.

Further reading

General

Essential Fatty Acids, 2014. Linus Pauling Institute Oregon State University. https://lpi.oregonstate.edu/mic/other-nutrients/essential-fatty-acids.

Omega Fatty Acids, 2019. https://ods.od.nih.gov/factsheets/Omega3FattyAcids-HealthProfessional/.

PDCAAS Protein score. 2019. https://en.wikipedia.org/wiki/Protein_Digestibility_Corrected_Amino_Acid_Score. [accessed June 2019].

The microbiome

Ercolini, D., Fogliano, V., 2018. Food design to feed the human gut microbiota. J. Agric. Food Chem. 66 (15), 3754—3758. https://doi.org/10.1021/acs.jafc.8b00456.

Human Microbiome Project. 2019. https://www.genome.gov/27549400/the-human-microbiome-project-extending-the-definition-of-what-constitutes-a-human/. [accessed Jan 2019],[accessed April 2019].

NIH MHP, 2015. Food Additives Alter Gut Microbes, Cause Diseases in Mice. https://www.nih.gov/news-events/nih-research-matters/food-additives-alter-gut-microbes-cause-diseases-mice.

Proteins of the future

Nadathur, S., 2017. Sustainable Protein Sources [VitalSource Bookshelf]. Retrieved from. https://bookshelf.vitalsource.com/#/books/9780128027769/.

NIH Human Microbiome Project (HMP) https://commonfund.nih.gov/hmp. [Update March].

Plant-Based Market Overview, 2019. Good Foods Overview. https://www.gfi.org/marketresearch.

Pulse Canada, 2019. http://www.pulsecanada.com/about-pulse-canada/what-is-a-pulse/.

Straight Talk about Soy, 2019. The Nutrition Source. https://www.hsph.harvard.edu/nutritionsource/soy/.

Clean meat

Cassaiday, L., 2018. Clean Meat. https://www.aocs.org/stay-informed/inform-magazine/featured-articles/clean-meat-february-2018.

Friedrich, B., 2016. "'Clean Meat': The 'clean Energy' of food." the Good Food Institute Blog. http://tinyurl.com/clean-meat.

Lab-Grown burger, 2013. BBC News. https://www.bbc.com/news/science-environment-23576143.

Review questions

1. Define the terms microbiome and microbiota.
2. What are the beneficial effects of the gut microbiome?
3. Define the terms prebiotic, probiotic and fiber.
4. Name three foods that are high in fiber.
5. How are short chain fatty acids beneficial to health?
6. What is the proposed effect of artificial sweeteners on the microbiome?
7. Define the terms macronutrient and micronutrient.
8. Which muscle protein is important to making processed meats?
9. What muscle protein is responsible for cooked meat flavor?
10. What is CLA and what food is it found in?
11. What fraction of milk protein is responsible for making cheese and yogurt?
12. Name the egg component essential to making mayonnaise.
13. What vitamins are found in eggs?
14. What are the major proteins and lipids of soybeans?
15. Name three antinutritional factors common to plant foods.
16. What is a pulse? Give three examples.
17. Is quinoa derived from plants, microorganisms, or insects?
18. Which protein-rich plant food source contains the highest level of protein?
19. What is Quorn and what is its source?
20. What is Spirulina and why is it green?

21. What is microcystin and where is it found?
22. Describe the advantages and disadvantages of algae and fungi as food sources.
23. Define the term plant-based meat and give an example.
24. What is leghemoglobin and why is it important to the impossible burger?
25. Define the term clean meat. What are its advantages and disadvantages?
26. What is the difference between plant-based meat and clean meat?
27. How soy milk is made?
28. Is it fair to use the term "milk" for products made from plant materials?
29. Define the term PDCAAS and its relevance to dietary proteins.
30. What is entomophagy?
31. Describe the advantages and disadvantages of insects as a food source?

Index

Note: 'Page numbers followed by "*f*" indicate figures and "*t*" indicate tables'.

Printed in the United States
By Bookmasters